P9-CBA-532

Fluid Mechanics for Engineering Technology

SECOND EDITION

Fluid Mechanics for Engineering Technology

IRVING GRANET, P.E.

Queensborough Community College
New York Institute of Technology

Prentice-Hall Inc., Englewood Cliffs, N.J. 07632

Library of Congress Cataloging in Publication Data

Granet, Irving.
Fluid mechanics for engineering technology.

Includes bibliographical references and index.
1. Fluid mechanics. I. Title.
TA357.G7 1981 620.1′06 80-18289
ISBN 0-13-322610-7

Editorial/production supervision and interior design by Mary Carnis
Cover Design: Wanda Lubelska
Manufacturing Buyer: Anthony Caruso

Printed in the United States of America

10 9 8 7 6 5 4

Prentice-Hall International, Inc., *London*
Prentice-Hall of Australia Pty. Limited, *Sydney*
Prentice-Hall of Canada, Ltd., *Toronto*
Prentice-Hall of India Private Limited, *New Delhi*
Prentice-Hall of Japan, Inc., *Tokyo*
Prentice-Hall of Southeast Asia Pte. Ltd., *Singapore*
Whitehall Books Limited, *Wellington, New Zealand*

This book is dedicated to my devoted wife, Arlene, whose love and forebearance made its completion possible.

Contents

Preface

Although some time has elapsed since the first edition of the book was written, all of the objectives stated at that time are still valid. It is a *textbook* designed to be used by a student who will be studying fluid mechanics at a community college, technical institute, or in bachelor of technology programs. With this objective in mind, I have avoided use of the calculus. However, as was noted in the first edition, the instructor has the discretion to present an alternative parallel development using calculus to increase the depth of presentation where it is deemed both warranted and feasible.

The widespread adoption of the first edition has led me to retain many of the features and arrangement of the original book. However, I have carefully considered the constructive suggestions made by users of the book and have incorporated many of them in this edition. Also, during the planning of this revision it became necessary to make a decision regarding the adoption of SI units. The generally slow adoption of the SI system of units by industry in the United States requires the technician to be familiar with both the English and SI systems at this time. Therefore, equations and problems are developed in both systems throughout the book. The first chapter of the book is devoted to introducing the student to systems of units and should prove to be valuable as a familiarization and as an introduction to the use of dimensional consistency in problem solving. The text contains over 500 problems, with complete solutions given for 134 of the problems as an integral part of the text.

Miss Cheryl Gorgoni, Mrs. Mae Shuman, and Mrs. Muriel Smith typed most of the manuscript and I am most grateful for their help in this undertaking. Professor Donald H. Wright of Suffolk County Community College reviewed the preliminary manuscript in detail making many suggestions which were subsequently incorporated into the final version. His efforts are greatly appreciated. I am also indebted to the many users of the book who took the time to make constructive suggestions. My colleagues at Queensborough Community College and New York Institute of Technology were most supportive of me during this time. My wife Arlene and our children, Ellen, Kenny, David, and Jeffrey, provided the love, patience, and understanding that made the successful completion of this undertaking possible.

Irving Granet
North Bellmore, New York

Systems of Units and Dimensional Consistency

1.1 INTRODUCTION

Fluid mechanics is the study of the behavior of fluids whether they are at rest or in motion; the study of fluids at rest is best known as *fluid statics* and the study of fluids in motion is termed *fluid dynamics*. In this book we use the term *fluid* to refer to both gases and liquids. To distinguish between a liquid and a gas, we note that while both will occupy the container in which they are placed, a liquid presents a free surface if it does not completely fill the container, but a gas will always fill the volume of the container in which it is placed. For gases it is important to take into account the change in volume that occurs when either the pressure or the temperature is changed, whereas in most cases it is possible to neglect the change in volume of a liquid when there is a change in pressure.

It is apparent that almost every part of our lives and the technology of modern life involves some dependence on and knowledge of the science of fluid mechanics. Whether we consider the flow of blood in the minute blood vessels of the human body or the motion of an aircraft or missile at speeds exceeding the velocity of sound, we need to utilize some branch of fluid mechanics to describe the motion. The literature of this subject is so vast that a brief description cannot adequately reflect its scope. At one time the subject was treated from a purely mathematical approach by one group of investigators and from an entirely empirical experimental approach by another group of investigators. In this text we use the modern technique of coordinating both approaches by supplementing theory with experiment.

Since all measurements as well as theoretical developments must explicitly state the units being used, we start our study with a discussion of systems of units.

1.2 THE SI SYSTEM OF UNITS

At the time of the French Revolution, the systems of weights and measures used throughout the world were an incoherent and almost hopeless jumble. International trade and the interchange of scientific information both suffered greatly because of this condition. French scientists and scholars of this era developed a rational system of weights and measures called the *metric system*, which was adopted by most countries of the world. In 1960, the General Conference of Weights and Measures extensively revised and simplified the older metric system and gave it the French title, *Système International d'Unités* (International System of Units), commonly abbreviated *SI*. The latest revisions and additions were made in an international conference in 1971, and work still continues on these standards.

For the engineer, the greatest confusion has been the units for mass and weight. The literature abounds with units such as slugs, pounds mass, pound force, poundal, kilogram force, kilogram mass, dyne, and so on. In the SI system, the base unit for *mass* (not weight or force) is the kilogram, which is equal to the mass of the international standard kilogram located at the International Bureau of Weights and Measures. It is used to specify the quantity of matter in a body. The mass of a body never varies, and it is independent of gravitational force.

The SI *derived* unit for force is the newton (N). The unit of force is defined from *Newton's second law of motion*: force is equal to mass times acceleration ($F = ma$). By this definition, 1 newton applied to a mass of 1 kilogram gives the mass an acceleration of 1 metre per second squared ($N = kg \cdot m/s^2$). The newton is used in all combinations of units that include force: pressure or stress (N/m^2), energy ($N \cdot m$), power ($N \cdot m/s = W$), and so on. By this procedure, the unit of force is not related to gravity as was the older kilogram force.

Weight is defined as a measure of gravitational force acting on a material object at a specified location. Thus, a constant mass has an approximate constant weight on the surface of the earth. The agreed standard value (standard acceleration) of gravity is 9.806 650 m/s². Figure 1.1 illustrates the difference between mass (kilogram) and force (newton).

The term "mass" or "unit mass" should be used only to indicate the quantity of matter in an object. The old practice of using weight in such cases should be avoided in engineering and scientific practice. The general relation that ties together mass (m) and weight (W) is

$$W = m \times g$$

where g is the local acceleration of gravity. In SI units, $g = 9.806$ m/s².

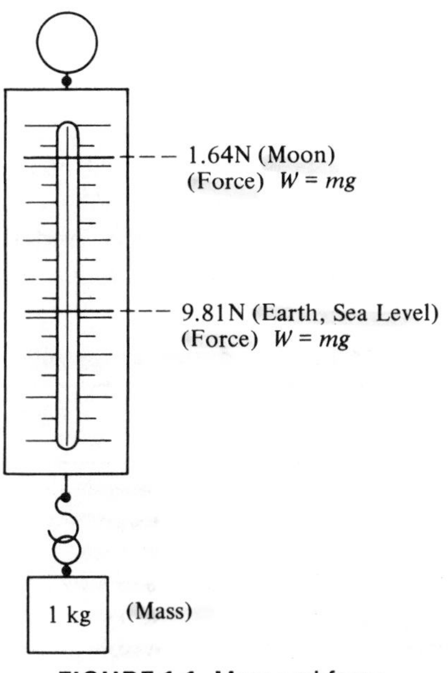

FIGURE 1.1 Mass and force.

ILLUSTRATIVE PROBLEM 1.1

One kilogram of lead is taken to the moon, where the local acceleration of gravity is one-sixth that of earth gravity. What is its mass on the moon, and how much does it weigh on the moon?

Solution

As shown in Figure 1.1, the body weighs 9.81 N on the earth. On the moon the mass will still be 1 kg, since the amount of matter in the body stays constant. However, since the local acceleration of gravity on the moon is one-sixth of the earth's gravity, it will weigh one-sixth of its earth's weight on the moon. Therefore,

$$\text{weight (moon)} = \tfrac{1}{6} \times 9.81 = 1.635 \text{ N}$$

The SI system consists of three classes of units:

1. Base units
2. Supplementary units
3. Derived units
   (a) With special names
   (b) Without special names

Table 1.1 gives the seven base units of the SI system. Several observations concerning this table should be noted. The unit of length is the metre (not meter), and the kilogram is a unit of mass, not weight. Also, symbols are never pluralized; never written with a period; and the use of upper- and lower-case symbols *must* be used as shown *without exception.*

TABLE 1.1 Base SI Units

Quantity	Name of Base SI Unit	Symbol
length	metre	m
mass	kilogram	kg
time	second	s
electric current	ampere	A
thermodynamic temperature	kelvin	K
amount of substance	mole	mol
luminous intensity	candela	cd

Reprinted with permission from *Strength of Materials for Engineering Technology*, 2nd edition, by Irving Granet, Reston Publishing Company, a Prentice-Hall Company, 11480 Sunset Hills Road, Reston, Va. 1980, p. 2.

Table 1.2 gives the supplementary units of the SI system. These units can be regarded as either base units or as derived units.

Table 1.3 gives the derived units (with and without symbols) often used in fluid mechanics. These derived units are formed by the algebraic combination of base and supplementary units. It is noted that where the name is named for a person, the first letter of the symbol appears as a capital (e.g.,

TABLE 1.2 Supplementary SI Units

Quantity	Supplementary SI Unit	Symbol
plane angle	radian	rad
solid angle	steradian	sr

Reprinted with permission from *Strength of Materials for Engineering Technology*, 2nd edition, by Irving Granet, Reston Publishing Company, a Prentice-Hall Company, 11480 Sunset Hills Road, Reston, Va., 1980, p. 2.

TABLE 1.3 Derived SI Units

Quantity	Name	Symbol	Formula	Expressed in Terms of Base Units
acceleration	acceleration	m/s^2	m/s^2	m/s^2
area	square metre	m^2	m^2	m^2
density	kilogram per cubic metre	—	kg/m^3	$kg \cdot m^{-3}$
energy or work	joule	J	$N \cdot m$	$m^2 \cdot kg \cdot s^{-2}$
force	newton	N	$m \cdot kg \cdot s^{-2}$	$m \cdot kg \cdot s^{-2}$
length	metre	m	m	m
mass	kilogram	kg	kg	kg
moment	newton-metre	$N \cdot m$	$N \cdot m$	$m^2 \cdot kg \cdot s^{-2}$
moment of inertia of area	—	m^4	m^4	m^4
plane angle	radian	rad	rad	rad
power	watt	W	J/s	$m^2 \cdot kg \cdot s^{-3}$
pressure or stress	pascal	Pa	N/m^2	$N \cdot m^{-2}$
rotational frequency	revolutions per second	rev. per sec.	s^{-1}	s^{-1}
temperature	degree celsius	°C	°C	1 °C = 1 K
time	second	s	s	s
torque (see moment)	newton-metre	$N \cdot m$	$N \cdot m$	$m^2 \cdot kg \cdot s^{-2}$
velocity (speed)	metre per second	metre per sec.	m/s	$m \cdot s^{-1}$
volume	cubic metre	—	m^3	m^3

Reprinted with permission from *Strength of Materials for Engineering Technology*, 2nd edition, by Irving Granet, Reston Publishing Company, a Prentice-Hall Company, 11480 Sunset Hills Road, Reston, Va., 1980, p. 2.

newton is N). Otherwise, the convention is to make the symbol lower-case.

In order for the SI system to be universally understood, it is most important that the symbols for the SI units and the conventions governing their use be strictly adhered to. Care should be taken to use the correct case for symbols, units, and their multiples (e.g., K for kelvin, k for kilo; m for milli, M for mega). As noted earlier, unit *names* are never capitalized except at the beginning of a sentence. SI unit *symbols* derived from proper names are written with the first letter in upper-case; all other symbols are written in lower-case. For example, m (metre), s (second), K (kelvin), Wb (weber). Also, unit names form their plurals in the usual manner. Unit symbols are always written in singular form: for example, 350 megapascals, or 350 MPa; 50 milligrams, or 50 mg. Since the unit symbols are standardized, the symbols should always be used in preference to the unit names. An exception is made when a number written out in words precedes the unit (e.g., seven metres, not seven m). Unit symbols are not followed by a period unless they occur at the end of a sentence and the numerical value associated with a symbol should be separated from that symbol by a space (e.g., 1.81 mm, *not* 1.81mm). The period is only to be used as a decimal marker. Since the comma is used by some countries as a decimal marker, the SI system does not use the comma. A space is used to separate large numbers in groups of threes starting from the decimal in either direction. Thus, 3 807 747.0 and 0.030 704 254 indicate this type of grouping. Notice that for numerical values of less than 1, the decimal point is preceded by a zero. For a number of four digits, the space can be omitted.

In addition, certain style rules should also be adhered to:

1. When a product is to be indicated, use a space between unit names (e.g., newton metre).
2. When a quotient is indicated, use the word "per" (e.g., metre per second).
3. When a product is indicated, use the word "square," "cubic," and so on (e.g., square metre).
4. In designating the product of units, use a centered dot (e.g., N·s, kg·m).
5. For quotients, use a solidus (/) or a negative exponent (e.g., m/s or $m \cdot s^{-1}$). The solidus (/) should not be repeated in the same expression unless ambiguity is avoided by using parentheses. Thus, one should use m/s^2 or $m \cdot s^{-2}$ but *not* m/s/s; also, use $m \cdot kg/(s^3 \cdot A)$ or $m \cdot kg \cdot s^{-3} \cdot A^{-1}$ but *not* $m \cdot kg/s^3/A$.

One of the features of the older metric system and the current SI system that is most useful is the fact that multiples and submultiples of the units are in terms of factors of 10. Thus, the prefixes given in Table 1.4 are used in conjunction with SI units to form names and symbols of multiples of SI units. Certain general rules apply to the use of these prefixes:

TABLE 1.4 Factors of Ten for SI Units

Prefix	Symbol		Factor
tera	T	10^{12}	1 000 000 000 000
giga	G	10^{9}	1 000 000 000
mega	M	10^{6}	1 000 000
kilo	k	10^{3}	1 000
hecto	h	10^{2}	100
deka	da	10^{1}	10
deci	d	10^{-1}	0.1
centi	c	10^{-2}	0.01
milli	m	10^{-3}	0.001
micro	μ	10^{-6}	0.000 001
nano	n	10^{-9}	0.000 000 001
pico	p	10^{-12}	0.000 000 000 001
femto	f	10^{-15}	0.000 000 000 000 001
atto	a	10^{-18}	0.000 000 000 000 000 001

Reprinted with permission from *Strength of Materials for Engineering Technology*, 2nd edition, by Irving Granet, Reston Publishing Company, a Prentice-Hall Company, 11480 Sunset Hills Road, Reston, Va., 1980, p. 5.

1. The prefix becomes part of the name or symbol with no separation (e.g., kilometre, megagram, etc.).
2. Compound prefixes should not be used: use GPa, not kMPa.
3. In calculations, use powers of 10 in place of prefixes.
4. Try to select a prefix where the numerical value will fall between 0.1 and 1000. This rule may be disregarded when it is better to use the same multiple for all items. It is also recommended that prefixes representing 10 raised to a power that is a multiple of 3 be used; (e.g., 100 mg, not 10 cg).
5. The prefix is combined with the unit to form a new unit, which can be provided with a positive or negative exponent. Therefore, mm^3 is $(10^{-3}\ m)^3$ or $10^{-9}\ m^3$.
6. Where possible, avoid the use of prefixes in the denominator of compound units. The exception to this rule is the prefix k in the base unit kg (kilogram).

There are certain units outside the SI system that may be used together with the SI units and their multiples. These are recognized by the International Committee for Weights and Measures as having to be retained because of their practical importance. These are listed in Table 1.5.

TABLE 1.5 Retained Common Units

Quantity	Name of Unit	Unit Symbol	Definition
time	minute	min	1 min = 60 s
	hour	h	1 h = 60 min = 3600 s
	day	d	1 d = 24 h = 86 400 s
plane angle	degree	°	$1° = 1/(\pi/180)$ rad
	minute	′	$1' = (1/60)° = 2.909 \times 10^{-4}$ rad
	second	″	$1'' = (1/60)' = 4.848 \times 10^{-6}$ rad
volume	litre	l	$l = 1\ dm^3 = 10^{-3} m^3$
mass	tonne	t	$1\ t = 1\ Mg = 10^3$ kg

Reprinted with permission from *Strength of Materials for Engineering Technology*, 2nd edition, by Irving Granet, Reston Publishing Company, a Prentice-Hall Company, 11480 Sunset Hills Road, Reston, Va., 1980, p. 6.

ILLUSTRATIVE PROBLEM 1.2

What is the weight of a body that has a mass of 10 kg?

Solution

The weight of the body will be its mass multiplied by the acceleration of gravity; that is,

$$w = mg$$

In terms of SI units,

$$w = 10\ \text{kg} \times 9.81\ \frac{\text{m}}{\text{s}^2}$$

$$w = 9.81\ \frac{\text{kg}\cdot\text{m}}{\text{s}^2}$$

and since

$$1\ \text{N} = 1\ \frac{\text{kg}\cdot\text{m}}{\text{s}^2}$$

then

$$w = 9.81\ \text{N}$$

98.10

It is almost universally agreed that when a new language is to be learned, the student should be completely immersed and made to "think" in the new language. This technique has been proven most effective by the Berlitz language schools and the Ulpan method of language teaching. A classic joke about this is of the American traveling in Europe who was amazed that two-year-old children were able to speak "foreign" languages. In dealing with the SI system, the student should not "think" in terms of customary units and then perform a mental conversion. It is better to learn to "think" in terms of the SI system, which will then become a second language. However, there will be times when it may be necessary to convert from customary U.S. units to SI units. To facilitate such conversions, Table 1.6 gives some commonly used conversion factors.

TABLE 1.6 Conversion Factors

Multiply	By	To Obtain
atmospheres	2.992×10^{1}	inches mercury (32 deg. F)
atmospheres	1.033×10^{4}	kilogram/sq metre
atmospheres (760 torr)	1.013×10^{2}	kilopascals
bars	9.869×10^{-1}	atmospheres
bars	1.000×10^{2}	kilopascals
British thermal units (Btu)	3.927×10^{-4}	horsepower-hours
British thermal units (Btu)	1.056	kilojoules
British thermal units (Btu)	2.928×10^{-4}	kilowatt-hours
British thermal units (Btu)	1.221×10^{-8}	megawatt-days
Btu/hr-square foot	3.153×10^{-4}	Watts/sq centimeter
Btu/hr-sq ft-deg. F	5.676×10^{-4}	W/sq cm-degree Celsius
Btu/minute	2.356×10^{-2}	horsepower
Btu/minute	1.757×10^{1}	Watts
calories	4.190	Joules
cubic feet	2.832×10^{-2}	cubic metres
cubic feet	2.832×10^{1}	litres
cubic feet/min	4.720×10^{-4}	cubic metres/sec
cubic metres	8.107×10^{-4}	acre-feet
cubic metres	3.531×10^{1}	cubic feet
cubic metres	2.642×10^{2}	gallons (US)
cubic metres/sec	2.119×10^{3}	cubic feet/min
cubic metres/sec	1.585×10^{4}	gallons/min
degrees Celsius	$(9/5)\ C + 32$	degrees Fahrenheit
degrees Fahrenheit	$5/9\ (F - 32)$	degrees Celsius

TABLE 1.6 (continued)

Multiply	By	To Obtain
feet	3.048×10^{-1}	metres
feet of H_2O (39.2 deg. F)	3.048×10^{2}	kilogram/sq metre
feet of H_2O (39.2 deg. F)	4.335×10^{-1}	pounds/sq inch
feet/sec	3.048×10^{-1}	metres/sec
foot-pound (force)	1.356	Joules
foot-pounds (force)/min	2.260×10^{-2}	Watts
gallons	3.785×10^{-3}	cubic metres
gallons/min	6.309×10^{-5}	cubic metres/sec
horsepower	4.244×10^{1}	British thermal units/min
horsepower	7.457×10^{-1}	kilowatts
horsepower-hours	2.547×10^{3}	British thermal units
horsepower-hours	7.457×10^{-1}	kilowatt-hrs
inches of H_2O (39.2 deg. F)	2.491×10^{-1}	kilopascals
inches mercury (32 deg. F)	3.342×10^{-2}	atmospheres
inches mercury (32 deg. F)	3.453×10^{2}	kilograms/sq metre
inches mercury (32 deg. F)	3.386	kilopascals
inches mercury (32 deg. F)	4.912×10^{-1}	pounds/sq inch
Joules	7.376×10^{-1}	foot-pounds (force)
Joules	1.000	Watt-seconds
Joules	2.387×10^{-1}	calories
kilograms	2.205	pounds
kilograms	1.102×10^{-3}	tons (short)
kilograms/cubic metre	6.243×10^{-2}	pounds/cubic foot
kilograms/square metre	9.678×10^{-5}	atmospheres
kilograms/square metre	3.281×10^{-3}	ft H_2O (at 39.2 deg. F)
kilograms/square metre	2.896×10^{-3}	inches mercury (32 deg. F)
kilograms/square metre	1.422×10^{-3}	pounds/sq inch
kilojoules	9.471×10^{-1}	British thermal units
kilopascals	4.015	inches H_2O (at 39.2 deg. F)
kilopascals	1.450×10^{-1}	pounds (force)/sq inch
kilopascals	2.953×10^{-1}	inches mercury (32 deg. F)
kilopascals	1.000×10^{-2}	bars
kilopascals	9.869×10^{-3}	atmospheres (760 torr)
kilowatts	1.341	horsepower
kilowatt-hours	3.413×10^{3}	British thermal units
kilowatt-hours	1.341	horsepower-hours
kilowatt-hours	4.167×10^{-5}	megawatt-days

TABLE 1.6 (continued)

litre	3.531×10^{-2}	cubic feet
megawatt-days	8.189×10^{7}	British thermal units
megawatt-days	2.400×10^{4}	kilowatt-hours
metres	3.281	feet
Newtons	2.248×10^{-1}	pounds (force)
pounds	4.536×10^{-1}	kilograms
pounds (force)	4.448	Newtons
pounds/cubic feet	1.602×10^{1}	kilograms/cu metre
pounds/square inch	2.307	ft H_2O (at 39.2 deg. F)
pounds/square inch	2.036	inches mercury (32 deg. F)
pounds/square inch	7.031×10^{2}	kilograms/sq metre
pounds/square inch	6.895	kilopascals
square feet	9.290×10^{-2}	square metres
square metres	2.471×10^{-4}	acres
square metres	1.076×10^{1}	square feet
tonnes	2.205×10^{3}	pounds
tons (short)	9.072×10^{2}	kilograms
Watts	5.688×10^{-2}	Btu/minute
Watts	4.427×10^{1}	foot-pounds (force)/minute
Watt-seconds	1.000	Joules
Watts/sq centimeter	3.171×10^{3}	Btu/hr-sq ft
Watts/sq cm-deg. C	1.762×10^{3}	Btu/hr-sq ft-deg. F

Reprinted with permission from *Strength of Materials for Engineering Technology*, 2nd edition, by Irving Granet, Reston Publishing Company, a Prentice-Hall Company, 11480 Sunset Hills Road, Reston, Va., 1980, p. 6.

ILLUSTRATIVE PROBLEM 1.3

Table 1.6 lists the conversion factor for obtaining cubic metres from cubic feet as 2.832×10^{-2}. Starting with the definition that 1 inch equals 2.54 cm (0.0254 m), derive this conversion factor.

Solution

We will use an approach in the solution of this problem which will be used throughout this book, namely, dimensional consistency. Units must be consistent, and we can use the familiar rules of algebra to manipulate units. Proceeding with this problem, we obtain

$$1 \text{ in.} = 0.0254 \text{ m}$$

or

$$1 = 0.0254 \frac{\text{m}}{\text{in.}} \tag{a}$$

Since 1 ft = 12 in.,

$$1 = 12 \frac{\text{in.}}{\text{ft}} \tag{b}$$

If we multiply equation (a) by equation (b), we obtain

$$1 = 0.0254 \frac{\text{m}}{\text{in.}} \times 12 \frac{\text{in.}}{\text{ft}}$$

Since the inches cancel, we have

$$1 = 0.0254 \times 12 \frac{\text{m}}{\text{ft}}$$

Notice that we can cube both sides, to obtain

$$(1)^3 = (0.0254 \times 12)^3 \frac{\text{m}^3}{\text{ft}^3} = 2.832 \times 10^{-2} \frac{\text{m}^3}{\text{ft}^3}$$

or

$$\text{ft}^3 = 2.832 \times 10^{-2} \text{ m}^3$$

Thus, multiplying cubic feet by 2.832×10^{-2} will yield cubic metres, the desired result. We could have obtained the same result in one chain-type calculation, as

$$\left(0.0254 \frac{\text{m}}{\text{in.}} \times 12 \frac{\text{in.}}{\text{ft}}\right)^3 = 2.832 \times 10^{-2} \frac{\text{m}^3}{\text{ft}^3}$$

The latter method is more commonly used.

ILLUSTRATIVE PROBLEM 1.4

Derive the conversion factor to convert cubic metres per second to gallons per minute if 1 gallon is a measure of volume that equals 231 in.^{3}

Solution

Using the chain-type calculation, we obtain

$$\frac{\text{m}^3}{\text{s}} \times \frac{60\ \text{s}}{\text{min}} \times \frac{1}{(0.0254\ \text{m/in.})^3} \times \frac{1}{231\ \text{in.}^3/\text{gal}} = 1.585 \times 10^4\ \frac{\text{gal}}{\text{min}}$$

Thus,

$$\frac{\text{m}^3}{\text{s}} \times 1.585 \times 10^4 = \frac{\text{gal}}{\text{min}} \qquad \text{which checks Table 1.6}$$

1.3 THE ENGLISH SYSTEM OF UNITS

After using both the English system of units and the SI system, the student will find that the English system of units is not as coherent or as easy to use as the SI system. The difficulty arises from the fact that mass and force can be defined independently of each other. This is due to the fact that two laws attributed to Newton—the law of universal gravitation and the second law of motion—each involve force and mass. We therefore find three basic systems of units to exist, consisting of various combinations of the *same* term to describe both force and weight. Table 1.7 is a simple listing of these combinations.

TABLE 1.7

Mass	*Force*
Pound	Pound
Pound	Poundal
Slug	Pound

At present, in the United States, the accepted consistent set of units commonly used in fluid mechanics is the pound (force), the slug (mass), the foot (length), and the second (time). For this set of units the slug is the derived unit that equals the mass that a force of 1 pound will accelerate at the rate of 1 foot per second per second. From Newton's second law of motion,

$$F = ma \tag{1.1}$$

we find that

$$m = \frac{F}{a} = \frac{\text{lb}}{\text{ft/sec}^2} = \frac{\text{lb}\cdot\text{sec}^2}{\text{ft}} \tag{1.2}$$

Thus, the units of the slug (mass) in the conventional English system are lb·sec²/ft.

If we now perform a simple experiment on earth that consists of dropping a weight (in a vacuum to eliminate the effects of the air on the body) and measuring its acceleration, we would obtain the acceleration, which from equation (1.1) is the same as the ratio of force (weight) to mass. It is generally accepted that the acceleration will be 32.174 ft/sec². Therefore, we obtain

$$\frac{w}{m} = 32.174 \tag{1.3}$$

or

$$m = \frac{w}{32.174} \tag{1.4}$$

If we compare equation (1.4) with the results obtained in Illustrative Problem 1.2, we can write as a general result, independent of the system of units,

$$w = mg \tag{1.5}$$

where $g = 9.806$ m/sec² or 32.174 ft/sec² on the surface of the earth, depending upon the system of units being used. For this book we will usually round off these numbers to 9.81 and 32.17, respectively.

ILLUSTRATIVE PROBLEM 1.5

What is the mass of a body that weights 20 lb? Assume that the local acceleration of gravity is 32.17 ft/sec².

Solution

Using equation (1.5), we obtain

$$m = \frac{w}{g} = \frac{20}{32.17} = \frac{\text{lb}}{\text{ft/sec}^2} = 0.622\,\frac{\text{lb}\cdot\text{sec}^2}{\text{ft}} = 0.622 \text{ slug}$$

(mass)

ILLUSTRATIVE PROBLEM 1.6

A mass of 10 slugs weighs 250 lb on a spring balance on an unknown planet. The spring balance was originally calibrated at a standard location on the earth. What is the value of the acceleration of gravity at this location on the unknown planet?

Solution

From equation (1.5),

$$g = \frac{w}{m} = \frac{250\ \text{lb}}{10\ \text{lb}\cdot\text{sec}^2/\text{ft}} = 25.0\ \frac{\text{ft}}{\text{sec}^2}$$

1.4 DIMENSIONAL CONSISTENCY

The physical sciences are concerned with the discovery and formulation of exact relations between the various quantities involved in recurring situations. Dimensional consistency or dimensional analysis is based upon the axiom that a general relationship may be established between two quantities only when these two quantities have the equivalent physical nature and are measured in the same units. Based upon this axiom, an equation must at least be dimensionally homogeneous (have the same dimensions on both sides of the equality) if it is to be generally valid. However, dimensional homogeneity does not guarantee that an equation is valid, nor does an equation with differing dimensions on both sides of the equality have necessarily to be invalid. Many empirical equations of this type (nonhomogeneous) exist either for convenience or to express experimental data over a limited range of the variables involved.

We have already used the principle of dimensional consistency when we established the physical units of N and slugs. In each case we express the symbols in an equation in terms of mass, force, length, and time (at times, temperature too) and using the method of algebraic cancellation of units, establish the unit or the conversion factor needed to obtained the desired unit.

ILLUSTRATIVE PROBLEM 1.7

As we will see in Chapter 2, the pressure at the base of a uniform column of liquid of density ρ kg/m³ is

$$p = \rho g h$$

where h is the height in metres. What is the corresponding unit for p, pressure?

Solution

Applying the dimensions of each unit,

$$p = \rho\left[\frac{\text{kg}}{\cancel{\text{m}^3}\,{}^{1}}\right] \times g\left[\frac{\cancel{\text{m}}}{\text{s}^2}\right] \times h\,(\cancel{\text{m}})$$

$$= \frac{\text{kg}}{\text{m}\cdot\text{s}^2}$$

If we note that pressure is force divided by area (newtons per metre²), $p = \text{N/m}^2$, where N is the unit of force. However, force,

$$F = m \cdot a = \frac{\text{kg}\cdot\text{m}}{\text{s}^2}$$

Therefore,

$$p = \frac{\text{kg}\cdot\cancel{\text{m}}}{\text{s}^2/\cancel{\text{m}^2}\,{}^{1}} = \frac{\text{kg}}{\text{m}\cdot\text{s}^2}$$

Inserting this into the equation for pressure,

$$\frac{\text{kg}}{\text{m}\cdot\text{s}^2} = \frac{\text{kg}}{\text{m}\cdot\text{s}^2}$$

and we find that the equation is dimensionally consistent. At this time the student should note that the unit of pressure, N/m^2, is defined to be the pascal (Pa).

ILLUSTRATIVE PROBLEM 1.8

Solve Illustrative Problem 1.7 in the English system of units. Assume that the desired pressure unit is lb/in.^2 (psi).

Solution

$$p = \rho g h$$

where ρ is in slugs/ft^3, g in ft/sec^2, and h in ft. Therefore,

$$p = \rho\left(\frac{\text{slugs}}{\cancel{\text{ft}^3}\,{}^{1}}\right) \times g\left(\frac{\cancel{\text{ft}}}{\text{sec}^2}\right) \times h\,(\cancel{\text{ft}})$$

$$= \frac{\text{slug}}{\text{ft}\cdot\text{sec}^2}$$

But we have already seen that the slug has the units lb·sec²/ft (Illustrative Problem 1.5). Using this value, we have

$$p = \frac{\text{slug}}{\text{ft}\cdot\text{sec}^2} = \frac{\text{lb}\cdot\text{sec}^2}{\text{ft}\cdot\text{ft}\cdot\text{sec}^2} = \frac{\text{lb}}{\text{ft}^2}$$

The unit for pressure that we obtain is lb/ft² (psf). Since we want to have as our pressure unit, lb/in.², we need to introduce the following conversion:

$$p\ (\text{psi}) = p\left(\frac{\text{lb}}{\text{ft}^2}\right) \times \frac{1}{(12\ \text{in./ft})^2} = p\left(\frac{\text{lb}}{\text{ft}^2}\right) \times \frac{1}{(12)^2\ \text{in.}^2/\text{ft}^2}$$

or

$$p\ (\text{psi}) = p\ (\text{psf}) \times \frac{1}{144}$$

This conversion is a very common conversion in the English system and will appear often in later chapters.

ILLUSTRATIVE PROBLEM 1.9

What net force is required to accelerate a mass of 15 kg at the rate of 30 m/s²?

Solution

Using Newton's second law of motion,

$$F = ma$$
$$= 15\ \text{kg} \times 30\ \frac{\text{m}}{\text{s}^2} = 450\ \frac{\text{kg}\cdot\text{m}}{\text{s}^2}$$

But the unit of force in the SI system is the newton, which dimensionally has the units kg·m/s² (see Illustrative Problem 1.2 and Table 1.3). Therefore,

$$F = 450\ \text{N}$$

1.5 CLOSURE

Although the student may have seen some of the material in previous courses, it is of the utmost importance that it be fully understood and mastered before proceeding. Be sure of all the units used in any equation and be equally sure of the meaning of all terms in the equation. The use of dimensional consistency is the key to obtaining meaningful results, not only in fluid mechanics but in all branches of technology. It should become a habit to put the dimensions into an equation to ensure that the desired result is obtained. The problems at the end of each chapter will help you to develop this habit.

PROBLEMS

1.1 A box has the dimensions $12 \times 12 \times 6$ in. How many cubic metres are there in this box?

1.2 A man runs a 1500-m race, sometimes called the "metric mile." What part of a mile did he run? (1 mi = 5280 ft.)

1.3 A car is operated at a speed of 50 mi/hr. How many feet per second does this correspond to?

1.4 What velocity in feet per second does 10 m/s correspond to?

1.5 Derive the multiplier in Table 1.6 to go from feet to metres.

1.6 Derive the multiplier in Table 1.6 to go from square feet to square metres.

1.7 Derive the multiplier in Table 1.6 to go from cubic feet to cubic metres.

1.8 A car travels at 64 km/h. How many miles per hour does this correspond to?

1.9 A barrel of oil contains 55 gal of oil. If there are 231 in.^{3} in a gallon, how many cubic metres are there in 10 barrels of oil?

1.10 A litre is a volume measurement that is equal to 1000 cm^3 (10^{-3} m^3). How many litres are there in 1 cubic foot?

1.11 Using the data of Problem 1.10 and the fact that 1 gallon is equal to 231 in.^{3}, determine the number of litres per gallon.

1.12 Table 1.6 gives the multiplier of cubic metres per second to be 2.119×10^3 to obtain cubic feet per minute. Derive this multiplier.

1.13 One gigapascal (GPa) is (a) 10^6 Pa; (b) 10^8 Pa; (c) 10^3 Pa; (d) 10^9 Pa; (e) none of these.

1.14 One megapascal (MPa) is (a) 10^5 Pa; (b) 10^4 Pa; (c) 10^6 Pa; (d) 10^7 Pa; (e) none of these.

1.15 A mass of 5 kg is placed on a planet whose gravitational force is 10 times that of the earth. What is its weight on this planet?

1.16 A mass of 10 slugs weighs 320 lb on a spring balance. The acceleration of gravity at this location is (a) 32.7 ft/sec^2; (b) 0.03125 ft/sec^2; (c) 320 ft/sec^2; (d) 32.0 ft/sec^2; (e) none of these.

1.17 A mass of 10 kg weights 90 N. The acceleration of gravity at this location is (a) $\frac{1}{9}$ m/s^2; (b) 9.0 m/s^2; (c) 90 m/s^2; (d) 10 m/s^2; (e) none of these.

1.18 A weight of 160.85 lb corresponds to a mass of (a) 0.2 slug; (b) 0.5 slug; (c) 7.5 slug; (d) 0.75 slug; (e) none of these.

1.19 What is the weight in newtons of a body whose mass is 12 kg?

1.20 What is the weight in newtons of a body whose mass is 8 slugs?

1.21 If a body has a mass of 200 g, what is its weight in newtons?

1.22 A body weighs 150 lb. What is its mass in slugs?

1.23 If a body has a mass of 25 slugs, it will weigh how many newtons?

1.24 At a location on earth where $g = 32.2$ ft/sec^2, a body weighs 161 lb. At another location where $g = 32.0$, the same body will weigh (a) 161 lb; (b) 162 lb; (c) 160 lb; (d) 163 lb; (e) none of these.

1.25 What is the weight of 10 kg at a location on earth where $g = 9.7$ m/s^2?

1.26 If the body of Problem 1.25 is moved to another location on earth where $g = 10$ m/s^2, what will its mass be?

1.27 A mass of 100 kg is hung from a spring in a local gravitational field where $g = 9.806$ m/s^2 and the spring is found to deflect 25 mm. If the same mass is taken to a planet where $g = 5.412$ m/s^2, how much will the spring deflect if its deflection is directly proportional to the force applied?

1.28 A balance scale is used to weigh a sample on the moon. If the value of g is one-sixth of earth's gravity and the "standard" weights (earth weights) add up to 20 lb, what is the mass of the body?

1.29 A balance scale is used to weigh a sample on the moon. If the value of g is one-sixth of earth's gravity and the "standard" weights (earth weights) add up to 100 N, what is the mass of the body?

1.30 Solve Problem 1.28 if a spring balance is used that was calibrated on earth and reads 20 lb.

1.31 Solve Problem 1.29 if a spring balance is used that was calibrated on earth and reads 100 N.

1.32 A net force of 10 N acts on a body whose mass is 0.5 kg. What is the acceleration of the body?

1.33 A net force of 25 lb acts on a body whose mass is 5 slugs. What is the acceleration of the body?

1.34 A chemist measures 200 cm^3. How many cubic metres is this?

1.35 A common measure of the quantity of matter used by pharmacists is the grain. There are 15 grains in 1 gram. How many milligrams are there in 5 grains of aspirin?

1.36 A physician prescribes a pill containing 670 mg of a medication. If the pharmacist has 5-grain, 10-grain, 15-grain pills, which one should she dispense? (See Problem 1.35.)

REFERENCES

1. *AISI Metric Practice Guide—SI Units and Conversion Factors for the Steel Industry*, American Iron and Steel Institute, Washington, D.C., 1975.
2. *ASME Orientation and Guide for Use of SI (Metric) Units*, 5th ed., American Society of Mechanical Engineers, New York, 1974.
3. *ASME Text Booklet—SI Units in Strength of Materials*, American Society of Mechanical Engineers, New York, 1975.
4. *Fundamentals of Fluid Mechanics* by J. A. SULLIVAN, Reston Publishing Co., Reston, Va, 1978.
5. *Applied Fluid Mechanics* by R. L. MOTT, 2nd ed., Charles E. Merrill Publishing Co., Columbus, Ohio, 1979.
6. *Fluid Mechanics* by V. L. STREETER and E. B. WYLIE, 7th ed., McGraw-Hill Book Company, New York, 1979.

Fluid Properties

2.1 INTRODUCTION

In Section 1.1, we decided that the general term "fluid" would be used to refer to both liquids and gases. To distinguish between a liquid and a gas, we also noted that although both will occupy the container in which they are placed, a liquid presents a free surface if it does not completely fill the container. A gas will always fill the volume of the container in which it is placed. For gases it is important to take into account the change in volume that occurs when either the pressure or temperature is changed, whereas in most cases it is possible to neglect the change in volume of a liquid when there is a change in pressure.

As a general concept applicable to all situations, we can simply define a *system* as a grouping of matter taken in any convenient or arbitrary manner. For the present we simply note that systems can have energy stored or transferred to or from them, and although we are at liberty to choose a system in an arbitrary manner, all forces and energies must be accounted for when deriving the equations governing the system and its motion.

Let us consider a given system and then ask ourselves how we can distinguish changes that occur to it. To answer this question, it is necessary for us to identify those *external characteristics* that enable us to both distinguish and evaluate system changes. Conversely, if the external characteristics of a system do not change, we should be able to infer that the system has not undergone any change. Some of the external characteristics that can be used to describe a system are *temperature*, *pressure*, and *volume*. These observable external characteristics of a system are called *properties*, and we note that *when all the properties of a system are reproduced at different times and we are unable to distinguish any difference in the system, the system is in the same state at both times.* For the remaining portions of this chapter, we consider those fluid properties that play principal roles in our study of fluid mechanics.

2.2 TEMPERATURE

The *temperature* of a system is a measure of the random motion of the molecules of the system. If there are different temperatures within the body (or bodies composing the system), the question arises as to how the temperature at a given location is measured and how this measurement is interpreted. Let us examine this question, since similar questions will also have to be considered when other properties of a system are studied. In air at room pressure and temperature there are approximately 2.7×10^{19} molecules per cubic centimeter. If we divide the cube whose dimensions are 1 cm on a side into smaller cubes each of whose sides is one thousandth of a centimeter, there will be about 2.7×10^{10} molecules in each of the smaller cubes, which is still an extraordinarily large number. Although we speak of temperature at a point, we really mean the average temperature of the molecules in the neighborhood of the point.

When measuring the temperature of a body with a thermometer, it should be noted that the thermometer measures only the temperature of its sensing head. For the thermometer to be able to measure the temperature of a system, it is necessary for both the thermometer and the system to be in thermal equilibrium. The measurement of temperature usually employs the measurement of some secondary characteristic (such as the height of a liquid in a bulb thermometer) when the system changes from one state to another. Other examples of physical effects that have been used to indicate temperature are the linear expansion of liquids and solids, the pressure change of a confined gas, the volume change of bodies, and the generation of voltages when dissimilar metals are placed in contact with each other. Although it is assumed that the student is familiar with the common temperature scales, a brief summary is given below.

The common scales of temperature are the Fahrenheit and the Celsius and are defined by using the ice point and the boiling point of water at atmospheric pressure. In the Celsius temperature scale, the interval between the ice point and the boiling point is divided into 100 equal parts. In addition, as shown in Figure 2.1, the Celsius ice point is zero and the Fahrenheit ice

°F	°C	
212	100	Atmospheric boiling point
32	0	Ice-point
−460	−273	Absolute zero

FIGURE 2.1

point is 32. The conversion from one scale to the other is directly derived from Figure 2.1 and results in the following relations:

$$°C = \tfrac{5}{9}(°F - 32) \tag{2.1}$$

$$°F = \tfrac{9}{5}(°C) + 32 \tag{2.2}$$

The ability to extrapolate to temperatures below the ice point and above the boiling point of water and to interpolate in these regions is provided by the *International Scale of Temperature*. This agreed-upon standard utilizes the boiling and melting points of different elements and establishes suitable interpolation formulas in the various temperature ranges between these elements. The data for these elements are given in Table 2.1.

TABLE 2.1

Element	*Melting or Boiling Point at 1 atm*	*Temperature*	
		°C	*°F*
Oxygen	Boiling	−182.97	−297.35
Sulfur	Boiling	444.60	832.28
Antimony	Melting	630.50	1166.90
Silver	Melting	960.8	1761.4
Gold	Melting	1063.0	1945.4
Water	Melting	0	32
	Boiling	100	212
Hydrogen	Boiling	−252.7	−422.86
Helium	Boiling	−268.9	−452.02
Nitrogen	Boiling	−195.8	−320.44

ILLUSTRATIVE PROBLEM 2.1

Determine the temperature at which the same value is indicated on both Fahrenheit and Celsius thermometers.

Solution

Using equation (2.1) and letting °C = °F, we obtain

$$°\text{F} = \tfrac{5}{9}(°\text{F} - 32)$$
$$\tfrac{4}{9}°\text{F} = -\tfrac{160}{9}$$
$$°\text{F} = -40$$

Therefore, both Celsius and Fahrenheit temperature scales indicate the same temperature at −40°.

By using the results of Illustrative Problem 2.1, it is possible to derive an alternative set of equations to convert from the Fahrenheit to the Celsius temperature scale. When this is done, we obtain

$$°\text{F} = \tfrac{9}{5}(40 + °\text{C}) - 40 \tag{2.3}$$
$$°\text{C} = \tfrac{5}{9}(40 + °\text{F}) - 40 \tag{2.4}$$

The symmetry of equations (2.3) and (2.4) makes them relatively easy to remember and use.

2.3 ABSOLUTE TEMPERATURE

Let us consider the case of a gas that is confined in a cylinder with a constant cross-sectional area by a piston that is free to move. If heat is now removed from the system, the piston will move down, but because of its weight, it will maintain a constant pressure on the gas. This procedure can be carried out for several gases, and if volume is plotted as a function of temperature, we shall obtain a family of straight lines that intersect at zero volume (Figure 2.2a). All of these lines intersect at one temperature. This unique temperature is known as the *absolute zero* temperature, and the accepted values on the Fahrenheit and Celsius temperature scales are −459.69° and −273.16°, respectively, with the values −460° and −273° used for most engineering calculations. Using these temperatures as being the zero of temperature, we arrive at two absolute temperature scales that are defined as:

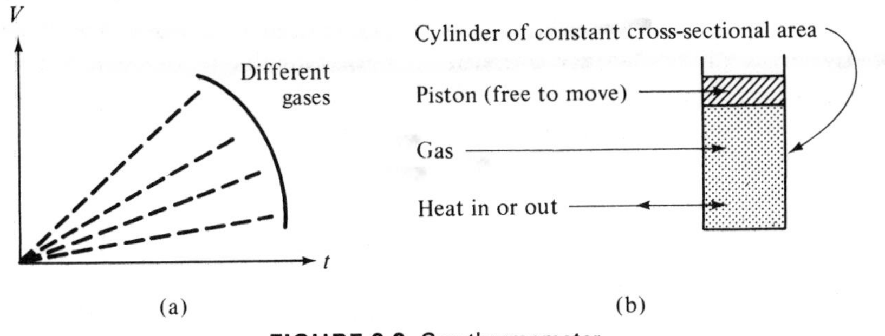

FIGURE 2.2 Gas thermometer.

$$\text{degrees Rankine } (°\text{R}) = °\text{F} + 460 \tag{2.5}$$

$$\text{degrees Kelvin (K)} = °\text{C} + 273 \tag{2.6}$$

It is also possible to define the absolute temperature scale independent of the properties of any substance, and the interested student is referred to texts on thermodynamics for this development.

ILLUSTRATIVE PROBLEM 2.2

Convert 68°F to °R and K.

Solution

To use equations (2.5) and (2.6), we need to know the given temperature of 68°F in °C. We obtain this by using equation (2.1):

$$°\text{C} = \tfrac{5}{9}(\text{F} - 32) = \tfrac{5}{9}(68 - 32) = 20$$

Therefore,

$$°\text{R} = 68 + 460 = 528$$

$$\text{K} = 20 + 273 = 293$$

As a check, note that $528/293 = 1.8$, which is $\frac{9}{5}$.

2.4 PRESSURE

When a gas is confined in a container, molecules of the gas strike the sides of the container, and these collisions with the walls of the container cause the molecules of the gas to exert a force on the walls. Liquids behave in a similar fashion. Although there is a difference in the way a force is applied to a fluid and to a solid, we can define the term *pressure* to equal the force applied to the object divided by the area that is perpendicular to the force. Therefore,

$$p = \frac{F}{A} \tag{2.7}$$

It is sometimes necessary to define the pressure at a point. To do this mathematically, it is necessary to consider the area to be steadily shrinking. We now define pressure at a point to be the normal (perpendicular) force per unit area as the area shrinks to zero. Mathematically, this statement is given as

$$p = \left(\frac{F}{A}\right)_{\lim A \to 0} \tag{2.8}$$

In common English engineering, pressure is expressed in units of pounds force per square inch (psi) or pounds force per square foot (psf). In SI units, pressure is expressed as newtons per square metre (N/m^2) or pascals (Pa). One important observation must be made at this time: *When a fluid is at rest, the pressure at any boundary exerted by the fluid* (*and on the fluid*) *will be perpendicular to the boundary.*

Most mechanical pressure gages read zero when open to the atmosphere and, therefore, measure the pressure of fluids *above* local atmospheric pressure. This pressure is called *gage pressure* and it is commonly designated in the English system as psig or psfg. *Absolute pressure* is the pressure in a system that would be measured from a base of a perfect vacuum at zero absolute pressure. Absolute pressure is designated as psia or psfa. Figure 2.3 shows the relationship between gage pressure and absolute pressure.

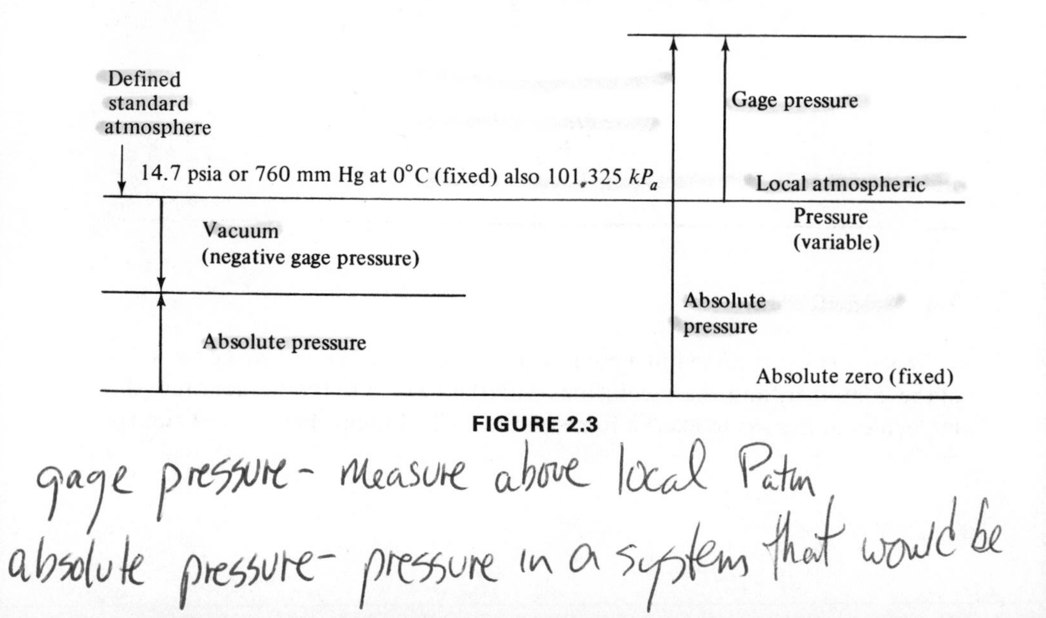

FIGURE 2.3

The relation of gage pressure to absolute pressure is readily written as:

$$\text{Absolute pressure } (p_a) = \text{Gage pressure } (p_g) + \text{Atmospheric pressure } (p_{atm}) \qquad (2.9)$$

ILLUSTRATIVE PROBLEM 2.3

A pressure gage reads 20 psig when the local atmospheric pressure is 14.5 psi. What is the absolute pressure corresponding to this reading?

Solution

From Figure 2.3 or equation (2.9),

$$p_a = 20 + 14.5 = 34.5 \text{ psia}$$

In SI units the equivalent of 14.696 (14.7) psia is 101.325 kPa. All pressures in the SI systems are absolute and the terms "gage" and "vacuum" are not used. A unit that is sometimes used in conjunction with the SI system is the *bar*. The bar is equal to 10^5 Pa. Although use of the bar does reduce the powers of 10 in a calculation, it is not a recommended unit.

When pressure is measured below atmospheric pressure, it is commonly called *vacuum*. By referring to Figure 2.3, we obtain the relation between vacuum and absolute pressure as

$$\text{Absolute pressure } (p_a) = \text{Atmospheric pressure } (p_{atm}) - \text{Vacuum } (p_g) \qquad (2.10)$$

ILLUSTRATIVE PROBLEM 2.4

If the atmospheric pressure is 14.7 psia and a vacuum gage reads a vacuum of 13.0 psig, what is the absolute pressure?

Solution

From Figure 2.3 or equation (2.10), we obtain

$$p_a = 14.7 - 13.0 = 1.7 \text{ psia}$$

The *Bourdon gage* is the device most commonly used to measure pressure commercially. This type of gage is shown in Figure 2.4 and it typically consists of a small-volume oval tube that is fixed at one end and is free at the other end to allow displacement under the deforming action of the pressure difference across the tube walls. In the most common model shown in Figure 2.4, a tube with an oval cross section is bent in a circular arc. Under pressure the tube tends to become circular, with a subsequent increase in the radius of the arc. By an almost frictionless linkage, the free end of the tube rotates a pointer over a calibrated scale to give a mechanical indication of pressure.

The reference pressure in the case containing the Bourdon tube is usually atmospheric, so that the pointer indicates gage pressures. Absolute pressures can be measured directly without evacuating the complete gage casing by biasing a sensing Bourdon tube against a reference Bourdon tube that is evacuated and sealed as shown in Figure 2.5. Bourdon gages are available for

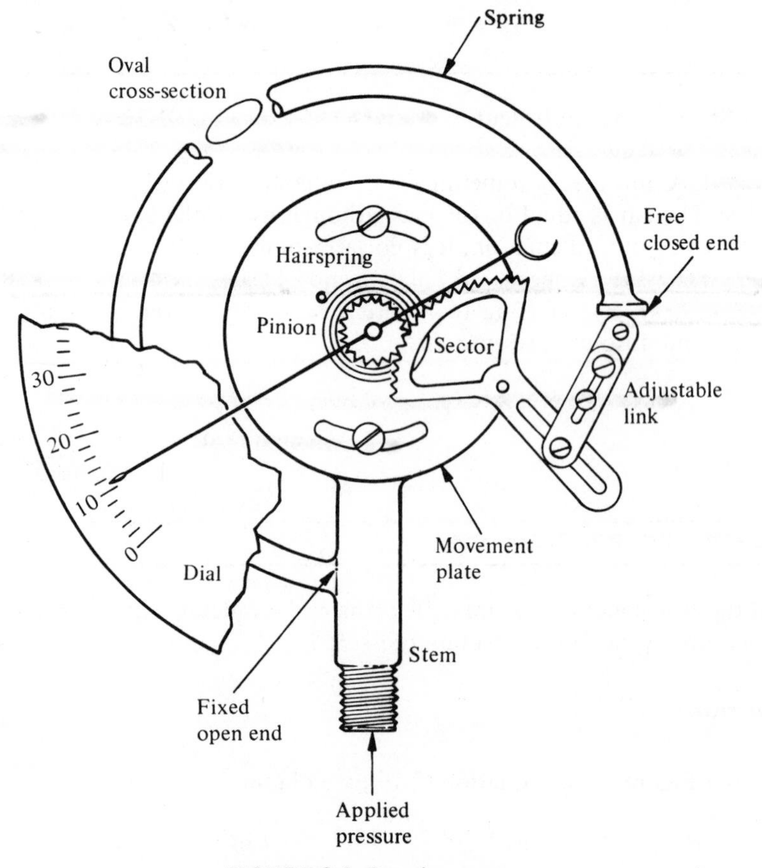

FIGURE 2.4 Bourdon gage.

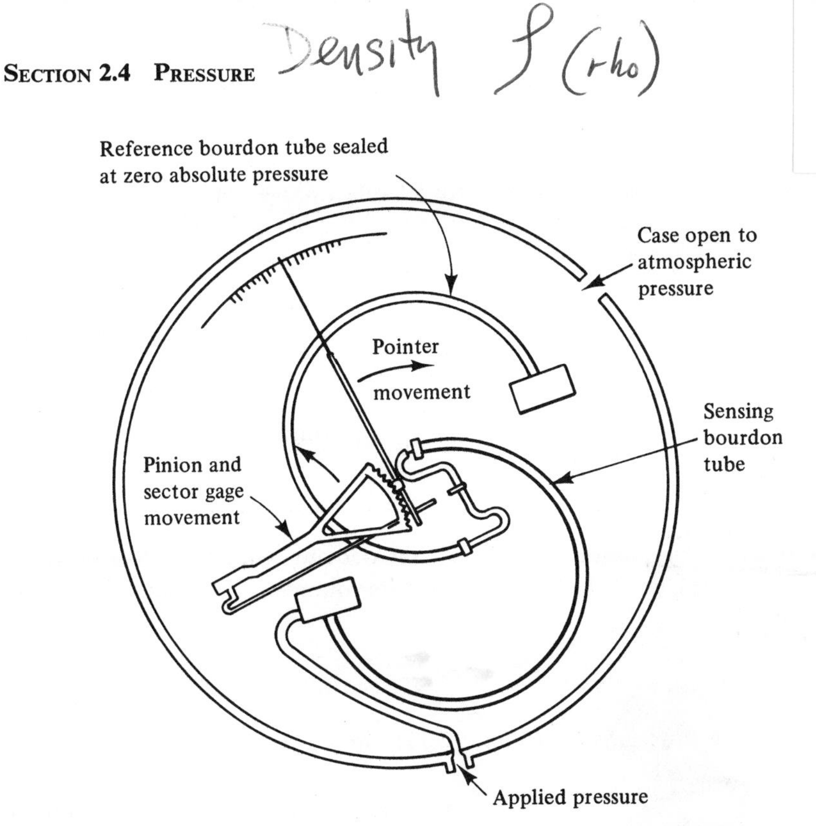

FIGURE 2.5 Bourdon gage for absolute pressure measurement.

a wide range of absolute, gage, and differential pressure measurements within a calibration uncertainty of 0.1% of the reading.

Before continuing our discussion of pressure, it is necessary for us to introduce four new terms: density, specific weight, specific volume, and specific gravity.

Density is defined as the mass per unit volume of a substance. If we denote the mass of a body as m and its volume as V, the density ρ (rho) is

$$\rho = \frac{m}{V} \tag{2.11}$$

In English units, where m is in slugs and V is in cubic feet, ρ has the units of slugs per cubic foot. In SI units, where m is in kg and V is in m^3, ρ has the units kg/m^3.

ILLUSTRATIVE PROBLEM 2.5

Convert 1 kg/m^3 to $slugs/ft^3$.

Solution

Using the methods of Chapter 1 and Table 1.6, we obtain

$$1\,\frac{\text{kg}}{\text{m}^3} \times 9.81\,\frac{\text{N}}{\text{kg}} \times 2.248 \times 10^{-1}\,\frac{\text{lb}}{\text{N}} \times \frac{1}{(3.281\ \text{ft/m})^3} \times \frac{1}{32.17\ \text{lb/slug}}$$

$$= 1.941 \times 10^{-3}\,\frac{\text{slug}}{\text{ft}^3}$$

Thus,

$$\frac{\text{kg}}{\text{m}^3} \times 1.941 \times 10^{-3} = \frac{\text{slug}}{\text{ft}^3}$$

Specific weight is defined as the weight per unit volume of a substance. Denoting the weight of the body as w and its volume as V, the specific weight γ (gamma) is

$$\gamma = \frac{w}{V} \tag{2.12}$$

If we compare equations (2.10) and (2.11), we find that

$$\frac{\gamma}{\rho} = \frac{w}{m} \tag{2.13}$$

But we have from equation (1.5) that $w/m = g$. Therefore,

$$\frac{\gamma}{\rho} = g \qquad \text{or} \qquad \gamma = \rho g \tag{2.14}$$

In English units w is in pounds, V is in cubic feet, and γ has the units pounds force per cubic feet. In SI units, w is in newtons, V is in m^3, and γ has the units N/m^3.

ILLUSTRATIVE PROBLEM 2.6

A cube has 6-in. sides. If the cube weighs 20 lb, what is its density and specific weight in both English and SI units?

Solution

The volume of the cube is

$$6\ \text{in.} \times 6\ \text{in.} \times 6\ \text{in.} = 216\ \text{in.}^3 = \frac{216\ \text{in.}^3}{(12\ \text{in./ft})^3} = 0.125\ \text{ft}^3$$

The specific weight,

$$\gamma = \frac{w}{v} = \frac{20 \text{ lb}}{0.125 \text{ ft}^3} = 160 \frac{\text{lb}}{\text{ft}^3}$$

Since $\gamma = \rho g$,

$$\rho = \frac{\gamma}{g} = \frac{160 \text{ lb/ft}^3}{32.17 \text{ ft/s}^2} = 4.97 \text{ slugs/ft}^3$$

In SI units,

$$V = 216 \text{ in.}^3 \times \left(0.0254 \frac{\text{m}}{\text{in.}}\right)^3 = 3.54 \times 10^{-3} \text{ m}^{-3}$$

The weight of 20 lb is

$$20 \text{ lb} \times 4.448 \frac{\text{N}}{\text{lb}} = 88.96 \text{ N} \qquad \text{(where 4.448 is from Table 1.6)}$$

Therefore,

$$\gamma = \frac{w}{V} = \frac{88.96 \text{ N}}{354 \times 10^{-3} \text{ m}^3} = 25\ 130 \text{ N/m}^3$$

Since $\rho = \gamma/g$,

$$\frac{25\ 130 \text{ N/m}^3}{9.81 \text{ N/kg}} = 2561.7 \text{ kg/m}^3$$

(As a check, using the results of Illustrative Problem 2.5, we obtain

$$2561.7 \frac{\text{kg}}{\text{m}^3} \times 1.941 \times 10^{-3} = 4.97 \text{ slugs/ft}^3)$$

Specific volume is defined as the reciprocal of specific weight or the volume per unit weight. Thus,

$$v = \frac{1}{\gamma} \tag{2.15}$$

and the units of specific weight are ft³/lb or m³/N.

The last of these four terms, *specific gravity*, is defined as the ratio of the specific weight or density of a substance to the specific weight or density of water at 4°C (39.2°F). From this definition we have

$$\text{Specific gravity (sg)} = \frac{\gamma_{\text{substance}}}{\gamma_{\text{water}} \text{ at } 4°\text{C}} = \frac{\rho_{\text{substance}}}{\rho_{\text{water}} \text{ at } 4°\text{C}} \tag{2.16}$$

For water at 4°C,

$$\gamma_w = 62.4 \frac{\text{lb}}{\text{ft}^3} = 9.81 \text{ kN/m}^3$$

$$\rho_w = 1.94 \frac{\text{slugs}}{\text{ft}^3} = 1000 \text{ kg/m}^3$$

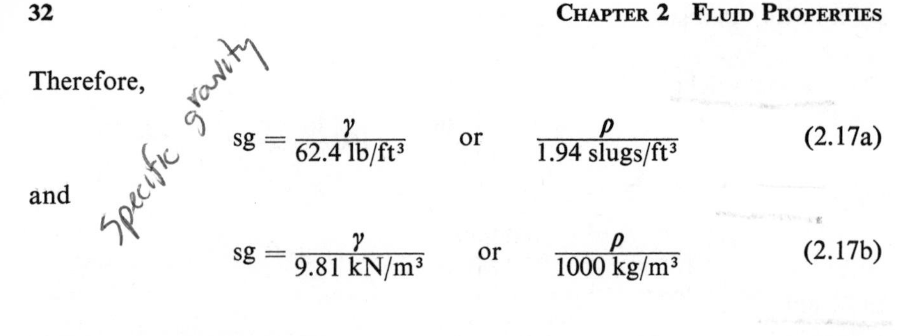

Therefore,

$$\text{sg} = \frac{\gamma}{62.4 \text{ lb/ft}^3} \quad \text{or} \quad \frac{\rho}{1.94 \text{ slugs/ft}^3} \tag{2.17a}$$

and

$$\text{sg} = \frac{\gamma}{9.81 \text{ kN/m}^3} \quad \text{or} \quad \frac{\rho}{1000 \text{ kg/m}^3} \tag{2.17b}$$

ILLUSTRATIVE PROBLEM 2.7

One gallon of an unknown substance weighs 7 lb. What is its specific gravity?

Solution

One gallon is 231 in.³ or

$$\frac{231 \text{ in.}^3}{1728 \text{ in.}^3/\text{ft}^3} = 0.1337 \text{ ft}^3$$

At 4°C, the weight of 1 gal of water is

$$62.4 \frac{\text{lb}}{\text{ft}^3} \times 0.1337 \text{ ft}^3 = 8.343 \text{ lb}$$

Therefore, the specific gravity, by definition, is

$$\text{sg} = \frac{7}{8.343} = 0.839$$

ILLUSTRATIVE PROBLEM 2.8

If 10 kg of a substance occupies a volume of 2×10^{-2} m³, what is its specific gravity?

Solution

The density of the substance is

$$\frac{10 \text{ kg}}{2 \times 10^{-2} \text{ m}^3} = 500 \text{ kg/m}^3$$

From equation (2.17),

$$sg = \frac{\rho}{1000} = \frac{500}{1000} = 0.5$$

Let us now return to the basic topic of this section—pressure—and address ourselves to the following problem. What will the pressure be at the base of a column of liquid due to the liquid if its specific weight is constant and equal to γ? Assume that the cross-sectional area is constant and equal to A and that the height of the fluid is h above the base of the column, as shown in Figure 2.6. From our definition of pressure as force divided by area, we must

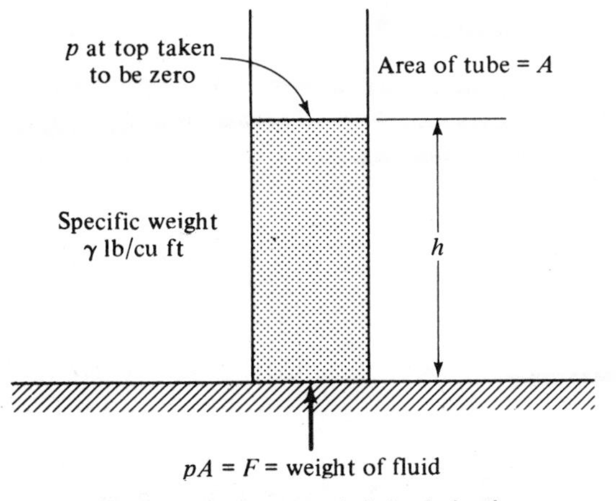

FIGURE 2.6 Pressure–height derivation.

first find the force being exerted on the base by the column of fluid. This force will be equal to the weight of the fluid, which is given by equation (2.12) as $w = \gamma V$. The volume of fluid is $V = Ah$ and the weight is $w = \gamma Ah$. This is the force exerted on the base. If we divide this force by the area of the base. we obtain the pressure:

$$p = \frac{\gamma Ah}{A} \tag{2.18}$$

$$p = \gamma h$$

Equation (2.18) will be used extensively in Chapter 3 and it should be thoroughly understood. Notice that as you go from the top of the column down, the pressure increases, or conversely as you go up from the bottom of the column the pressure decreases. These changes in pressure are directly proportional to the changes in elevation. Also note that we have assumed that γ, the specific weight, is constant. If distance is measured up from the base, equation (2.18) should be written as

$$p = -\gamma h \tag{2.18a}$$

The most common fluid used to measure pressure differences, atmospheric pressure, and vacuum pressure is mercury. At approximately room temperature, the density of mercury is very nearly 13.6 g/cm³. It will usually be assumed that the density of mercury is 13.6 g/cm³ unless otherwise stated. The effect of temperature on the specific weight of mercury is given in Appendix 2-1.

ILLUSTRATIVE PROBLEM 2.9

A glass tube open at the top has a 1-in. high column of mercury in it. If the density of mercury is 13.6 g/cm³ = 13 600 kg/m³, determine the pressure at the base of the column in psi and Pa.

Solution

The specific weight,

$$\gamma = \rho g = 13{,}600 \frac{\text{kg}}{\text{m}^3} \times 9.81 \frac{\text{N}}{\text{kg}} = 133{,}416 \text{ N/m}^3$$

Therefore,

$$p = \gamma h = 133{,}416 \frac{\text{N}}{\text{m}^3} \times 1 \text{ in.} \times 0.0254 \frac{\text{m}}{\text{in.}} = 3388.8 \frac{\text{N}}{\text{m}^2} = 3388.8 \text{ Pa}$$

In terms of psi,

$$3388.8 \frac{\text{N}}{\text{m}^2} \times 2.248 \times 10^{-1} \frac{\text{lb}}{\text{N}} \times \frac{1}{(3.281 \text{ ft/m})^2} \times \frac{1}{(12 \text{ in./ft})^2} = 0.491 \text{ psi}$$

Thus, from Illustrative Problem 2.9 a column height of 1 in. of mercury is equivalent to a pressure of 0.491 psi. Notice that this enables us to express pressures in terms of equivalent heights of fluids. We will find this of great utility in Chapter 3.

In vacuum work it is common to express the absolute pressure in a vacuum chamber in terms of millimeters of mercury. Thus, a vacuum may be expressed as 10^{-5} mm Hg. If this is expressed in pounds force per square inch, it would be equivalent to 0.000000193 psi, with the assumption that the density of mercury (for vacuum work) is 13.6 g/cm³. Recently, the term *torr* has entered the technical literature. A torr is defined as 1 mm Hg. Thus, 10^{-5} torr is the same as 10^{-5} mm Hg. Another unit of pressure used in vacuum work is the micron. A *micron* is defined as one thousandth of 1 mm Hg, so that 10^{-3} mm Hg is equal to 1 micron.

ILLUSTRATIVE PROBLEM 2.10

A mercury manometer (vacuum gage) reads 26.5 in. of vacuum when the local barometer reads 30.0 in. Hg at standard temperature. Determine the absolute pressure in psia.

Solution

$$p = (30.0 - 26.5) \text{ in. Hg vacuum}$$

and from the preceding problem,

$$p = (30 - 26.5)0.491 = 1.72 \text{ psia}$$

If the solution is desired in psfa, it is necessary to convert from square inches to square feet. Since there are 12 in. in 1 ft, there will be 144 in.^{2} in 1 ft^2. Thus, the conversion to psfa requires that psia be multiplied by 144. This conversion is often necessary to keep the dimensions of equations consistent in English units.

2.5 SURFACE TENSION

When a liquid has a free surface, there is a discontinuity that takes place at the surface since there is a liquid on one side of the surface and a different liquid and gas on the other side of the surface. For example, consider a drop of liquid in air. Within the drop, each molecule of liquid experiences forces due to the other liquid molecules. At the surface of the drop the liquid molecules experience forces due to both the air and the liquid molecules. If the drop is to maintain its shape, the attraction of the liquid molecules must exceed the attraction between the liquid molecules and air molecules. Effectively, this is the same as if the liquid were enclosed by a stretchable membrane under tension. The force of this *surface tension* is always tangent to the interface. A liquid is said to wet a surface in contact with it if the attraction of the molecules for the surface exceeds the attraction of the molecules for each other; on the other hand, the liquid is classified as a nonwetting liquid if the attraction of other liquid molecules for each other is greater than their attraction to the surface.

Quantitatively, surface tension can be defined as the force of molecular attraction per unit length of free surface. It should be noted that surface tension is a function of both the liquid and the surface in contact with the liquid. Denoting surface tension by σ, we have, in general,

$$\sigma = \frac{F}{L} \tag{2.19}$$

Since the forces of attraction between molecules decreases with increasing temperature, it is found that surface tension also decreases with increasing temperature. Table 2.2 gives the surface tension of some fluids in contact with air at 68°F.

TABLE 2.2 Surface Tension of Common Liquids in Contact with Air at 68°F†‡

	Surface Tension, σ	
	lb/ft	*N/m*
Alcohol, ethyl	0.00153	2.23×10^{-2}
Benzene	0.00198	2.89×10^{-2}
Carbon tetrachloride	0.00183	2.67×10^{-2}
Kerosene	0.0016–0.0022	$2.34\text{–}3.21 \times 10^{-2}$
Water	0.00498	7.27×10^{-2}
Mercury		
In air	0.0352	5.14×10^{-1}
In water	0.0269	3.93×10^{-1}
In vacuum	0.0333	4.86×10^{-1}
Oil		
Lubricating	0.0024–0.0026	$3.50\text{–}3.79 \times 10^{-2}$
Crude	0.0016–0.0026	$2.34\text{–}3.79 \times 10^{-2}$

†Reproduced with permission from *Fluid Mechanics* by V.L. Streeter, McGraw-Hill Book Company, New York, 1958, p. 14. SI units added.

‡See Tables B.6 and B.7 in Appendix B for other data on surface tension.

With the foregoing in mind, let us refer to Figure 2.7, which shows the contact angle relations for the solid–liquid–vapor interfaces for both wetting and nonwetting liquids. The subscripts *vs*, *vl*, and *ls* denote, respectively, the vapor–solid interface, the vapor–liquid interface, and the liquid–solid interface. For the liquid to be in equilibrium with the solid surface, the surface tension forces at the solid–liquid–vapor interfaces must be in balance parallel to the solid surface. Therefore,

$$\sigma_{vs} = \sigma_{ls} + \sigma_{vl} \cos\theta \tag{2.20}$$

where θ is the contact angle between the fluid and the surface. When $(\sigma_{vs} -$

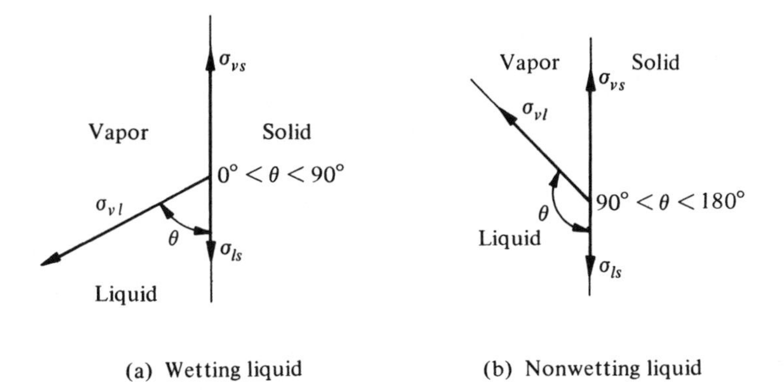

FIGURE 2.7 Contact angles.

$\sigma_{ls})/\sigma_{vl}$ is between zero and unity, the contact angle lies between 0 and 90° and the liquid is called a *wetting* liquid. If $(\sigma_{vs} - \sigma_{ls})/\sigma_{vl}$ lies between 0 and -1, the contact angle will be between 90 and 180° and the liquid is termed *nonwetting*. The magnitude of the contact angle θ is dependent on the three surface tension forces.

Let us calculate the rise h shown in Figure 2.8a for the case of a wetting fluid in a circular tube. The basic consideration here is that the weight of the column of height h must be supported by an equal force in the vertical direction, where the source of this supporting force must be the vertical component of the surface tension. Therefore, the force down (equal to the weight of the column of height h) is $\gamma h(\pi d^2/4)$, where γ is the specific weight of the fluid and d is the inside diameter of the tube. The total force due to the surface tension is the surface tension multiplied by the perimeter of the tube: $\sigma\pi d$. However, since we only desire the vertical component of this force, it is

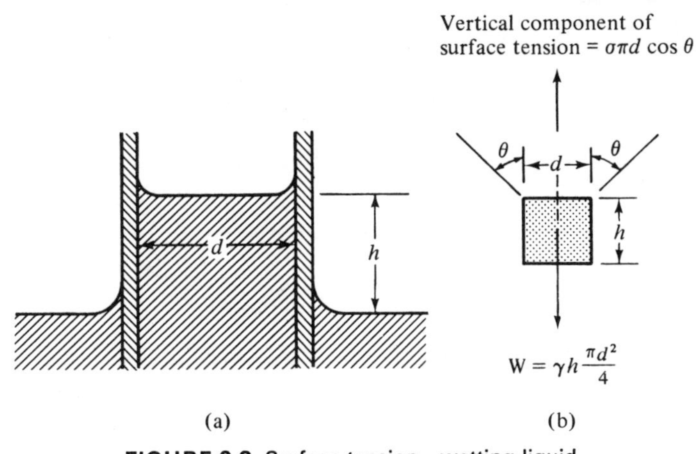

FIGURE 2.8 Surface tension—wetting liquid.

necessary to multiply the total force by $\cos\theta$. Equating the two vertical forces,

$$\gamma h \frac{\pi d^2}{4} = \sigma \pi d \cos\theta$$

or

$$h = \frac{4\sigma\cos\theta}{\gamma d} = \frac{4\sigma\cos\theta}{\rho g d} \tag{2.21}$$

Water in contact with glass and air has a contact angle whose value is essentially zero, while mercury in contact with glass and air has a contact angle of 129°.

If we consider two infinite parallel plates instead of a circular tube, the equation of equilibrium becomes (for a unit depth)

$$\gamma h d = 2\sigma\cos\theta \tag{2.22}$$

or

$$h = \frac{2\sigma\cos\theta}{\gamma d} = \frac{2\sigma\cos\theta}{\rho g d} \tag{2.22a}$$

Notice that the rise in the tube is twice the rise between the parallel plates.

ILLUSTRATIVE PROBLEM 2.11

A glass tube that has an inside diameter of 1.5×10^{-3} m is placed in a fluid whose contact angle is 10°. If the surface tension for this liquid in contact with glass is 7.3×10^{-2} N/m, determine the rise of the liquid in the tube above the normal liquid level. Assume that the density of the fluid is 800 kg/m³.

Solution

From equation (2.21), with $\gamma = \rho g = 800 \times 9.81$ N/m³,

$$h = \frac{4\sigma\cos\theta}{\gamma d} = \frac{4 \times 7.3 \times 10^{-2} \times \cos 10}{800 \times 9.81 \times 1.5 \times 10^{-3}} = 0.0244 \text{ m} = 2.44 \text{ cm} = 0.96 \text{ in.}$$

2.6 COMPRESSIBILITY

When a liquid is confined, it will offer resistance to any external agency that tends to change its shape. This resistance to change in shape is called the *compressibility* of the fluid, and the change in volume is measured as a function of the original volume and the applied pressure. Mathematically, the *modulus of elasticity* or *bulk modulus* of the fluid at a point is defined by

$$\beta = -V\left(\frac{\Delta p}{\Delta V}\right)_{\lim \Delta V \to 0} \tag{2.23}$$

where V is the original volume, ΔV is the change in volume, Δp is the change in pressure, and the notation $\lim \Delta V \longrightarrow 0$ indicates the value at a point as the volume decreases toward zero. The negative sign accounts for the fact that as the pressure increases, the volume decreases. It is usually assumed that the bulk modulus is insensitive to pressure but does vary with temperature. For water the bulk modulus is approximately 300,000 psi. Since steel has a modulus of elasticity of 30 million psi it is seen that water is relatively compressible. The physical properties of water are given in Tables 2.3 and 2.4 as a function of temperature. Figure 2.9 gives the variation of the bulk modulus of water with pressure and temperature.

TABLE 2.3 Physical Properties of Water in English Units†

Temp, °F	Specific Weight γ, lb/ft^3	Density ρ, slugs/ft^3	Viscosity μ, lb·s/ft^2 $10^5\mu=$	Kinematic Viscosity ν, ft^2/s $10^5\nu=$	Surface Tension σ, lb/ft $100\sigma=$	Vapor-pressure Head p_v/γ, ft	Bulk Modulus of Elasticity β lb/in^2 $10^{-3}\beta=$
32	62.42	1.940	3.746	1.931	0.518	0.20	293
40	62.43	1.940	3.229	1.664	0.514	0.28	294
50	62.41	1.940	2.735	1.410	0.509	0.41	305
60	62.37	1.938	2.359	1.217	0.504	0.59	311
70	62.30	1.936	2.050	1.059	0.500	0.84	320
80	62.22	1.934	1.799	0.930	0.492	1.17	322
90	62.11	1.931	1.595	0.826	0.486	1.61	323
100	62.00	1.927	1.424	0.739	0.480	2.19	327
110	61.86	1.923	1.284	0.667	0.473	2.95	331
120	61.71	1.918	1.168	0.609	0.465	3.91	333
130	61.55	1.913	1.069	0.558	0.460	5.13	334
140	61.38	1.908	0.981	0.514	0.454	6.67	330
150	61.20	1.902	0.905	0.476	0.447	8.58	328
160	61.00	1.896	0.838	0.442	0.441	10.95	326
170	60.80	1.890	0.780	0.413	0.433	13.83	322
180	60.58	1.883	0.726	0.385	0.426	17.33	313
190	60.36	1.876	0.678	0.362	0.419	21.55	313
200	60.12	1.868	0.637	0.341	0.412	26.59	308
212	59.83	1.860	0.593	0.319	0.404	33.90	300

†Reprinted from *Fluid Mechanics* by V. L. Streeter and E. B. Wylie, 7th ed., 1979, p. 534, by permission of McGraw-Hill Book Company, New York.

***TABLE 2.4 Physical Properties of Water in SI Units*†**

Temp, °C	Specific Weight γ, N/m^3	Density ρ, kg/m^3	Viscosity μ, $N \cdot s/m^2$ $10^3\mu=$	Kinematic Viscosity ν, m^2/s $10^6\nu=$	Surface Tension σ, N/m $100\sigma=$	Vapor-pressure Head p_v/γ, m	Bulk Modulus of Elasticity β N/m^2 $10^{-7}\beta=$
0	9805	999.9	1.792	1.792	7.62	0.06	204
5	9806	1000.0	1.519	1.519	7.54	0.09	206
10	9803	999.7	1.308	1.308	7.48	0.12	211
15	9798	999.1	1.140	1.141	7.41	0.17	214
20	9789	998.2	1.005	1.007	7.36	0.25	220
25	9779	997.1	0.894	0.897	7.26	0.33	222
30	9767	995.7	0.801	0.804	7.18	0.44	223
35	9752	994.1	0.723	0.727	7.10	0.58	224
40	9737	992.2	0.656	0.661	7.01	0.76	227
45	9720	990.2	0.599	0.605	6.92	0.98	229
50	9697	988.1	0.549	0.556	6.82	1.26	230
55	9679	985.7	0.506	0.513	6.74	1.61	231
60	9658	983.2	0.469	0.477	6.68	2.03	228
65	9635	980.6	0.436	0.444	6.58	2.56	226
70	9600	977.8	0.406	0.415	6.50	3.20	225
75	9589	974.9	0.380	0.390	6.40	3.96	223
80	9557	971.8	0.357	0.367	6.30	4.86	221
85	9529	968.6	0.336	0.347	6.20	5.93	217
90	9499	965.3	0.317	0.328	6.12	7.18	216
95	9469	961.9	0.299	0.311	6.02	8.62	211
100	9438	958.4	0.284	0.296	5.94	10.33	207

†Reprinted from *Fluid Mechanics* by V. L. Streeter and E. B. Wylie, 7th ed., 1979, p. 535, by permission of McGraw-Hill Book Company, New York.

ILLUSTRATIVE PROBLEM 2.12

Determine the percentage change in the volume of water if its pressure is increased by 30,000 psi.

Solution

From equation (2.23),

$$\beta = -V\left(\frac{\Delta p}{\Delta V}\right)$$

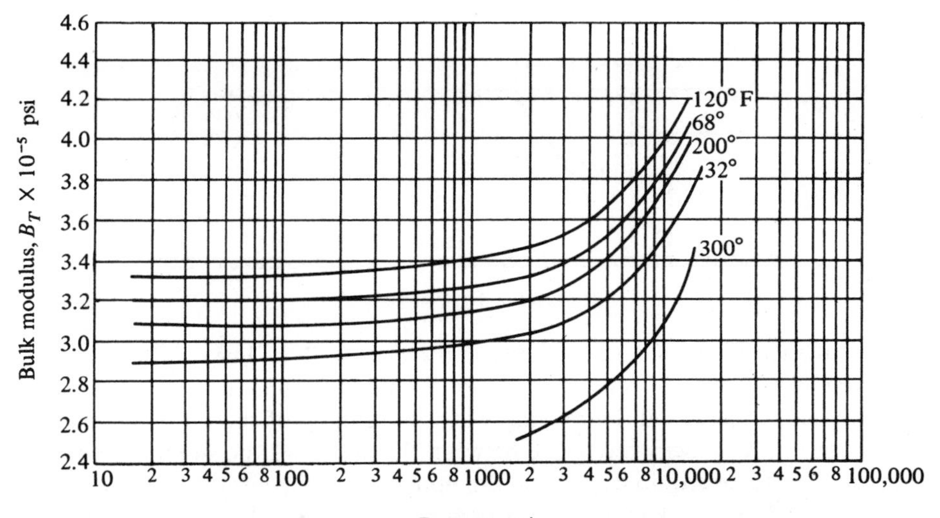

FIGURE 2.9 Variation of the bulk modulus of water with pressure and temperature. (Reproduced with permission from *Fluid Mechanics* by Arthur G. Hansen, John Wiley & Sons, Inc., New York, 1967, p. 487.)

or since the bulk modulus of water is 300,000 psi, the change in volume is $\Delta V/V$, or

$$\frac{\Delta V}{V} = -\left(\frac{\Delta p}{\beta}\right) = \frac{-30,000}{300,000} = -0.1 \quad \text{or} \quad -10\%$$

2.7 VISCOSITY

Because of the attractive forces between their molecules, fluids have the ability to resist forces tending to change the shape of a body of fluid. The stresses within an elemental volume of fluid are analogous to those within an elastic body. In an elastic body the stresses are proportional to the change of shape of the body (i.e., stress is proportional to strain). In the case of fluids, the attractive forces between molecules are weaker, so stress is assumed to be proportional to the time rate of change of shape with respect to distance; the constant of proportionality is called *viscosity*. This assumption was introduced by Newton, who set the force acting tangentially on a unit surface between adjacent layers of fluid (i.e., the stress) equal to a constant (i.e., viscosity) times the velocity gradient in a direction perpendicular to the layers.

Consider the situation shown in Figure 2.10. The upper plate is moving to the right with a velocity V relative to the lower plate. If the viscosity is con-

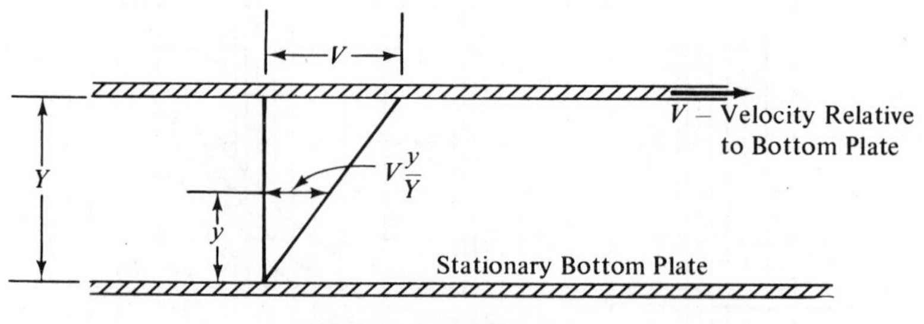

FIGURE 2.10 Viscosity.

stant, the variation of velocity between the plates will be linear. The particle in contact with the bottom plate will not be moving, while the particle touching the upper plate will be moving with the velocity of the upper plate, V. As shown, the velocity anywhere between the plates will be simply given by Vy/Y. The force required to move the upper plate at constant velocity relative to the bottom plate is proportional to the area of the plates, inversely proportional to the distance separating the plates, and directly proportional to the relative velocity of the plates. Thus,

$$F \propto \frac{V}{Y}A \qquad \text{or} \qquad \frac{F}{A} \propto \frac{V}{Y} \tag{2.24}$$

By definition, stress is force divided by area. In this case the stress is a shear stress in that the fluid acts as if each layer were composed of thin sheets tending to move (shear) with respect to each other. The shear stress, τ, is therefore

$$\tau \propto \frac{V}{Y} \tag{2.25}$$

Combining equations (2.24) and (2.25) and expressing the proportionality of equation (2.25) as an equality,

$$\frac{F}{A} = \tau = \mu\frac{V}{Y} \tag{2.26}$$

The proportionality constant μ is known as the *coefficient of viscosity*, *dynamic viscosity*, the *absolute viscosity*, or simply the *viscosity of a fluid*. If the viscosity varies with y due to temperature or other local conditions, μ can be written as

$$\mu = \tau\left(\frac{\Delta y}{\Delta V}\right)_{\lim \Delta V \to 0} \tag{2.27}$$

where μ may vary from location to location.

Since shear stress has the dimensions of force divided by area, and velocity has the dimensions of length per unit time, viscosity must have the dimension of force per unit area multiplied by time. The units of viscosity are therefore pound seconds per square foot, gram seconds per square centimeter, poundal seconds per square foot, dyne seconds per square centimeter, or dyne seconds per square foot. Of these units, dyne seconds per square centimeter is given the name *poise* and it is found that the viscosity of water at 68.4°F is 1 cP or one hundredth of 1 P. The viscosity of most liquids and gases is essentially independent of pressure except for very high pressures, but as temperature increases, the viscosity of all liquids decreases, while the viscosity of all gases increases, as can be seen from Figure 2.11. Table 2.5 gives some convenient conversion factors for viscosity. In SI units,

$$\text{P} \times 10^{-1} = \frac{\text{N} \cdot \text{s}}{\text{m}^2} \qquad \text{and} \qquad \text{cP} \times 10^{-3} = \frac{\text{N} \cdot \text{s}}{\text{m}^2}$$

TABLE 2.5

1 P	= 0.0672 poundal sec/ft^2
1 P	= 0.00209 lb sec/ft^2
1 poundal sec/ft^2	= 14.88 P
1 lb sec/ft^2	= 478.8 P
1 cm^2/s	= 0.001076 ft^2/sec
1 ft^2/sec	= 929 cm^2/s

ILLUSTRATIVE PROBLEM 2.13

Derive the conversion factor

$$1 \text{ P} = 0.00209 \frac{\text{lb sec}}{\text{ft}^2}$$

Solution

It is indeed unfortunate that this type of conversion mixes units. To solve a problem of this type, it is first necessary to define all terms. A poise is a dyne second per square centimeter and the dyne is given in the centimeter–gram–second system as a gram centimeter per second squared. Therefore,

$$1 \text{ P} = \frac{1 \text{ dyne s}}{\text{cm}^2} = \frac{1 \text{ dyne s}}{\text{cm}^2} \times \frac{1}{1 \text{ ft}^2/(12 \times 2.54)^2 \text{ cm}^2}$$

$$= \frac{1 \text{ dyne sec}}{\text{ft}^2}$$

$$1 \text{ dyne} = \frac{1 \text{ g cm}}{\text{s}^2} = \frac{1 \text{ g cm}}{\text{s}^2} \times \frac{1}{980 \text{ cm/s}^2} \times \frac{1}{454 \text{ g/lb}} = \text{lb}$$

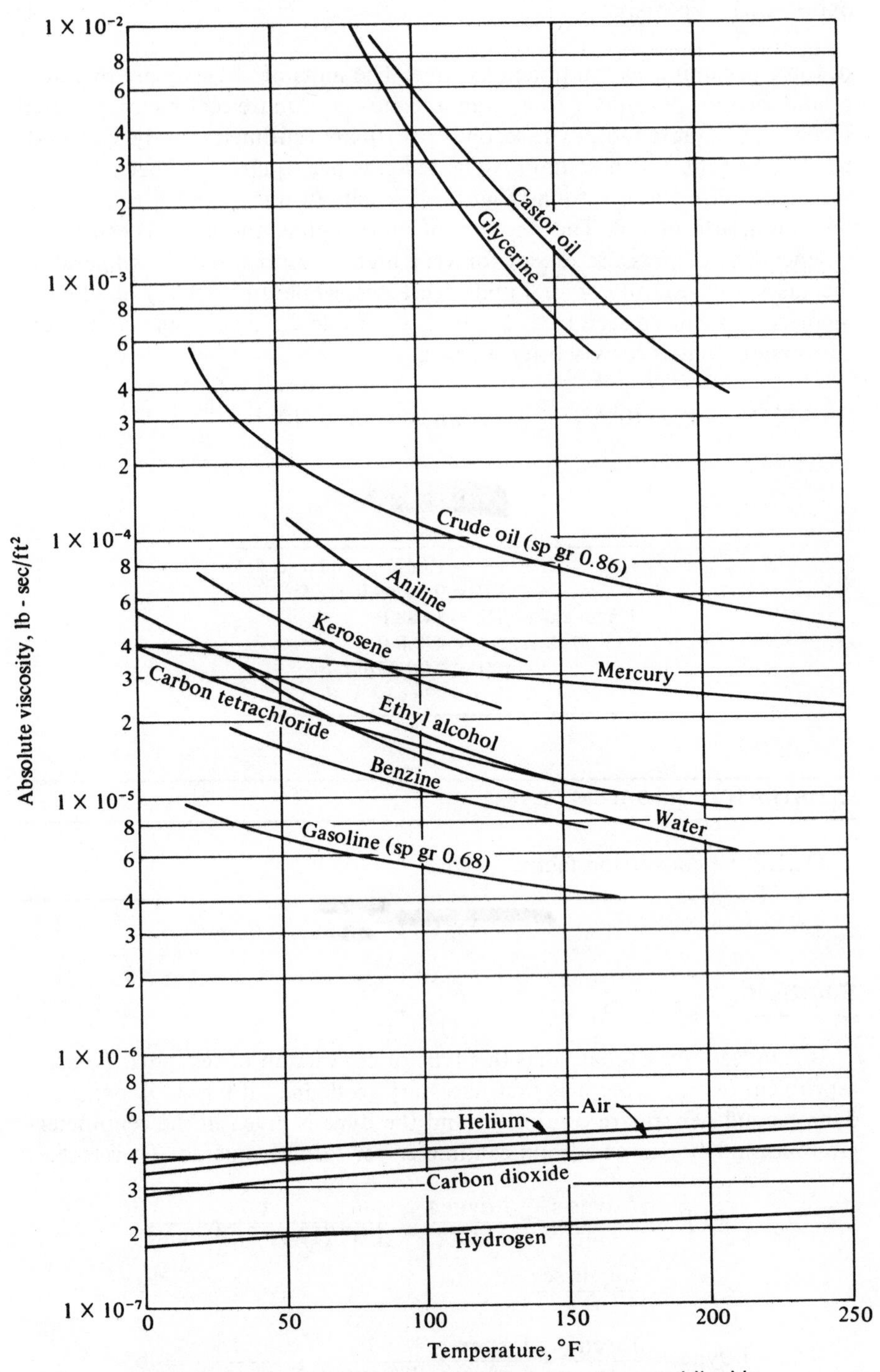

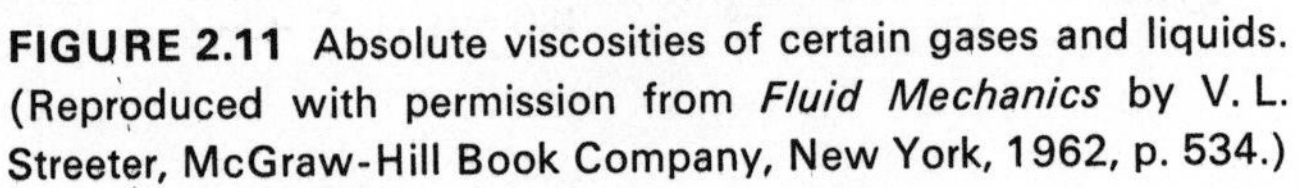

FIGURE 2.11 Absolute viscosities of certain gases and liquids. (Reproduced with permission from *Fluid Mechanics* by V. L. Streeter, McGraw-Hill Book Company, New York, 1962, p. 534.)

Therefore,

$$1\ \text{P} = \frac{1}{1/(12 \times 2.54)^2} \times \frac{1}{980 \times 454} = \frac{\text{lb sec}}{\text{ft}^2}$$

$$= 0.00209\ \frac{\text{lb sec}}{\text{ft}^2}$$

ILLUSTRATIVE PROBLEM 2.14

The viscosity of mercury at 68°F is 1.58×10^{-2} P, that is, 1.58×10^{-3} N·s/m^2. Determine the force necessary to maintain a relative velocity of 2 m/s between two plates that are separated by 0.1 m and whose area is 0.1 m^2. Consider only viscous effects.

Solution

From equation (2.26),

$$\frac{F}{A} = \mu \frac{V}{Y}$$

Therefore,

$$F = \mu A \frac{V}{Y} = \frac{1.58 \times 10^{-3}\ \text{N}\cdot\text{s/m}^2 \times 0.1\ \text{m}^2 \times 2\ \text{m/s}}{0.1\ \text{m}} = 0.00316\ \text{N}$$

There are many methods of measuring viscosity or relative quantities that can be compared to viscosity. In general, the devices used to make viscosity measurements (known as *viscometers* or *viscosimeters*) fall into three categories: *rotational*, *falling sphere*, or *flow-type* (capillary) units. Because of the effects of walls, concentricity, finite size, and so on, it is almost impossible to analytically obtain the constants of a given device. In almost every instance, the units are calibrated against liquids of known viscosity. However, it is interesting to briefly consider a sphere that is falling freely in a still, viscous fluid.

If a spherical object falls freely in a still viscous fluid, it will reach a velocity relative to the fluid that remains constant and is known as the terminal velocity. At this time we can consider the body to be in equilibrium and treat it as a problem in fluid statics, where the body is subjected to three forces: its weight, the buoyant force due to displaced fluid, and the force due to the relative motion of the sphere (drag) and the fluid. These forces are shown in Figure 2.12. Mathematically,

$$F_{\text{drag}} + F_{\text{buoyant force}} = W \qquad (2.28)$$

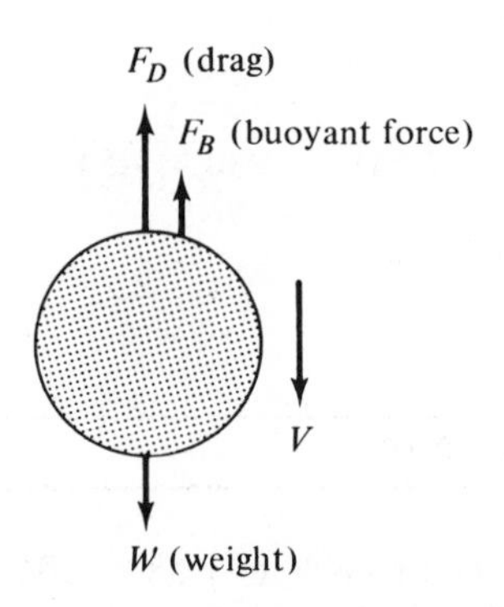

FIGURE 2.12 Free-body diagram of a freely falling sphere.

If a sphere is totally submerged in a fluid flowing past it with a relative velocity V, it can be shown that the force exerted on the sphere due to this relative motion arises predominantly from viscous effects on the surface of the sphere. The resultant force on the sphere is given by equation (2.29), which was originally derived by Sir G. G. Stokes in 1851:

$$F = 6\pi r_o \mu V \tag{2.29}$$

where r_o is the outside radius of the sphere, μ the viscosity of the fluid, and V the relative velocity between the sphere and the undisturbed fluid. Using equations (2.28) and (2.29), we obtain

$$6\pi r_o \mu V + \tfrac{4}{3}\pi r_o^3 \gamma_F = \tfrac{4}{3}\pi r_o^3 \gamma_S \tag{2.30}$$

where γ_S is the specific weight of the sphere and γ_F the specific weight of the fluid. Simplifying yields

$$V = \frac{2}{9}\frac{r_o^2}{\mu}(\gamma_S - \gamma_F) \tag{2.31}$$

Equation (2.31) is known as *Stokes' law*, and it is frequently used in problems for the determination of small particle sizes. It is also useful as a method of obtaining the viscosity of a fluid by measuring the terminal velocity of a falling sphere either optically, mechanically, or electrically.

ILLUSTRATIVE PROBLEM 2.15

A spherical metal ball is allowed to fall freely in 68°F water. If the ball is made of aluminum that weighs 168.8 lb/ft³ and is 0.01 in. in diameter, determine its terminal velocity.

Solution

At 68°F water has a viscosity of 1 cP. From Illustrative Problem 2.13, we have 1 P = 0.00209 lb sec/ft^2. Therefore, 1 cP = 2.09×10^{-5} lb sec/ft^2. Using 62.4 for the specific weight of water and $r_o = 0.01/(2 \times 12) = 0.000417$ ft,

$$V = \frac{2}{9} \frac{r_o^2}{\mu}(\gamma_S - \gamma_F)$$

Therefore,

$$V = \frac{2}{9} \frac{(0.000417)^2}{2.09 \times 10^{-5}}(168.8 - 62.4)$$

and

$$V = 0.197 \text{ ft/sec}$$

It is sometimes convenient to combine viscosity and density; viscosity divided by density is known as *kinematic viscosity*. In the metric system the name *stoke* (or one hundredth of its value, the *centistoke*) is used to designate kinematic viscosity; a centistoke is 1 cP divided by a density of 1 g/cm^3. Figure 2.13 shows the kinematic viscosity of certain gases and liquids, where the gases are at atmospheric pressure. By definition,

$$\nu = \frac{\mu}{\rho} \tag{2.32}$$

In the SI system,

$$\nu = \frac{\text{N}\cdot\text{s/m}^2}{\text{N}\cdot\text{s}^2/\text{m}^4} = \frac{\text{m}^2}{\text{s}}$$

The centistoke can be converted to SI units by noting that ν, centistokes $\times\ 10^{-6} = \nu$, SI.

2.8 CLOSURE

We have concerned ourselves with identifying and discussing those properties that will be necessary to establish the state of a given system. It is particularly important to understand the physical significance and units of those properties since our future studies will be based upon the use of them. By now it will be noticed that we are extensively utilizing the method of dimensional consistency that was discussed in Chapter 1. Two of the key elements needed for the successful completion of this study are now in place,

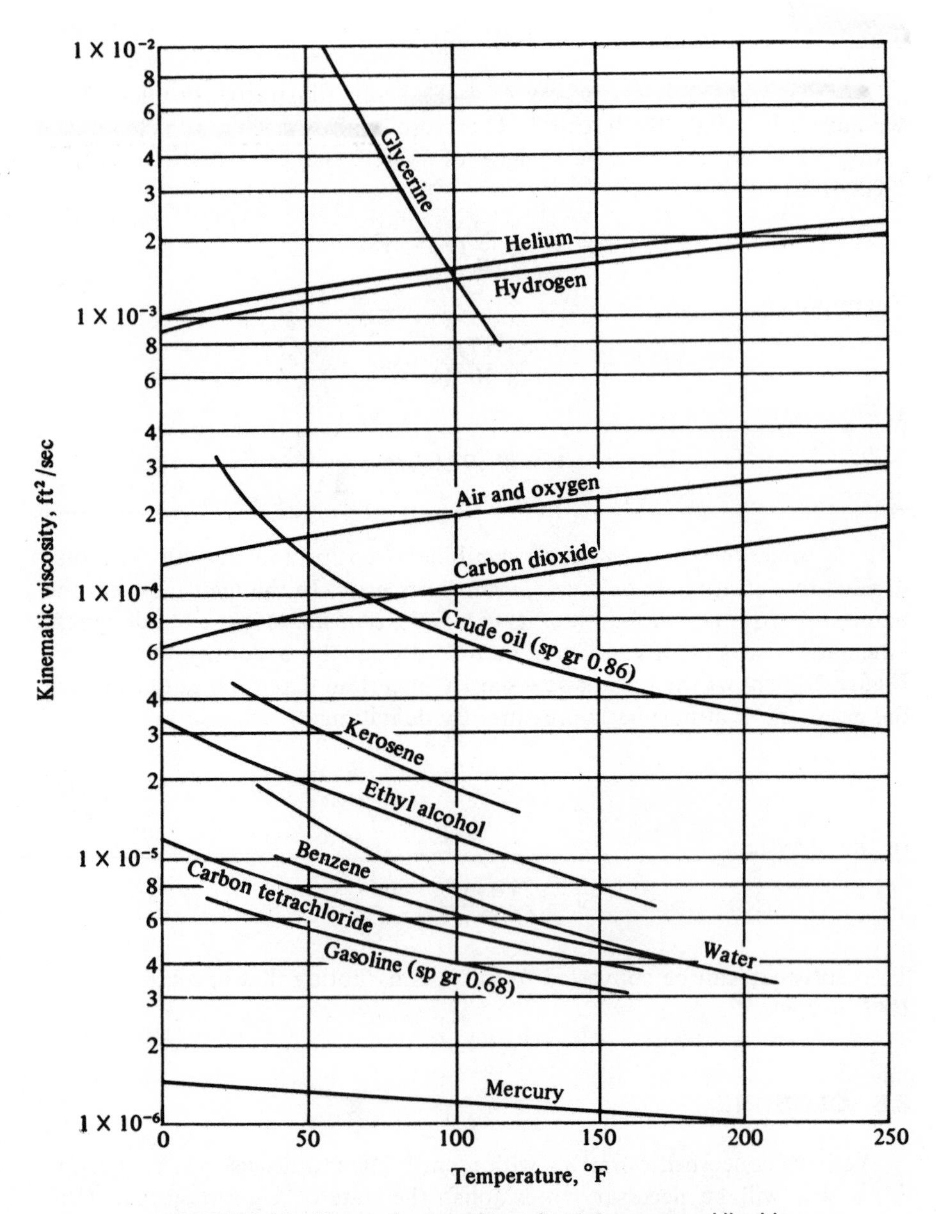

FIGURE 2.13 Kinematic viscosities of certain gases and liquids. (Reproduced with permission from *Fluid Mechanics* by V. L. Streeter, McGraw-Hill Book Company, New York, 1962, p. 535.)

that is, dimensional consistency and an understanding of the properties of fluids. In succeeding chapters the student will see them applied to the solving of problems in fluid mechanics.

Because the technically trained person will be involved at some time with the measurement of pressure, Appendix 2-1, "Pressure Measurements," has been placed at the end of this chapter for ready reference.

PROBLEMS

2.1 Convert 20, 40, and 60°C to equivalent degrees Fahrenheit.

2.2 Change 0, 10, and 50°F to equivalent degrees Celsius.

2.3 Convert 500°R, 500K, 600K, and 650°R to degrees Celsius.

2.4 An arbitrary temperature scale is proposed in which 20° is assigned to the ice point and 75° is assigned to the boiling point. Derive an equation relating this scale to the Celsius scale.

2.5 For the temperature scale proposed in Problem 2.4, what temperature corresponds to absolute zero?

2.6 Derive a relation between degrees Rankine and degrees Kelvin, and based upon the results, show that (°C + 273)1.8 = °F + 460.

2.7 A Fahrenheit and Celsius thermometer are used to measure the temperature of a fluid. If the Fahrenheit reading is 1.5 times that of the Celsius reading, what are both readings?

2.8 A new thermometer scale on which the ice point of water at atmospheric pressure would correspond to a marking of 200 and the boiling point of water at atmospheric pressure would correspond to a marking of −400 is proposed. What would be the reading of this new thermometer if a Fahrenheit thermometer placed in the same environment read 80°F?

2.9 A skin diver descends to a depth of 60 ft in fresh water. What is the pressure on his body? The specific weight of fresh water can be taken as 62.4 lb/ft^3.

2.10 A skin diver descends to a depth of 25 m in a salt lake, where the density is 1026 kg/m^3. What is the pressure on this person's body at this depth?

2.11 If a bourdon gage reads 25 psi, what is the absolute pressure in Pa?

2.12 A column of fluid is 1 m high. The fluid has a density of 2500 kg/m^3. What is the pressure at the base of the column?

2.13 A column of fluid is 25 in. high. If the specific weight of the fluid is 60.0 lb/ft^3, what is the pressure in psi at the base of the column?

2.14 Determine the density and specific volume of the contents of a 10-ft^3 tank if the contents weigh 250 lb.

2.15 A tank contains 500 kg of a fluid. If the volume of the tank is 0.5 m^3, what is the density of the fluid and what is the specific volume?

2.16 Calculate the specific weight of a fluid if a gallon of it weighs 20 lb. Also calculate its density.

2.17 The specific gravity of a fluid is 1.2. Calculate its density and its specific weight in SI units.

2.18 A fluid has a density of 0.90 g/cm^3. What is its specific weight in both the English and SI systems?

2.19 It is proposed by gasoline dealers to sell gasoline by the litre. If gasoline has a density of 1.3 slugs/ft^3, what is the weight of 60 litres of gasoline?

2.20 A household oil tank can hold 275 gal of oil. If oil has a specific weight of 8800 N/m^3, how many pounds of oil will there be in a full tank?

2.21 Convert 14.696 psia to 101 325 KPa.

2.22 A pressure gage indicates 25 psi when the barometer is at a pressure equivalent to 14.5 psia. Compute the absolute pressure in psi and feet of mercury when the specific gravity of mercury is 13.0 gm/cm^3.

2.23 The same as Problem 2.22. The barometer stands at 750 mm Hg and its specific gravity is 13.6 gm/cm^3.

2.24 A vacuum gage reads 8 in. Hg when the atmospheric pressure is 29.0 in. Hg. If the specific gravity of mercury is 13.6 gm/cm^3, compute the absolute pressure in psi.

2.25 A vacuum gage reads 10 in. Hg when the atmospheric pressure is 30 in. Hg. Assuming the density of mercury to be 13 595 kg/m^3, determine the pressure in Pa.

2.26 A tube whose diameter is 0.001 ft is placed in a mercury pool. If the surface tension of mercury is 35×10^{-3} lb/ft, determine the depression of the mercury in the tube below the surface of the mercury pool. Assume that $\theta = 129°$ and that the specific weight of mercury is 13.6 times that of water.

2.27 If instead of the tube in Problem 2.26, two infinite parallel plates are used, determine the depression of the mercury between the plates below the surface of the pool.

2.28 Solve Problem 2.26 for water having a surface tension of 5×10^{-3} lb/ft.

2.29 Solve Problem 2.27 for water having a surface tension of 5×10^{-3} lb/ft.

2.30 If a fluid has a surface tension of 2.9×10^{-3} N/m and a specific gravity of 0.85, how high will it rise in a capillary tube having a diameter of 2×10^{-4} m if the wetting angle is 25°?

2.31 Solve Problem 2.30 for two parallel plates.

2.32 A sphere of graphite (soot) having a specific weight of 138 lb/ft³ falls in still air. If the diameter is 0.5 mm, what will be its terminal velocity? Assume that $\mu_{air} = 3.73 \times 10^{-7}$ lb sec/ft² and $\gamma_{air} = 0.075$ lb/ft³.

2.33 If the specific weight of the air had been neglected in Problem 2.32, would there be a significant difference in the solution?

2.34 A spherical steel ball is dropped in oil. If the viscosity of the oil is 2 P, its specific weight is 8500 N/m³, and the diameter of the ball is 0.1 cm, determine the terminal velocity of the ball. The specific weight of steel is 74 870 N/m³.

2.35 Derive the conversion from poise to poundal seconds per square foot. Check the result with Table 2.4. (1 P = 0.0672 poundal sec/ft².)

2.36 Two plates separated by 3 in. are placed in oil with a viscosity of 133×10^{-6} lb sec/ft². If the plates are parallel and have an area of 5 ft², determine the relative velocity (considering only viscous effects) if a force of 0.1 lb is applied to one of the plates parallel to the orientation of the plates.

2.37 Determine the bulk modulus of a fluid if it undergoes a 1% change in volume when subjected to a pressure of 10,000 psi.

REFERENCES

1. *Fluid Mechanics for Engineers* by P. S. BARNA, Butterworth & Co. (Publishers) Ltd., London, 1957.

2. *Thermodynamics and Heat Power* by Irving Granet, 2nd ed., Reston Publishing Co., Reston, Va., 1980.

3. *Elementary Theoretical Fluid Mechanics* by K. BRENKERT, Jr., John Wiley & Sons, Inc., New York, 1960.

4. *Mechanics of Fluids* by G. MURPHY, International Textbook Company, Scranton, Pa, 1942.

5. *Engineering Applications of Fluid Mechanics* by J. C. HUNSAKER and B. G. RIGHTMIRE, McGraw-Hill Book Company, New York, 1947.

6. *Fluid Mechanics* by R. C. BINDER, 4th ed., Prentice-Hall, Inc., Englewood Cliffs, N.J., 1962.
7. *Thermodynamics of Fluid Flow* by N. A. HALL, Prentice-Hall, Inc., Englewood Cliffs, N.J., 1951.
8. *Hydraulics* by R. L. DAUGHERTY, 4th ed., McGraw-Hill Book Company, New York, 1937.
9. *Fluid Mechanics* by V. L. STREETER, 3rd ed., McGraw-Hill Book Company, New York, 1958; 1962; 7th ed., 1979.
10. *Fundamentals of Hydro- and Aero-Mechanics* by L. PRANDTL and O. G. TIETJENS, McGraw-Hill Book Company, New York, 1934.
11. *Elementary Fluid Mechanics* by J. K. VENNARD, John Wiley & Sons, Inc., New York, 1961.
12. *A Textbook of Fluid Mechanics* by J. R. D. FRANCES, Edward Arnold (Publishers) Ltd., London, 1958.
13. *Applied Fluid Mechanics* by ROBERT L. MOTT, 2nd ed., Charles E. Merrill Publishing Co., Columbus, Ohio, 1979.
14. *Fundamentals of Fluid Mechanics* by J. A. SULLIVAN, Reston Publishing Co., Reston, Va., 1978.

Pressure Measurements[1]

A2.1 INTRODUCTION

Pressure is a property of a system and as a fundamental parameter it is of the utmost importance that it be measured accurately. In the text of Chapter 2 pressure was defined to be the normal force exerted by a fluid on a surface. It would be more correct to qualify this definition and for the present restrict it to fluids at rest. However, no definition is really useful unless and until it can be converted into measurable characteristics. Figure A2.1 illustrates the basic definition of pressure and the elementary static concepts upon which the entire field of pressure measurement is based.

There is unfortunately a confusing number of units used to express pressure. While all of these units are based upon the definition of a normal force per unit area, their usage has entered the technical literature based upon particular applications. Table A2.1 gives conversion factors between some of

[1]The material in this appendix is taken with permission from "Pressure and Its Measurement" by R. P. Benedict, *Electro-Technology* (New York), **80**, October 1967, p. 69 et sequi.

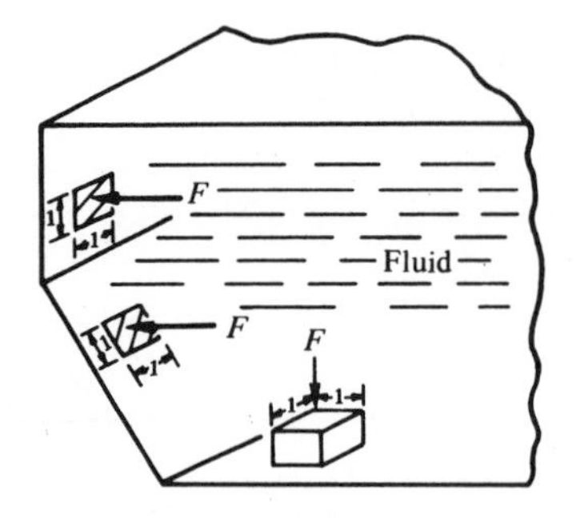

Pressure is the normal force F (lb/in.^{2}, lb/ft^2, dynes/cm^2) exerted on a unit area of a surface bounding a fluid.

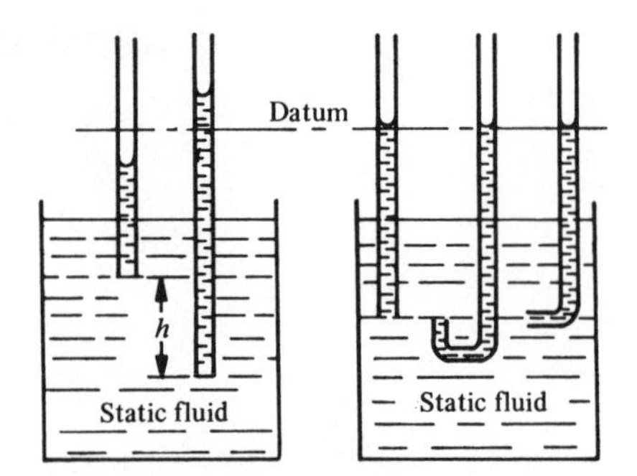

Fluid pressure varies with depth, but it is the same in all directions at a given depth.

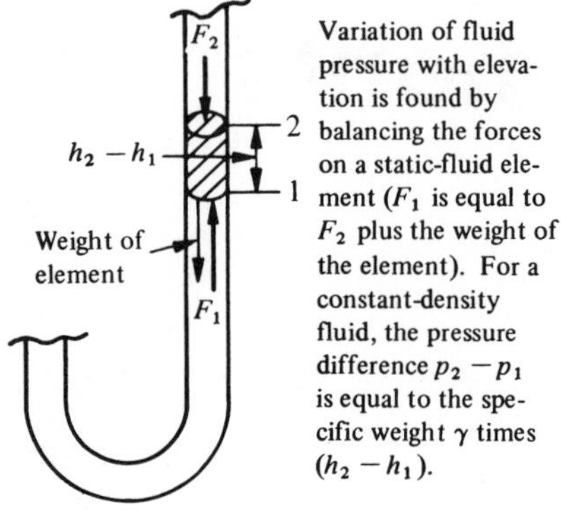

Variation of fluid pressure with elevation is found by balancing the forces on a static-fluid element (F_1 is equal to F_2 plus the weight of the element). For a constant-density fluid, the pressure difference $p_2 - p_1$ is equal to the specific weight γ times $(h_2 - h_1)$.

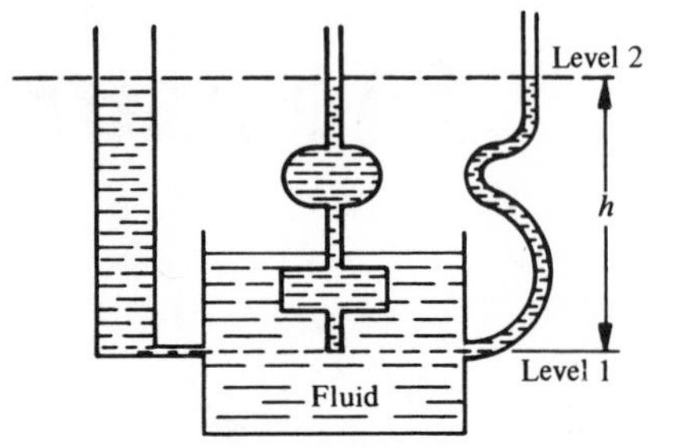

Pressure is independent of the shape and size of the vessel. The pressure difference between level 1 and level 2 is always $p_1 - p_2 = \gamma h$, where γ is the specific weight of the constant-density fluid in the vessel.

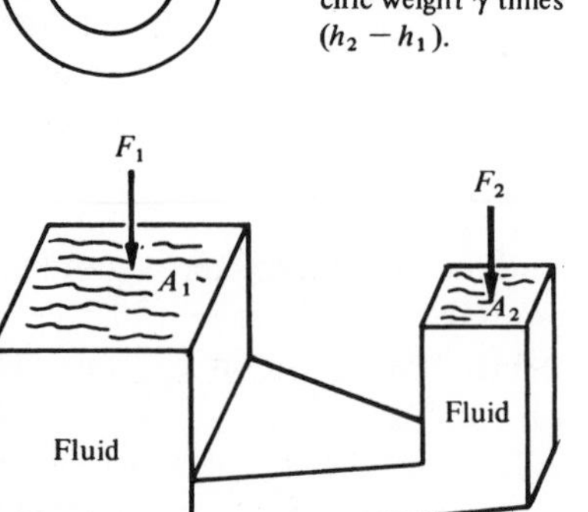

Constant pressure transmission in a confined fluid can be used to multiply force by the relation $p = F_1/A_1 = F_2/A_2$.

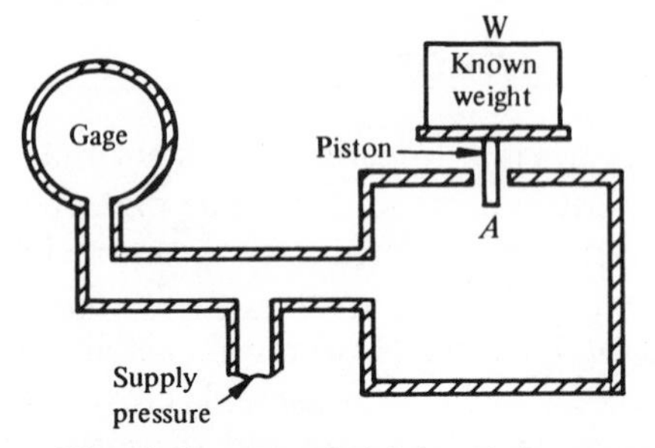

When the known weight is balanced, the gage pressure is $p = W/A$; this is the basic principle of deadweight testing.

FIGURE A2.1 Basic pressure-measurement concepts.

the commonly used units of pressure. For convenience, there are two approximations that can be used to give the student a "feel" for some of these units:

$$1 \text{ mm} = 1000 \text{ microns} \simeq 0.04 \text{ in. Hg}$$

TABLE A2.1 Pressure-Unit Conversion Factors

Pressure Unit	psi	in. H_2O	in. Hg	Atmospheres	Microbars	mm Hg	Microns
1 psi	1.000	27.730	2.0360	6.8046×10^{-2}	68,947.6	51.715	51,715.0
1 in. H_2O (68°F)	0.036063	1.000	0.073424	2.4539×10^{-3}	2,486.4	1.8650	1,865.0
1 in. Hg (32°F)	0.49115	13.619	1.000	3.3421×10^{-2}	33,864.0	25.400	25,400.0
1 atm	14.69595	407.513	29.9213	1.000	1.01325×10^{6}	760.000	7.6000×10^{5}
1 microbar (dynes /cm^2)	1.4504×10^{-5}	4.0218×10^{-4}	2.9530×10^{-5}	9.8692×10^{-7}	1.000	7.5006×10^{-4}	0.75006
1 mm Hg (32°F)	0.019337	0.53620	0.03937	1.3158×10^{-3}	1,333.2	1.000	1,000.0
1 micron (32°F)	1.9337×10^{-5}	5.3620×10^{-4}	3.9370×10^{-5}	1.3158×10^{-6}	1.3332	0.0010	1.000

and

$$1 \text{ micron} = 0.001 \text{ mm} \simeq 0.00004 \text{ in. Hg}$$

In performing any measurement it is necessary to have a standard of comparison in order to calibrate the measuring instrument. In Section A2.2 five pressure standards currently used as the basis for all pressure-measurement work will be discussed. Table A2.2 summarizes these standards, the pressure range in which they are used, and their accuracy.

TABLE A2.2 Characteristics of Pressure Standards

Type	Range	Accuracy
Dead-weight piston gage	0.01 to 10,000 psig	0.01 to 0.05 percent of reading
Manometer	0.1 to 100 psig	0.02 to 0.2 percent of reading
Micro-manometer	0.0002 to 20 in. H_2O	1 percent of reading to 0.001 in. H_2O
Barometer	27 to 31 in. Hg	0.001 to 0.03 percent of reading
McLeod gage	0.01 micron to 1 mm Hg	0.5 to 3 percent of reading

A2.2 DEAD-WEIGHT PISTON GAGE

The dead-weight, free-piston gage consists of an accurately machined piston inserted into a close-fitting cylinder. Masses of known weight are loaded on one end of the free piston, and pressure is applied to the other end until enough force is developed to lift the piston–weight combination. When the piston is floating freely between the cylinder limit stops, the gage is in equilibrium with the unknown system pressure. The dead-weight pressure can then be defined as

$$P_{dw} = \frac{F_e}{A_e} \tag{A2.1}$$

where F_e, the equivalent force of the piston–weight combination, depends on such factors as local gravity and air buoyancy, and A_e, the equivalent area of the piston–cylinder combination, depends on such factors as piston–cylinder clearance, pressure level, and temperature.

A fluid film provides the necessary lubrication between the piston and cylinder. In addition, the piston—or, less frequently, the cylinder—may be rotated or oscillated to reduce friction even further. Because of fluid leakage, system pressure must be continuously trimmed upward to keep the piston–weight combination floating. This is often achieved by decreasing system volume using a pressure volume apparatus (as shown in Figure A2.2). As long as the piston is freely balanced, system pressure is defined by equation (A2.1).

Corrections must be applied to the indication of the dead-weight piston gage, p_i, to obtain the accurate system pressure, P_{dw}. The two most important corrections concern air buoyancy and local gravity. According to Archimedes' principle, the air displaced by the weights and the piston exerts a

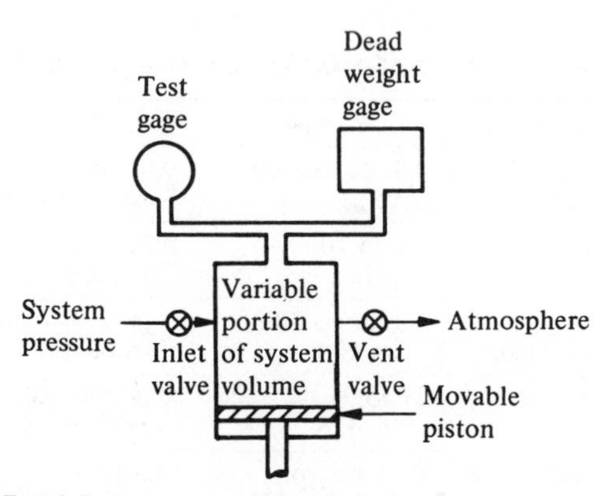

FIGURE A2.2 Pressure volume regulator to compensate for fluid leakage in a dead-weight gage.

buoyant force that causes the gage to indicate too high a pressure. The correction term for this effect is the ratio of two specific weights,

$$C_b = -\frac{\gamma_{air}}{\gamma_{weights}}$$

Whenever gravity at the measurement site varies from the standard sea-level value of 32.1740 ft/sec² because of latitude or altitude variations, a gravity correction term must be applied;

$$\begin{aligned} C_g &= \frac{g_{local}}{g_{std}} - 1 \\ &= -2.637 \times 10^{-3} \cos 2\phi - 9.6 \times 10^{-8} h - 5 \times 10^{-5} \end{aligned} \tag{A2.2}$$

where ϕ is latitude in degrees and h is altitude above sea level in feet.

The corrected dead-weight piston gage pressure is therefore

$$P_{dw} = p_i(1 + C_b + C_g) \tag{A2.3}$$

The effective area of the dead-weight piston gage is usually taken as the mean of the cylinder and piston areas, but temperature affects this dimension. The effective area increases between 13 and 18 ppm/°F for commonly used materials, and a suitable correction for this effect may also be applied.

A2.3 MANOMETER

The manometer will be discussed later in this book as a means of measuring pressure. However, for accurate measurements there are certain factors that must be taken into account. Therefore, for the sake of continuity and completeness a brief discussion of the manometer will be included in this section, and those corrections most necessary for accurate use of this instrument will also be discussed.

A manometer can simply consist of a transparent tube in the form of an elongated U, partially filled with a suitable liquid. Mercury and water are the most commonly used manometric fluids because detailed information is available on their specific weights.

The fluid whose pressure is unknown is applied to the top of one of the tubes of the manometer, while a reference fluid pressure is applied to the other tube, as shown in Figure A2.3. In the steady state, the difference between the unknown pressure and the reference pressure is balanced by the weight/unit area of the displaced manometer liquid, so that

$$\Delta p_{mano} = \gamma_m \, \Delta h_e \tag{A2.4}$$

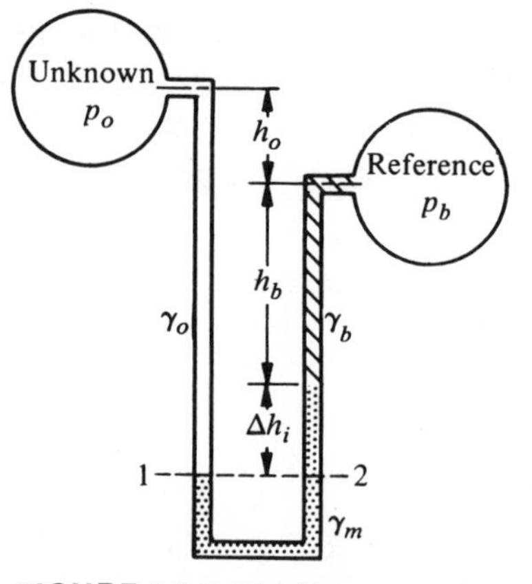

FIGURE A2.3 U-tube manometer.

where γ_m, the corrected specific weight of the manometer fluid, depends on such factors as temperature and local gravity, and h_e, the equivalent manometer fluid height, depends on such factors as scale variations with temperature, relative specific weights, heights of the fluids involved, and capillary effects. This equation holds as long as manometer fluid displacement is constant.

The value of specific weight γ_m corrected for local gravity is

$$\gamma_c = \gamma_{s.t}(1 + C_g) \tag{A2.5}$$

where C_g is the correction term given in equation (A2.3) and γ_c is the corrected value of specific weight for any fluid whose specific weight is $\gamma_{s.t}$ at standard gravity (32.1740 ft/sec^2) and a given Fahrenheit temperature t.

The variation of specific weight with temperature is described by empirical relationships; for mercury,

$$\gamma_{s.t} = \frac{0.491154}{1 + 1.01(t - 32) \times 10^{-4}} \tag{A2.6}$$

and for water,

$$\gamma_{s.t} = \frac{62.2523 + 0.978476 \times 10^{-2}t - 0.145 \times 10^{-3}t^2 + 0.217 \times 10^{-6}t^3}{1728} \tag{A2.7}$$

where $\gamma_{s.t}$ is in pounds per cubic inch. These relationships have proved satisfactory for accurate manometer measurements; some typically useful values are listed in Table A2.3.

One important correction needed to find the equivalent manometer fluid height h_e is associated with the relative weights and heights of the fluids

TABLE A2.3 Specific Weight of Mercury and Water (At Standard Gravity Value of 32.1740 ft/sec²)

	Specific weight ($\gamma_{s.t}$)	
Temperature (°F)	**Hg**	**H_2O**
32	0.491154	0.036122
36	0.490956	0.036126
40	0.490757	0.036126
44	0.490559	0.036124
48	0.490362	0.036120
52	0.490164	0.036113
56	0.489966	0.036104
60	0.489769	0.036092
64	0.489572	0.036078
68	0.489375	0.036062
72	0.489178	0.036045
76	0.488981	0.036026
80	0.488784	0.036005
84	0.488588	0.035983
88	0.488392	0.035958
92	0.488196	0.035932
96	0.488000	0.035905
100	0.487804	0.035877

involved. This hydraulic correction is

$$C_h = 1 + \frac{\gamma_b}{\gamma_m}\left(\frac{h_b}{\Delta h_i}\right) - \frac{\gamma_a}{\gamma_m}\left(\frac{h_a + h_b}{\Delta h_i} + 1\right) \qquad \text{(A2.8)}$$

The quantities in this equation are depicted in Figure A2.3.

The second main correction needed to determine h_e concerns capillary effects. The shape of the interface between two fluids at rest depends on the relative gravity and cohesion and adhesion forces between the fluids and their containing walls. In water–air–glass combinations, the crescent shape of the liquid surface (called the meniscus) is concave upward, and the water is said to wet the glass. In this situation, adhesive forces dominate, and water in a tube will be elevated by capillary action. For mercury–air–glass combinations, cohesive forces dominate, the meniscus is concave downward, and the mer-

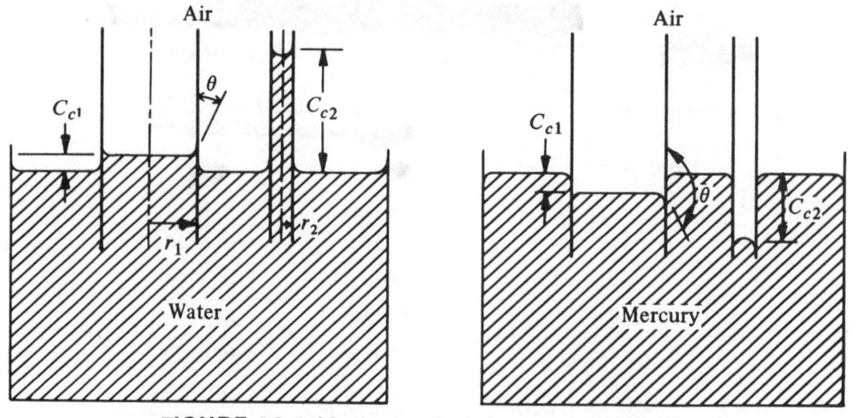

FIGURE A2.4 Meniscus effects in water and mercury.

cury level in a tube will be depressed by capillary action, as shown in Figure A2.4.

Based upon the discussion of surface tension in Chapter 2, the capillary correction factor for manometers can be written as

$$C_c = \frac{2 \cos \theta_m}{\gamma_m} \left(\frac{\sigma_{a-m}}{r_a} - \frac{\sigma_{b-m}}{r_b} \right) \tag{A2.9}$$

where θ_m is the angle of contact between the manometer fluid and the glass, σ_{a-m} and σ_{b-m} are the surface-tension coefficients of the manometer fluid (m) with respect to the fluids (a and b) above it, and r_a and r_b are the radii of the tubes containing fluids a and b. Typical values for these capillary variables are given in Table A2.4. For mercury manometers, C_c is positive when the larger capillary effect occurs in the tube in which the manometer fluid is highest. For water manometers under the same conditions, C_c is negative.

TABLE A2.4 Capillary Effects

Combination	Surface Tension σ dynes/cm	Surface Tension σ lb/in.	Contact Angle θ (deg)
Mercury, vacuum, glass	480	2.74×10^{-3}	140
Mercury, air, glass	470	2.68×10^{-3}	140
Mercury, water, glass	380	2.17×10^{-3}	140
Water air, glass	73	0.416×10^{-3}	0

When the same fluid is applied to both legs of a standard U-tube manometer, the capillary effect is often neglected. This can be done because the tube bores are approximately equal, and capillarity in one tube counterbalances that in the other. However, capillary effect can be extremely important and must always be considered in manometer-type instruments.

To minimize the effect of a variable meniscus, which can be caused by dirt, the method of approaching equilibrium, tube bore, and so on, the tubes are always tapped before reading and the measured liquid height is always based on readings taken at the center of the meniscus in each leg of the manometer. To reduce the capillary effect itself, the use of large-bore tubes (over $\frac{3}{8}$-in. diameter) is most effective.

When both hydraulic and capillary corrections are taken into account, the equivalent manometer fluid height is

$$\Delta h_e = C_h \, \Delta h_i \pm C_c \tag{A2.10}$$

where Δh_i is the indicated height shown in Figure A2.3.

A2.4 MICROMANOMETER

While the manometer is a useful and convenient tool for pressure measurements it is unfortunately limited when making low-pressure measurements. To extend its usefulness in the low-pressure range, micromanometers have been developed that have extended the useful range of low-pressure manometer measurements to pressures as low as 0.0002 in. H_2O.

One type of micromanometer is the so-called Prandtl-type micromanometer in which capillary and meniscus errors are minimized by returning the meniscus of the manometer liquid to a null position before measuring the applied pressure difference. As shown in Figure A2.5, a reservoir, which forms one side of the manometer, is moved vertically to locate the null position. This position is reached when the meniscus falls within two closely scribed marks on the near-horizontal portion of the micromanometer tube. Either the reservoir or the inclined tube is then moved by a precision lead-screw arrangement to determine the micromanometer liquid displacement (Δh), which corresponds to the applied pressure difference. The Prandtl-type micromanometer is generally accepted as a pressure standard within a calibration uncertainty of 0.001 in. H_2O.

Another method for minimizing capillary and meniscus effects in manometry is to measure liquid displacements with micrometer heads fitted with adjustable, sharp index points. Figure A2.6 shows a manometer of this type; the micrometers are located in two connected transparent containers. In some commercial micromanometers, contact with the surface of the manometric liquid may be sensed visually by dimpling the surface with the index point, or

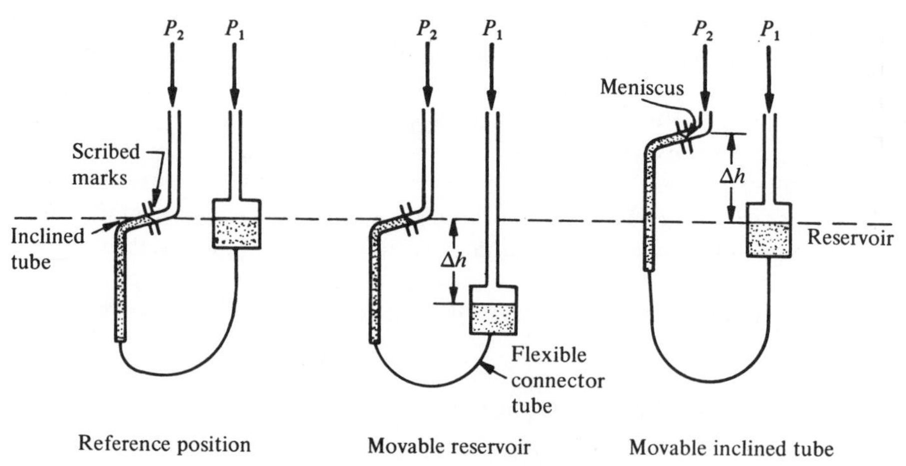

FIGURE A2.5 Two variations of the Prandtl-type micromanometer.

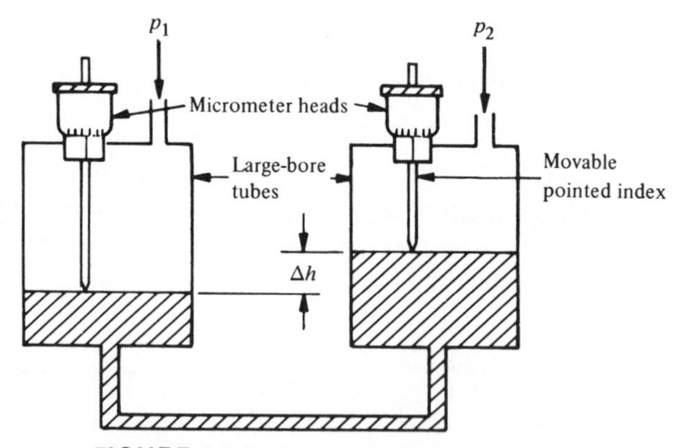

FIGURE A2.6 Micrometer-type manometer.

even by electrical contact. Micrometer-type micromanometers also serve as pressure standards within a calibration uncertainty of 0.001 in. H_2O.

An extremely sensitive high-response micromanometer uses air as the working fluid and thus avoids all the capillary and meniscus effects usually encountered in liquid manometry. In this device, shown in Figure A2.7, the reference pressure is mechanically amplified by centrifugal action in a rotating disk. The disk speed is adjusted until the amplified reference pressure just balances the unknown pressure. This null position is recognized by observing the lack of movement of minute oil droplets sprayed into a glass indicator tube. At balance, the air micromanometer yields the applied pressure difference

$$\Delta p_{\text{micro}} = Kpn^2 \tag{A2.11}$$

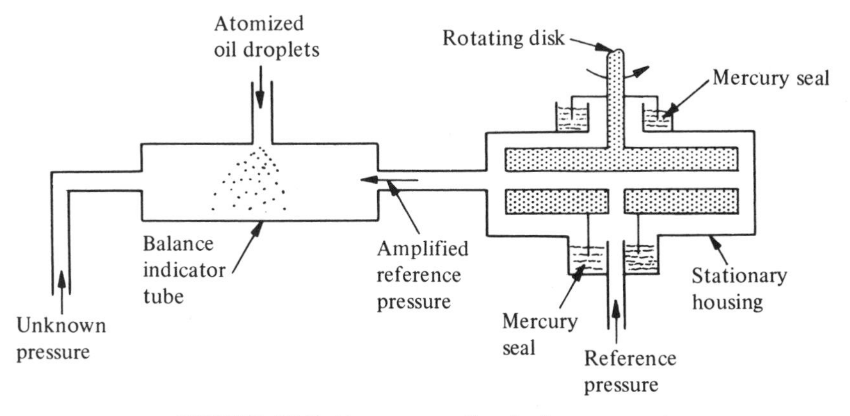

FIGURE A2.7 Air-type centrifugal micromanometer.

where p is the reference air density, n the rotational speed of the disk, and K a constant that depends on disk radius and annular clearance between the disk and the housing. Measurements of pressure differences as small as 0.0002 in. H_2O can be made with this type of micromanometer within an uncertainty of 1%.

A word on the reference pressures employed in manometry is pertinent at this point in the discussion. If atmospheric pressure is used as a reference, the manometer yields gage pressures. Because of the variability of air pressure, gage pressures vary with time, altitude, latitude, and temperature. If, however, a vacuum is used as reference, the manometer yields absolute pressures directly, and it may serve as a barometer. In any case, the absolute pressure is always equal to the sum of the gage and ambient pressures; by ambient pressure we mean the pressure surrounding the gage, which is usually atmospheric pressure.

A2.5 BAROMETERS

The reservoir or cistern barometer consists of a vacuum-reference mercury column immersed in a large-diameter ambient-vented mercury column that serves as a reservoir. The most common cistern barometer in general use is the Fortin type, in which the height of the mercury surface in the cistern can be adjusted. The operation of this instrument can best be explained with reference to Figure A2.8.

The datum-adjusting screw is turned until the mercury in the cistern makes contact with the ivory index, at which point the mercury surface is aligned with zero on the instrument scale. Next, the indicated height of the mercury column in the glass tube is determined. The lower edge of a sighting ring is lined up with the top of the meniscus in the tube. A scale reading ıd a

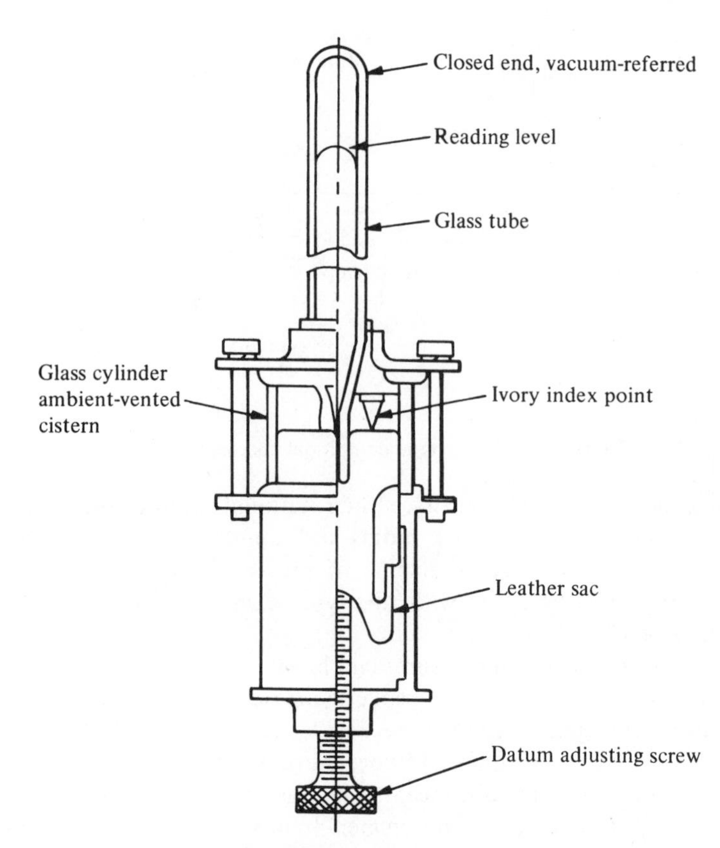

FIGURE A2.8 Fortin barometer.

vernier reading are taken and combined to yield the indicated mercury height at the barometer temperature.

Since atmospheric pressure on the mercury in the cistern is exactly balanced by the weight per unit area of the mercury column in the glass tube,

$$p_{\text{baro}} = \gamma_{\text{Hg}} h_{to} \tag{A2.12}$$

The referenced specific weight of mercury, γ_{Hg}, depends on such factors as temperature and local gravity; the referenced height of mercury, h_{to}, depends on such factors as thermal expansion of the scale and the mercury.

When the scale zero of the Fortin barometer is adjusted to agree with the mercury level in the cistern, the correct height of mercury at temperature t will be

$$h_t = h_{ti}[1 + S(t - t_s)] \tag{A2.13}$$

where S is the linear coefficient of thermal expansion of the scale, h_{ti} the indicated height of mercury, and t_s the temperature at which the scale was calibrated. Whenever t is greater than reference temperature t_o, the correct height of mercury at t will be greater than the height of mercury at t_o:

$$h_t = h_{to}[1 + m(t - t_o)] \tag{A2.14}$$

where m is the cubical coefficient of thermal expansion of mercury and h_{to} the referenced height of the mercury column.

A temperature correction factor can be defined by

$$C_t = h_{to} - h_{ti} \tag{A2.15}$$

From equations (A2.13) and (A2.14),

$$C_t = \left[\frac{S(t - t_s) - m(t - t_o)}{1 + m(t - t_o)}\right] h_{ti} \tag{A2.16}$$

When standard values of $S = 10.2 \times 10^{-6}/°\text{F}$, $m = 101 \times 10^{-6}/°\text{F}$, $t_s = 62°\text{F}$, and $t_o = 32°\text{F}$ are substituted in equation (A2.16),

$$C_t = -\left[\frac{9.08(t - 28.63)10^{-5}}{1 + 1.01(t - 32)10^{-4}}\right] h_{ti} \tag{A2.17}$$

This temperature correction is zero at $t = 28.63°\text{F}$ for all values of h_{ti}. It can be approximated with very little loss of accuracy as

$$C_t \simeq -9(t - 28.6)10^{-5} h_{ti} \tag{A2.18}$$

The uncertainty introduced in h_{to} by equation (A2.18) is always less than 0.001 in. Hg for values of h_{ti} from 28.5 to 31.5 in. Hg and values of t from 60 to 100°F.

The specific weight γ_{Hg} in equation (A2.12) must be based upon the local value of gravity and the reference temperature t_o; $\gamma_{\text{Hg}} = \gamma_{s,to}(1 + C_g)$, where C_g is the gravity correction term given by equation (A2.2) and $\gamma_{s,to}$ is 0.491154 lb/in.^{3} Hg when $t_o = 32°\text{F}$.

When the barometer is read at an elevation other than that of the test site, an altitude correction factor must be applied to the local absolute barometric pressure. The altitude correction factor may be obtained from the pressure–height relationship to be derived in Chapter 3. When this is done, we obtain

$$C_z = p_{\text{baro}}[e^{(Z_{\text{baro}} - Z_{\text{site}})/RT} - 1] \tag{A2.19}$$

where Z_{baro} and Z_{site} are altitudes in feet, R is the gas constant (53.35 ft/°R), and T is the absolute temperature in degrees Rankine.

Other factors may also contribute to the uncertainty of h_{ti}. Proper illumination is essential to define the location of the crown of the meniscus. Precision meniscus sighting under optimum viewing conditions can approach ± 0.001 in. With proper lighting, contact between the ivory index and the mercury surface in the cistern can be detected to much better than ± 0.001 in.

To keep the uncertainty in h_{ti} within 0.01% ($\simeq$ 0.003 in. Hg), the mercury temperature must be known within $\pm 1°$F. Scale temperature need not be known to better than $\pm 10°$F for comparable accuracy. Uncertainties caused by nonequilibrium temperature conditions can be avoided by installing the barometer in a uniform temperature room.

The barometer tube must be vertically aligned for accurate pressure determination. This is accomplished by a separately supported ring encircling the cistern; adjustment screws control the horizontal position.

Depression of the mercury column in commercial barometers is accounted for in the initial calibration setting at the factory. The quality of the barometer is largely determined by the bore of the glass tube. Barometers with a bore of $\frac{1}{4}$ in. are suitable for readings of about 0.01 in., whereas barometers with a bore of $\frac{1}{2}$ in. are suitable for readings down to 0.002 in.

A2.6 McLEOD GAGE

The *McLeod gage* is used in making low-pressure measurements. This instrument is shown in Figure A2.9 to consist of glass tubing arranged so that a sample of gas at unknown pressure can be trapped and then isothermally compressed by a rising mercury column. This amplifies the unknown pressure and allows measurement by conventional manometric means. All the mercury is initially contained in the area below the cutoff level. The gage is first exposed to the unknown gas pressure, p_1; the mercury is then raised in tube A beyond the cutoff, trapping a gas sample of initial volume $\bar{V}_1 = \bar{V} + ah_c$. The mercury is continuously forced upward until it reaches the zero level in the reference capillary B. At this time the mercury in the measuring capillary C reaches a level h, where the gas sample is at its final volume, $\bar{V}_2 = ah$, and at the amplified pressure, $p_2 = p_1 + h$. Then

$$p_1\bar{V}_1 = p_2\bar{V}_2 \tag{A2.20}$$

$$p_1 = \frac{ah^2}{\bar{V}_1 - ah} \tag{A2.21}$$

If $ah \ll \bar{V}_1$, as is usually the case,

$$p_1 = \frac{h^2}{\bar{V}_1} \tag{A2.22}$$

The larger the volume ratio $(\bar{V}_1/\bar{V}_2)$, the greater will be the amplified pressure p_2 and the manometer reading h. Therefore, it is desirable that measuring

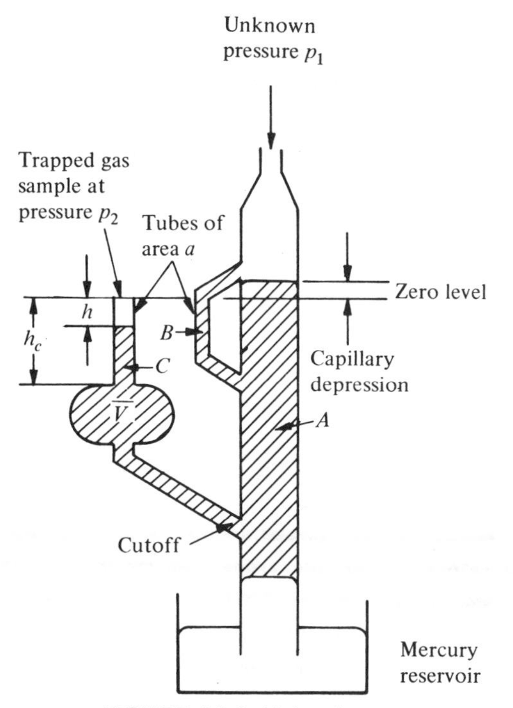

FIGURE A2.9 McLeod gage.

tube C have a small bore. Unfortunately, for tube bores under 1 mm, the compression gain is offset by reading uncertainty caused by capillary effects.

Reference tube B is introduced to provide a meaningful zero for the measuring tube. If the zero is fixed, equation (A2.22) indicates that manometer indication h varies nonlinearly with initial pressure p_1. A McLeod gage with an expanded scale at the lower pressures exhibits a higher sensitivity in this region. The McLeod pressure scale, once established, serves equally well for all the permanent gases (those whose critical pressure is appreciably below room temperature).

There are no corrections to be applied to the McLeod gage reading, but certain precautions should be taken. Moisture traps must be provided to avoid taking any condensable vapors into the gage. Such vapors occupy a larger volume at the initial low pressures than they occupy in the liquid phase at the high reading pressures. Thus, the presence of condensable vapors always causes pressure readings to be too low. Capillary effects, although partially counterbalanced by using a reference capillary, can still introduce significant uncertainties, since the angle of contact between mercury and glass can vary $\pm 30°$. Finally, since the McLeod gage does not give continuous readings, steady-state conditions must prevail for the measurements to be useful.

The mercury piston of the McLeod gage can be actuated in a number of ways. A mechanical plunger may force the mercury up tube *A*, while a partial vacuum over the mercury reservoir holds the mercury below the cutoff until the gage is charged. The mercury can then be raised by bleeding dry gas into the reservoir. There are also several types of swivel gages where the mercury reservoir is located above the gage zero during charging. A 90° rotation of the gage causes the mercury to rise in tube *A* by the action of gravity. In a variation of the McLeod principle, the gas sample may be compressed between two mercury columns, thus avoiding the need for a reference capillary and a sealed-off measuring capillary.

A2.7 PRESSURE TRANSDUCERS

In the earlier sections of this appendix we discussed five pressure standards that can be used for either calibration or the measurement of pressure in static systems. In this section we discuss some commonly used devices that are used for making measurements using an elastic element to convert fluid energy to mechanical energy. Such a device is known as a *pressure transducer*. Examples of mechanical pressure transducers having elastic elements only are dead-weight free-piston gages, manometers, Bourdon gages, bellows, and diaphragm gages.

Electrical transducers have an element that converts this displacement to an electrical signal. Active electrical transducers generate their own voltage or current output as a function of displacement. Passive transducers require an external signal. The piezoelectric pickup is an example of an active electrical pressure transducer. Electric elements employed in passive electrical pressure transducers include strain gages, slide-wire potentiometers, capacitance pickups, linear differential transformers, and variable-reluctance units.

The Bourdon gage was described briefly earlier in Chapter 2. In this transducer the elastic element is a small-volume tube that is fixed at one end but free at the other end to allow displacement under the deforming action of the pressure difference across the tube walls. In the most common model, shown in Figure A2.10, a tube with an oval cross section is bent in a circular arc. Under pressure the tube tends to become circular, with a subsequent increase in the radius of the arc. By an almost frictionless linkage, the free end of the tube rotates a pointer over a calibrated scale to give a mechanical indication of pressure.

The reference pressure in the case containing the Bourdon tube is usually atmospheric, so that the pointer indicates gage pressures. Absolute pressures can be measured directly without evacuating the complete gage casing by biasing a sensing Bourdon tube against a reference Bourdon tube, which is evacuated and sealed as shown in Figure A2.11. Bourdon gages are available

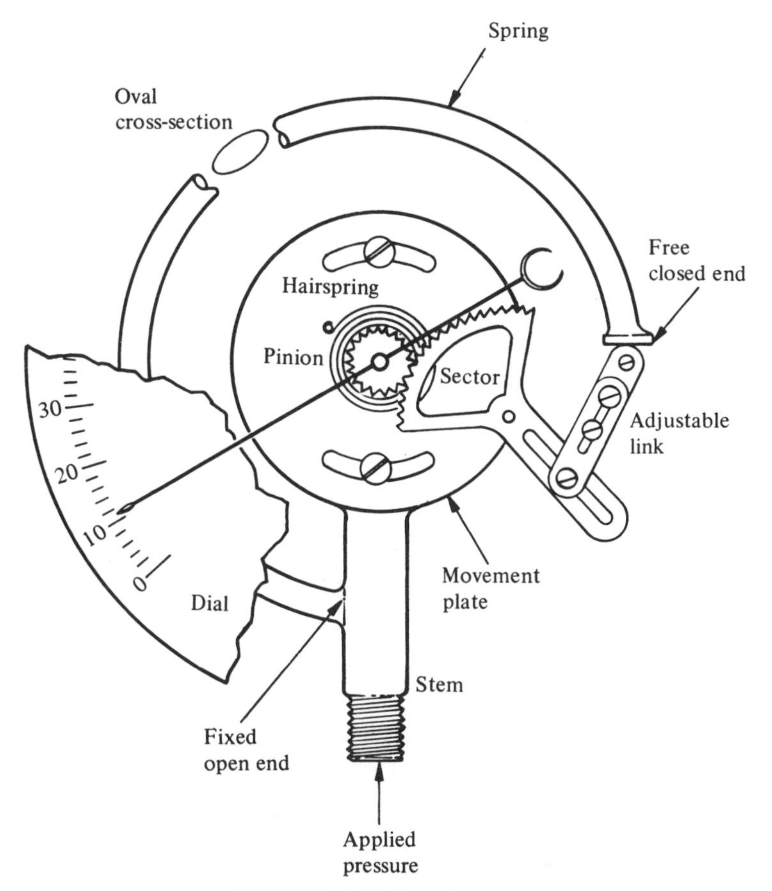

FIGURE A2.10 Bourdon gage.

for wide ranges of absolute, gage, and differential pressure measurements within a calibration uncertainty of 0.1% of the reading.

Another common elastic element used in pressure transducers is the bellows, shown as a gage element in Figure A2.12.

In one arrangement, pressure is applied to one side of a bellows and the resulting deflection is partially counterbalanced by a spring. In a differential arrangement, one pressure is applied to the inside of one sealed bellows, the other pressure is led to the inside of another sealed bellows, and the pressure difference is indicated by a pointer.

A final elastic element to be mentioned because of its widespread use in pressure transducers is the diaphragm. One such arrangement is shown in Figure A2.13. Such elements may be flat, corrugated, or dished plates; the choice depends on the strength and amount of deflection desired. In high-precision instruments, a pair of diaphragms is used back to back to form an

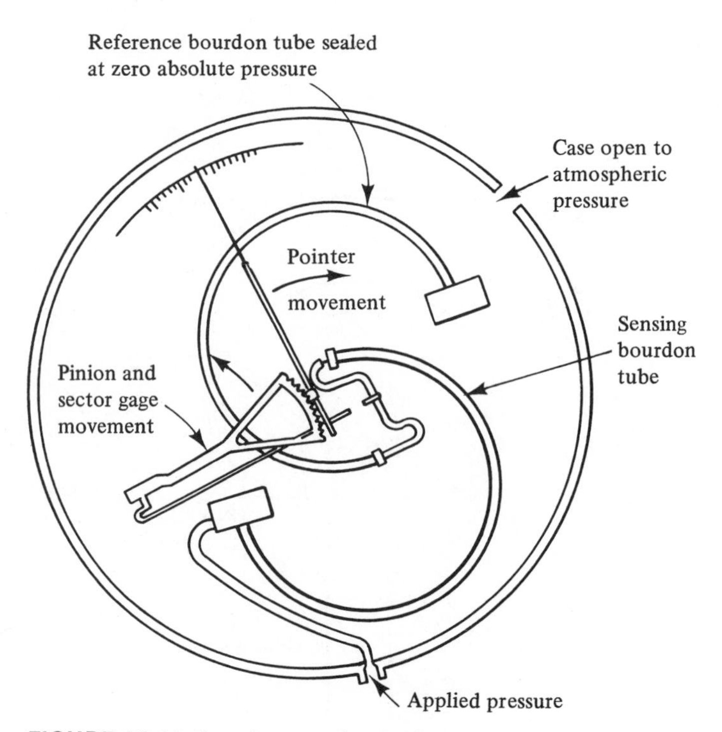

FIGURE A2.11 Bourdon gage for absolute pressure measurement.

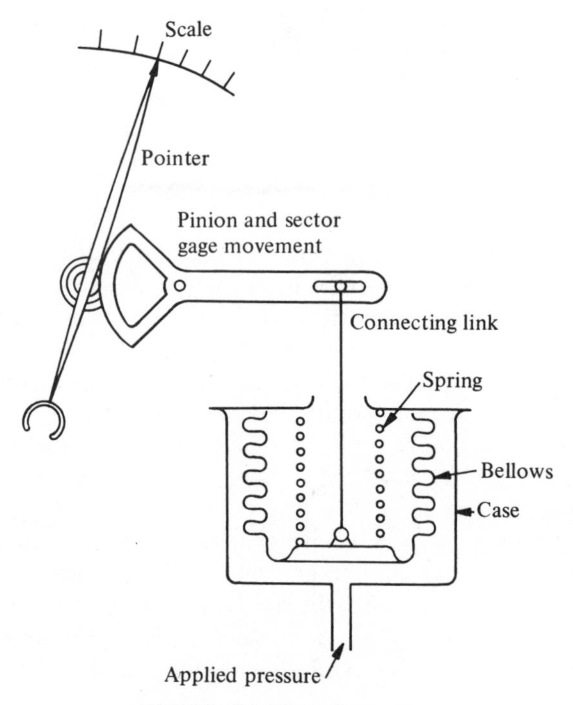

FIGURE A2.12 Bellows gage.

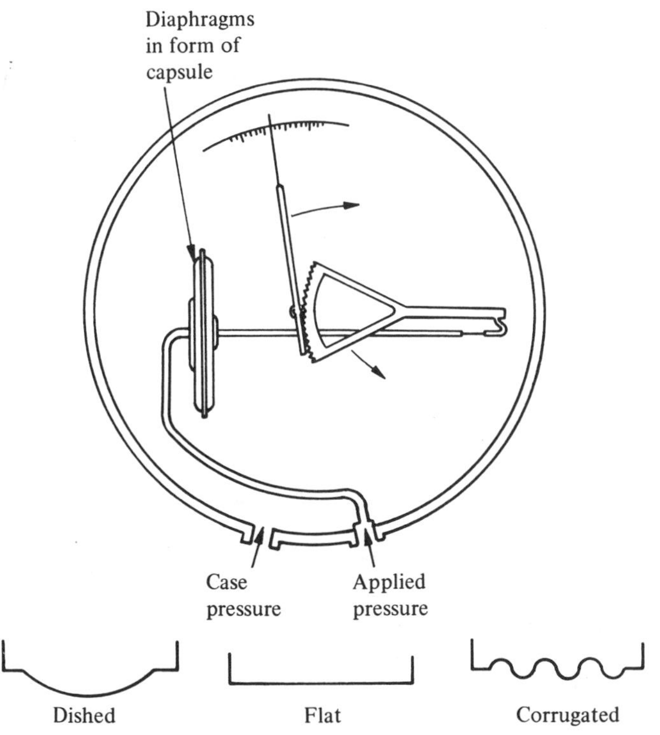

FIGURE A2.13 Diaphragm-based pressure transducer.

elastic capsule. One pressure is applied to the inside of the capsule; the other pressure is external. The calibration of this differential transducer is relatively independent of pressure magnitude.

Thus far we have discussed a few mechanical pressure transducers. In many applications it is more convenient to use transducer elements that depend on the change in the electrical parameters of the element as a function of the applied pressure. The only active electrical pressure transducer in common use is the piezoelectric transducer. Sound-pressure instrumentation makes extensive use of piezoelectric pickups in such forms as hollow cylinders and disks. Piezoelectric pressure transducers are also used in measuring rapidly fluctuating or transient pressures. These transducers are difficult to calibrate by static procedures. In a recently introduced technique called electrocalibration, the transducer is calibrated by electric field excitation rather than by physical pressure. The most common passive electrical pressure transducers are the variable-resistance types.

The strain gage is probably the most used pressure transducer element. Strain gages operate on the principle that the electrical resistance of a wire varies with its length under load. In unbonded strain gages, four wires run between electrically insulated pins located on a fixed frame and other pins

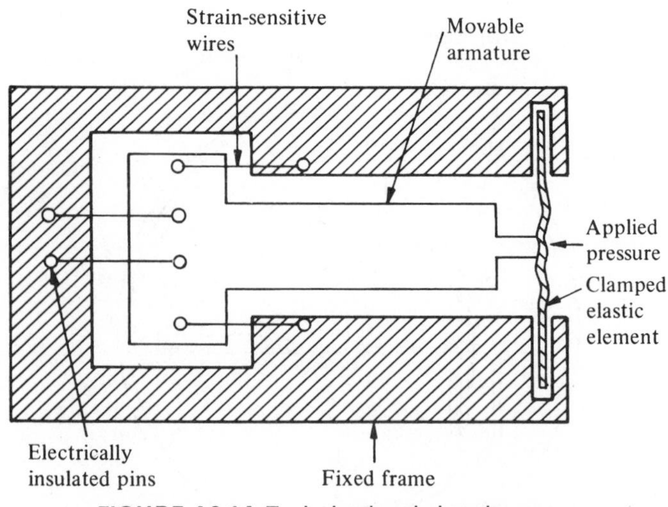

FIGURE A2.14 Typical unbonded strain gage.

located on a movable armature, as shown in Figure A2.14. The wires are installed under tension and form the legs of a bridge circuit. Under pressure the elastic element (usually a diaphragm) displaces the armature, causing two of the wires to elongate while reducing the tension in the remaining two wires. The resistance change causes a bridge imbalance proportional to the applied pressure.

The bonded strain gage takes the form of a fine wire filament, set in cloth, paper, or plastic and fastened by a suitable cement to a flexible plate, which takes the load of the elastic element. This is shown in Figure A2.15. Two similar strain gage elements are often connected in a bridge circuit to balance

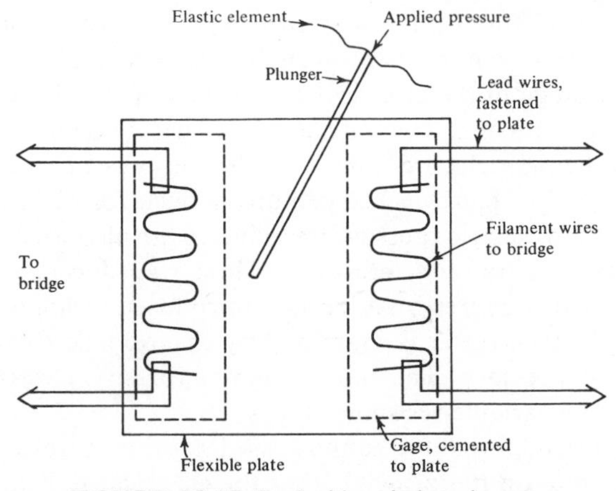

FIGURE A2.15 Typical bonded strain gage.

unavoidable temperature effects. The nominal bridge output impedance of most strain gage pressure transducers is 350 Ω, nominal excitation voltage is 10 V (ac or dc), and natural frequency can be as high as 50 Hz. Transducer resolution is infinite, and the usual calibration uncertainty of such gages is within 1% of full scale.

Many other forms of electrical pressure transducers are in use in industry, and only a few of these will be discussed briefly. These elements fall under the categories of potentiometer, variable-capacitance, linear variable differential transformer (LVDT), and variable-reluctance transducers. The potentiometer types are those that operate as variable resistance pressure transducers. In one arrangement, the elastic element is a helical Bourdon tube, while a precision wire-wound potentiometer serves as the electric element. As pressure is applied to the open end of the Bourdon tube, it unwinds, causing the wiper (connected directly to the closed end of the tube) to move over the potentiometer.

In the variable-capacitance pressure transducer, the elastic element is usually a metal diaphragm that serves as one plate of a capacitor. Under an applied pressure the diaphragm moves with respect to a fixed plate. By means of a suitable bridge circuit, the variation in capacitance can be measured and related to pressure by calibration.

The electric element in an LVDT is made up of three coils mounted in a common frame to form the device shown in Figure A2.16. A magnetic core centered in the coils is free to be displaced by a bellows, Bourdon, or diaphragm elastic element. The center coil, the primary winding of the transformer, has an ac excitation voltage impressed across it. The two outside coils form the secondaries of the transformer. When the core is centered, the induced voltages in the two outer coils are equal and 180° out of phase; this

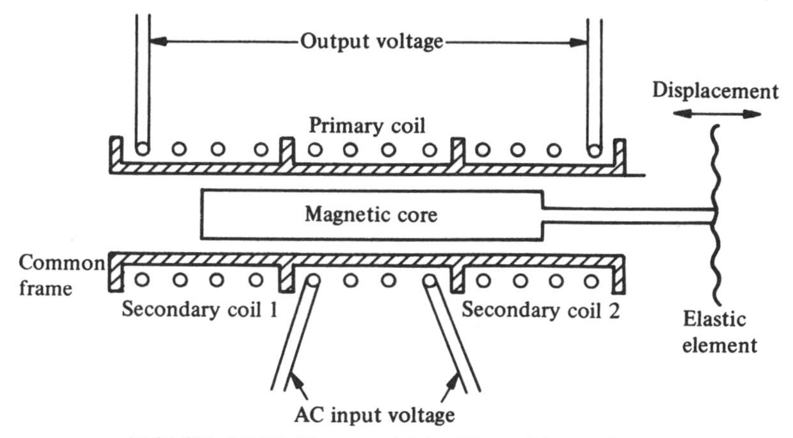

FIGURE A2.16 Linear variable differential transformer.

represents the zero-pressure position. However, when the core is displaced by the action of an applied pressure, the voltage induced in one secondary increases, while that in the other decreases. The output-voltage difference varies essentially linearly with pressure for the small core displacements allowed in these transducers. The voltage difference is measured and related to the applied pressure by calibration.

The last electric element that will be discussed is the variable-reluctance pressure transducer. The basic element of the variable-reluctance pressure transducer is a movable magnetic vane in a magnetic field. In one type the elastic element is a flat magnetic diaphragm located between two magnetic output coils. Displacement of the diaphragm changes the inductance ratio between the output coils and results in an output voltage proportional to pressure.

In this appendix the pressure elements that have been discussed were primarily for the measurement of pressures in systems that are at rest and in which the fluid is in equilibrium. The measurement of pressure in systems in which the fluid is moving requires the further definition of pressure as well as an understanding of the pressure gradients, turbulence, viscous effects, and time-dependent effects that occur in such systems. The usual elements for making these measurements are the static or wall tap, static tubes, aerodynamic probes, Pitot tubes, Pitot-static tubes, and so on. The principles of these devices will be discussed elsewhere in this text. The student desiring further information is referred to the article by R. P. Benedict cited earlier in this appendix or the book by Benedict (*Fundamentals of Temperature, Pressure and Flow Measurements* by R. P. Benedict, John Wiley & Sons, Inc., New York, 1969) for an extensive bibliography on this subject.

Fluid Statics

3.1 INTRODUCTION

The study of fluids at rest or having no velocity with respect to an observer in a gravitational field is known as *fluid statics*. Since the fluid will be at rest with respect to an observer, there will be no relative motion between adjacent fluid layers and viscosity will not enter into problems in fluid statics. These considerations enable us to treat fluids at rest mathematically, and the results of these mathematical studies will be very accurate for engineering purposes. For most engineering purposes, if a fluid is at rest with respect to a system that is moving with a uniform translation relative to the earth, it can be treated as if it were at rest since it will have no acceleration with respect to the earth.

To solve problems in fluid statics it is only necessary to use the principle of equilibrium of bodies that is developed in mechanics. Thus, for a body (or element of mass) to be in equilibrium, it is necessary that the sum of the external forces and moments acting on the body be zero. This principle, combined with a knowledge of the fluid density, enables one to readily analyze problems in fluid statics.

3.2 PRESSURE RELATIONSHIPS

In Chapter 2 the pressure at the base of a uniform column of liquid was derived as a function of the specific weight of the liquid and the height of the liquid column. It is useful at this time to again consider this same problem from a slightly different viewpoint. Consider a cylinder of fluid having a height h and a specific weight γ. The height of the column will be measured as positive in the vertical direction, as shown in Figure 3.1. We now apply the con-

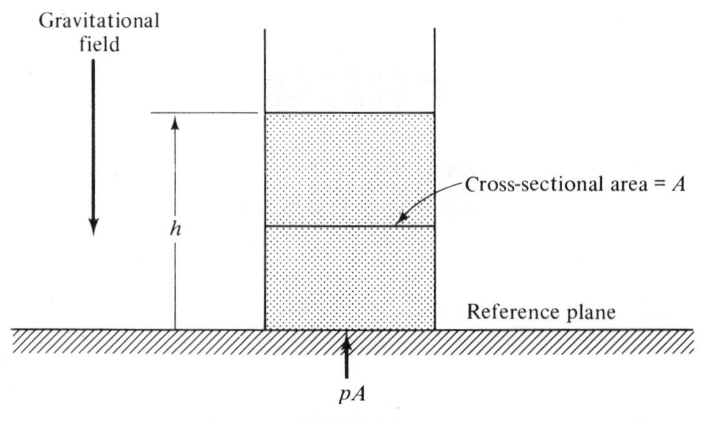

FIGURE 3.1

dition of equilibrium that the sum of the forces in the vertical direction at the base of the column must be zero. Since we have the weight acting down as,

$$W = \gamma h A \tag{3.1}$$

then,

$$pA = \gamma h A \tag{3.2}$$

or

$$p = \gamma h \tag{3.2a}$$

Since we are measuring height, h, to be positive in the "up" direction, it is necessary to introduce a minus sign into equation (3.2a) to account for this, since weight increases as we proceed "down" the column. Equation (3.2a) is therefore rewritten as

$$p = -\gamma h \tag{3.2b}$$

which corresponds to equation (2.18a). Note that equation (3.2a) is for the pressure due to the liquid column only; the negative sign is to be interpreted to mean that pressure decreases as we go up the column of liquid, and γ was taken to be constant.

Equations (3.2) can also be interpreted by referring to Figure 3.2, which

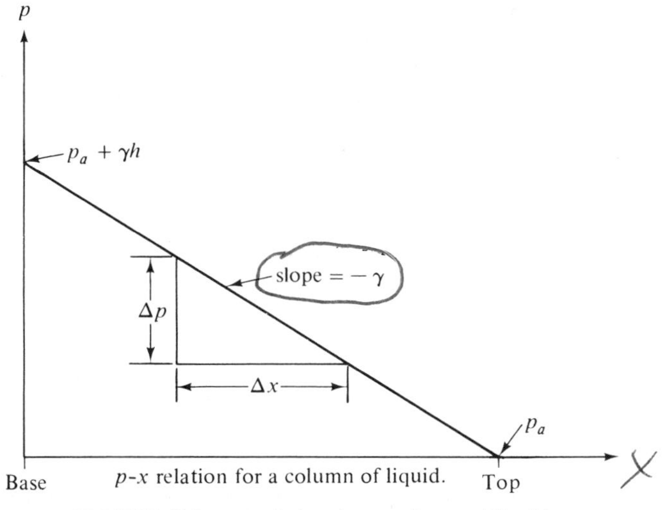

FIGURE 3.2 *p–x* relation for a column of liquid.

is a plot of pressure against height. As can be seen, γ is simply the constant slope, and from Figure 3.2 we can write directly

$$p = p_a + \gamma h - \gamma x = p_a + \gamma(h - x) \tag{3.3}$$

where p_a is the pressure at the free surface on top of the column. Equations (3.2) represent the fundamental relations among pressure, specific weight, and column height. As stated earlier, the negative sign indicates that the pressure decreases as one goes up the column. Note that the pressure in a column can be specified by stating feet of a fluid of a given density. In the nomenclature of hydraulics, this was known as a "head," and this nomenclature is still used today. Thus, 3 ft of water represents the head of water or a gage pressure of $3 \times 62.4 = 187.2$ psfg or 1.3 psig.

If the liquid does not have a constant density (say due to temperature gradients or to pressure effects), it is necessary to evaluate equations (3.2) taking into account the fact that γ is a variable. This can be done by either using the methods of calculus or by graphically solving the equation. One way of evaluating this equation graphically is to evaluate the mean or average value of γ for a specified variation of γ as a function of height. Illustrative Problem 3.1 demonstrates this procedure.

ILLUSTRATIVE PROBLEM 3.1

If γ varies linearly from γ_1 at the bottom of a column of fluid to γ_2 at the top of the column, determine the pressure at the bottom of the column.

Determine Pressure at bottom of column

Solution

A plot of γ against height is shown in Figure 3.3. By definition, the "average" is that value of γ that when multiplied by the base of the figure yields the area under the curve of the function. Therefore, the area of the triangle must

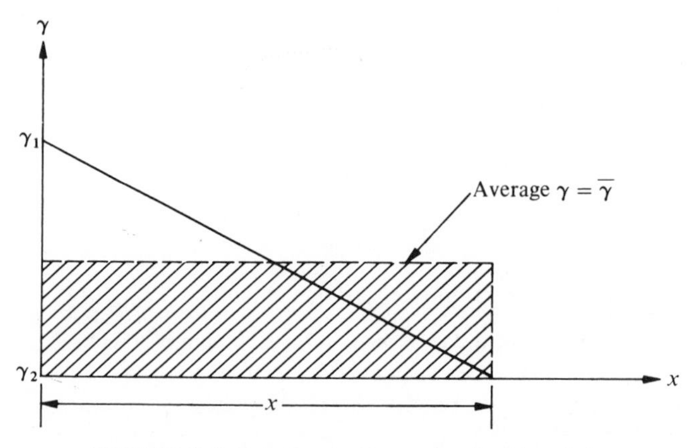

FIGURE 3.3 Solution to Illustrative Problem 3.1.

equal the area of the shaded rectangle shown in the figure. Since the area of a triangle is half the base multiplied by the altitude,

$$\frac{x(\gamma_1 + \gamma_2)}{2} = \bar{\gamma}x$$

or

$$\bar{\gamma} = \frac{\gamma_1 + \gamma_2}{2}$$

Therefore, the average γ, $\bar{\gamma}$, is simply the arithmetic mean of γ_1 and γ_2 *for this case only*. This corresponds to a uniform column of liquid of specific weight $(\gamma_1 + \gamma_2)/2$ in a tube.

At this time let us examine Figure 3.4, where four different containers hold the same specific weight fluid and are open to the atmosphere at the top. At the top, level A, the pressure is atmospheric. At some other level, level B, which is a distance x below level A, the pressure in *each* of the containers will be the *same*, since $p_B = p_A + \gamma x$. The variation in pressure is only a function of the specific weight of the fluid and the vertical depth. It is independent of the size, shape, or orientation of the container. This is also shown in Figure

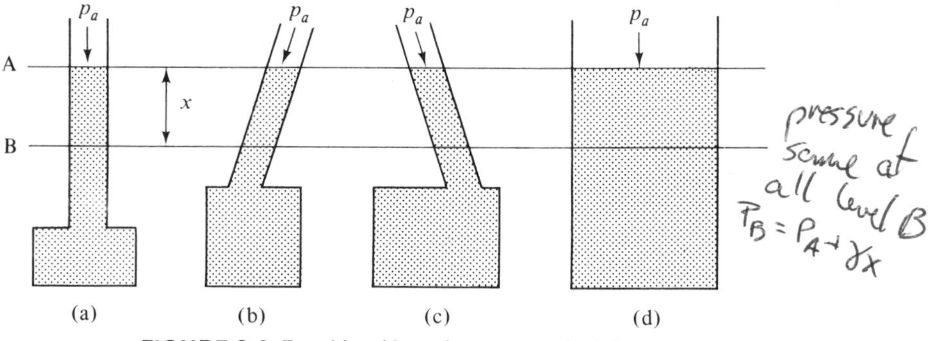

FIGURE 3.4 Equal level/equal pressure principle.

3.5, where we see a tank having several different sections. Notice that $p_1 = p_2 = p_3$; that is, a line of constant height is also a line of constant pressure. This fact has been termed the *equal level/equal pressure principle* and forms the basis for the measurement of pressure using a barometer or manometer.

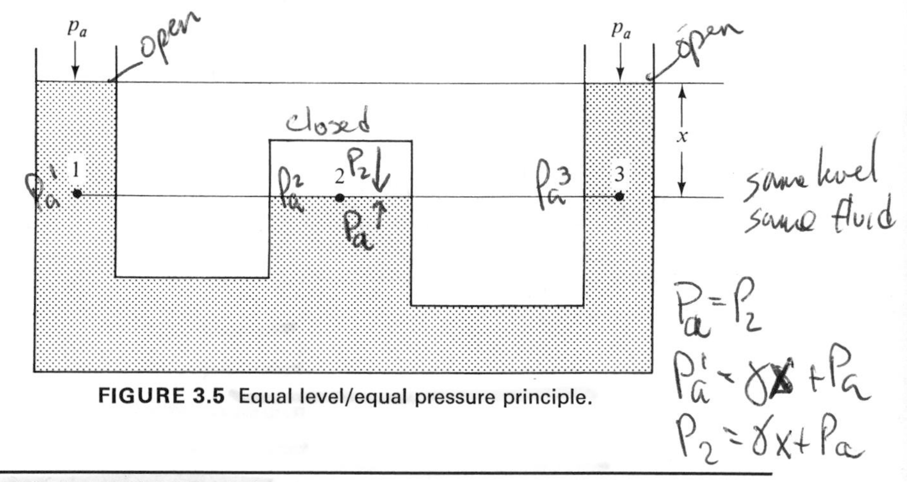

FIGURE 3.5 Equal level/equal pressure principle.

ILLUSTRATIVE PROBLEM 3.2

A tube is filled with mercury (specific gravity = 13.6) and is inverted into a well of mercury as shown in Figure 3.6. Neglecting the pressure of the vapor in the tube above the mercury, how high will the column of mercury be in the tube above the level in the well?

Solution

At level A in both the tube and the well, the pressure must be the same and for this problem it will be p_a, atmospheric pressure. If we proceed from the

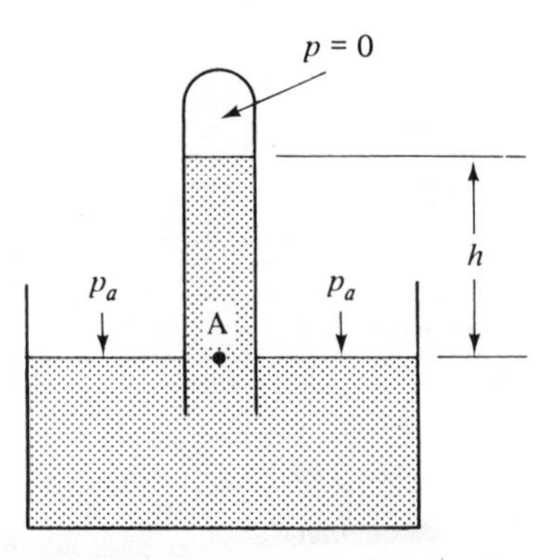

FIGURE 3.6 Illustrative Problem 3.2.

top of the tube down to level A, we have $p_A = 0 + \gamma h$. But this must equal p_a. Therefore,

$$p_a = 0 + \gamma h = \gamma h$$

and

$$h = \frac{p_a}{\gamma}$$

If atmospheric pressure is 101 325 Pa,

$$h = \frac{101\ 325\ \text{N/m}^2}{1000\ \text{kg/m}^3 \times 9.806\ \text{N/kg} \times 13.6} = 0.760\ \text{m} = 760\ \text{mm}$$

We can generalize this result in the following manner:

$$1\ \text{standard atmosphere} = \begin{cases} 14.7\ \text{psia} \\ 2116\ \text{psfa} \\ 29.92\ \text{in. Hg} \\ 760\ \text{mm Hg} \\ 101{,}325\ \text{Pa} \\ 10.34\ \text{m of water} \\ 33.91\ \text{ft of water} \\ 1.01\ \text{bars} \end{cases}$$

(The material in Appendix 2-1 should be reviewed at this time, especially the section on the barometer.)

There is one case of variable density that is of special use. Consider a condition where the local atmospheric pressure at sea level is known. If an airplane is at 1000 ft above sea level, what would the pressure be? Conversely, if the reading of a barometer is known at sea level and some unknown elevation, what is the elevation? To answer this problem we shall make the assumption that the air temperature remains constant and apply equation (3.2b), rewritten to account for the variation in γ, as

$$\frac{\Delta p}{\Delta x} = -\gamma \tag{3.3}$$

The relation that governs the pressure, temperature, and density for an *ideal gas* (air can be so considered for our purposes) is

$$pv = RT \quad \text{or} \quad \frac{p}{\gamma} = RT \tag{3.4}$$

where p is the pressure in psfa, R is a constant for a gas (53.3 for air in these engineering units), and T is the absolute temperature in degrees Rankine. In SI units, p is in Pa, T is in degrees Kelvin, γ is in N/m^3, and R for air is 29.24 in these units. Inserting γ from equation (3.4) into equation (3.3),

$$\frac{\Delta p}{\Delta x} = \frac{-p}{RT} \tag{3.5}$$

or

$$\frac{\Delta p}{p} = \frac{-\Delta x}{RT} \tag{3.6}$$

To obtain a solution to equation (3.6), it is necessary to sum this equation between the required pressure limits. Graphically, this process of summation can be illustrated by referring to Figure 3.7. Plotting $1/p$ against p gives us the coordinates of this graph. By selecting a value of Δp as shown, the shaded area gives us $\Delta p/p$. Therefore, the sum of the $\Delta p/p$ values is the area under the curve between the limits of p_1 and p_2. By the methods of calculus, it can be shown that this area is $\ln (p_2/p_1)$, where $\ln x = \log_e x = 2.3026 \log_{10} x$.

Returning to equation (3.6) and noting that R and T are constants for this problem, we can easily evaluate the right-hand side graphically by plotting x against $1/RT$, as shown in Figure 3.8. For all values of Δx, $1/RT$ is a constant and the curve is simply a rectangle whose area is x/RT. Thus, equation (3.4) becomes

$$\ln \frac{p_1}{p_1} = -\frac{x}{RT}$$

or

$$\ln \frac{p_1}{p_2} = \frac{x}{RT} \tag{3.7}$$

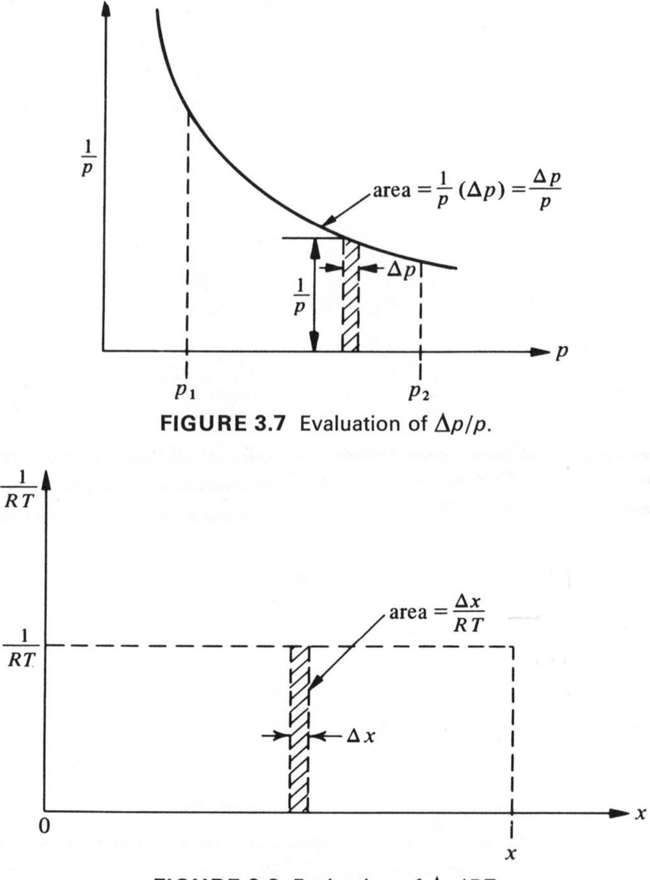

FIGURE 3.7 Evaluation of $\Delta p/p$.

FIGURE 3.8 Evaluation of $\Delta x/RT$.

The readings of a barometer are directly proportional to the local atmospheric pressure and therefore the ratio p_1/p_2 can be replaced by the barometer readings b_1 and b_2 as the ratio b_1/b_2. Equation (3.7) becomes

$$\ln \frac{b_1}{b_2} = \frac{x}{RT} \tag{3.8}$$

where x is the difference in elevation and b_1 and b_2 are the barometer pressures at sea level and elevation, respectively.

ILLUSTRATIVE PROBLEM 3.3

A barometer reads 760 mm Hg at sea level and 750 mm Hg at some other elevation. If the air temperature can be taken to be constant and equal to 20°C (293 K), determine the unknown elevation.

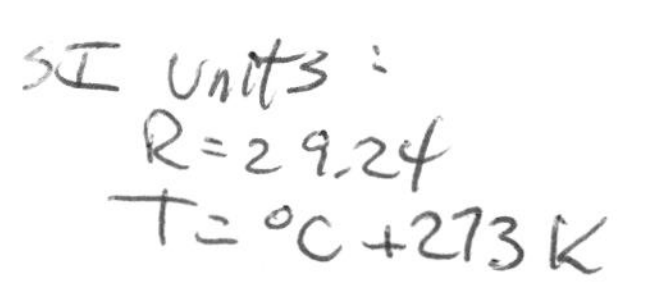

Solution

Using equation (3.8), we obtain

$$\ln \frac{760}{750} = \frac{x}{29.24 \times 293}$$

Therefore,

$$x = 29.24 \times 293 \ln \frac{760}{750}$$

$$= 113.48 \text{ m} \quad (372 \text{ ft})$$

The student should note that for small elevation differences the assumption of constant temperature is quite good. For large elevation differences there is a large change in temperature and equation (3.7) is no longer applicable.

For convenience and for standardization of performance data, a *standard atmosphere* has been defined; it is given in Table 3.1. It will be noted from

TABLE 3.1 The ICAO Standard Atmosphere†

Altitude (ft)	Temperature (°F)	Pressure (psia)	Specific Weight (lb/ft³)	Density (slug/ft³)	Viscosity × 10^7 (lb sec/ft²)
0	59.00	14.696	0.07648	0.002377	3.737
5,000	41.17	12.243	0.06587	0.002048	3.637
10,000	23.36	10.108	0.05643	0.001756	3.534
15,000	5.55	8.297	0.04807	0.001496	3.430
20,000	—12.26	6.759	0.04070	0.001267	3.325
25,000	—30.05	5.461	0.03422	0.001066	3.217
30,000	—47.83	4.373	0.02858	0.000891	3.107
35,000	—65.61	3.468	0.02367	0.000738	2.995
40,000	—69.70	2.730	0.01882	0.000587	2.969
45,000	—69.70	2.149	0.01481	0.000462	2.969
50,000	—69.70	1.690	0.01165	0.000364	2.969
55,000	—69.70	1.331	0.00917	0.000287	2.969
60,000	—69.70	1.049	0.00722	0.000226	2.969
65,000	—69.70	0.826	0.00568	0.000178	2.969
70,000	—69.70	0.650	0.00447	0.000140	2.969
75,000	—69.70	0.512	0.00352	0.000110	2.969
80,000	—69.70	0.404	0.00277	0.000087	2.969
85,000	—65.37	0.318	0.00216	0.000068	2.997
90,000	—57.20	0.252	0.00168	0.000053	3.048
95,000	—49.05	0.200	0.00131	0.000041	3.099
100,000	—40.89	0.160	0.00102	0.000032	3.150

†Reproduced with permission from *Elementary Fluid Mechanics* by J. K. Vennard, McGraw-Hill Book Company, New York, 1961, p. 548.

this table that the temperature is not constant and that the equations previously derived are valid only for small changes in elevation.

If h is the height above sea level in feet and the temperature is assumed to decrease linearly, it can be approximated by

$$T = (519 - 0.00357h)°\text{R} \tag{3.9}$$

At, 35,000 ft, the temperature becomes −67°F and is assumed to be constant above this level.

ILLUSTRATIVE PROBLEM 3.4

A barometer reads 760 mm Hg at sea level and 20°C. It is taken up a hill to a location where the elevation is 300 m above sea level. If the temperature stays constant, what is the expected reading of the barometer?

Solution

Applying equation (3.8), we obtain

$$\ln \frac{b_1}{b_2} = \frac{x}{RT}$$

$$\ln \frac{760}{b_2} = \frac{300}{29.24 \times (20 + 273)}$$

Since $\ln (760/b_2) = \ln 760 - \ln b_2$,

$$\ln 760 - \ln b_2 = \frac{300}{29.24(20 + 273)}$$

$$6.63332 - \ln b_2 = 0.03502$$

$$\ln b_2 = 6.59830$$

$$b_2 = 733.8 \text{ mm}$$

3.3 PRESSURE MEASUREMENT—MANOMETRY[1]

Since pressure is a property of a system, its measurement is both desirable and necessary. In addition, as will be seen in later chapters, a knowledge of the pressure in various portions of a system is necessary to the determination

[1]This student is referred to Appendix 2.1 for a more detailed discussion of pressure measurement.

of the flow and the thermodynamic state of a system. The simplest and most widely used mechanical pressure gage is the Bourdon gage. As is repeated in Figure 3.9, this gage consists simply of a bent tube closed on one end with a mechanism to indicate the movement of the closed end. The tube is usually noncircular in cross section and pressure is applied at the open end. When a pressure in excess of atmospheric pressure is applied to the open end, the closed end will move as the tube tends to straighten out (similar to a New Year's Eve blowout noisemaker) and the reading of the gage is proportional to the displacement of the tube. This type of gage can readily be calibrated and is rugged and easy to use. It is equally capable of measuring pressures above and below local atmospheric pressure. The student should note that the device illustrated measures gage pressure only. It is possible to evacuate the case of the gage and thereby create a gage capable of measuring absolute pressures.

If the gage is connected at an elevation different from the elevation where the pressure is to be measured, it is necessary to make a correction to the reading of the gage. Referring to Figure 3.9, it will be seen that if the gage is mounted above a pipe, the gage will read *low* by an amount equal to γh. The

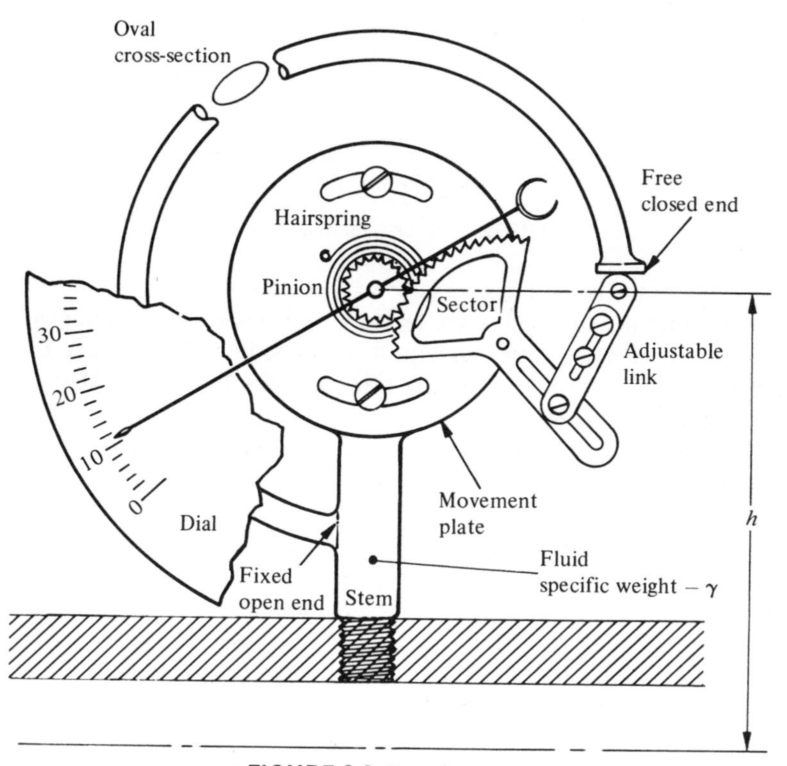

FIGURE 3.9 Bourdon gage.

specific weight γ should be obtained at the temperature of the fluid in the connecting line to the gage since this may be different from the temperature in the pipe. For commercial usage, this correction is usually negligible.

We have already shown that the pressure at the base of a column of liquid is simply a function of the height of the column and the specific weight of the liquid. Therefore, the height of a column of liquid of known specific weight can be and is used to measure pressure and pressure differences. Instruments that utilize this principle are known as manometers and the study of these pressure-measuring devices is known as manometry. By properly arranging a manometer and selecting the fluid judiciously, it is possible to measure extremely small pressures, very large pressures, and pressure differences. A simple manometer is shown in Figure 3.10, where the right arm is exposed to

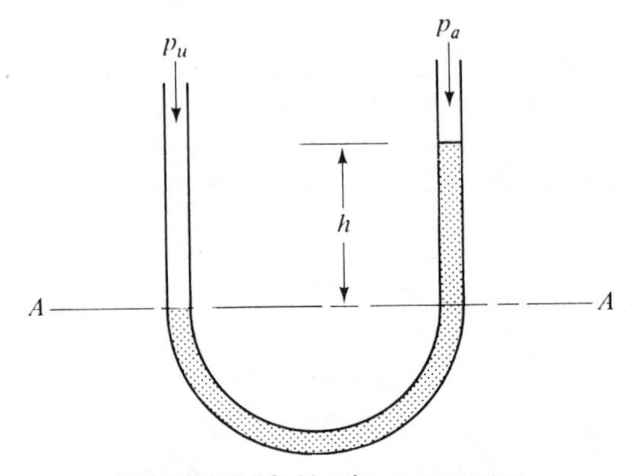

FIGURE 3.10 U-tube manometer.

the atmosphere while the left arm is connected to the unknown pressure. As shown, the fluid is depressed in the left arm and raised in the right arm until no unbalanced pressure forces remain. It has already been demonstrated that the pressure at a given level in either arm must be the same so that we can select any level as a reference and write a relation for the pressure. Actually, it is much easier and more convenient to select the interface between the manometer fluid and the unknown fluid as a common reference level. In Figure 3.10 the pressure at elevation AA is the same in both arms of the manometer based upon the equal level/equal pressure principle. Starting with the open manometer arm (right), we have atmospheric pressure p_a acting on the fluid. As one proceeds down the arm the pressure increases until we arrive at level AA, where the pressure in the right arm is $p_a + \gamma h$. This must be equal to the unknown pressure in the connected arm, p_u. Therefore,

$$p_u = p_a + \gamma h \tag{3.10a}$$

or

$$p_u - p_a = \gamma h \tag{3.10b}$$

ILLUSTRATIVE PROBLEM 3.5

The tank shown in Figure 3.11 is filled with water ($\gamma = 9810\ N/m^3$). The manometer on the side of the tank has air trapped in it as the tank is filled. After filling of the tank, the level of the water in the manometer is found to be 1 m below the level in the tank. Determine the pressure of the trapped air and the pressures at levels *B* and *C*.

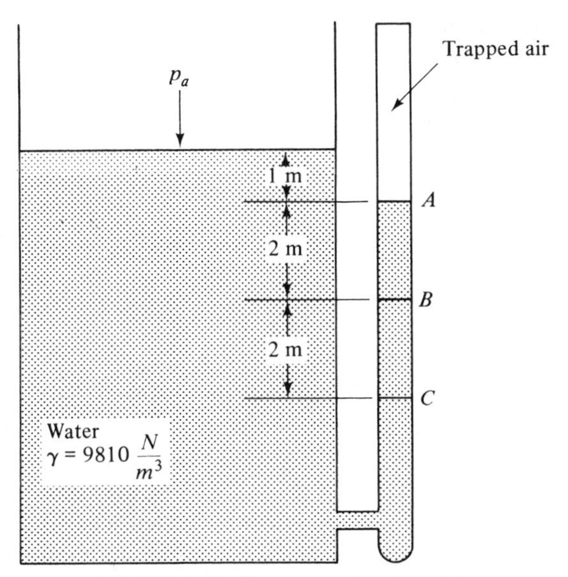

FIGURE 3.11 Illustrative Problem 3.5.

Solution

Starting at the open part of the tank, the pressure at this level is p_a. If we now proceed down the tank to level *A*, the pressure will be $p_a + \gamma h = p_a + 9810(1)$ N. This must also be the pressure of the trapped air, since level *A* is also the interface between the air and the water in the manometer. If we now proceed to level *B*, which is 3 m below the exposed surface of the tank, $p_B = p_a + \gamma h = p_a + 9810(3) = p_a + 29\,430$ N. At level *C* the pressure is $p_a + 9810(5) = p_a + 49\,050$ N. Notice that at each common level, the pressure in the tank and manometer must be the same.

ILLUSTRATIVE PROBLEM 3.6

A manometer is connected to a tank as shown in Figure 3.12. Determine the pressure in the tank if atmospheric pressure is 100 kPa.

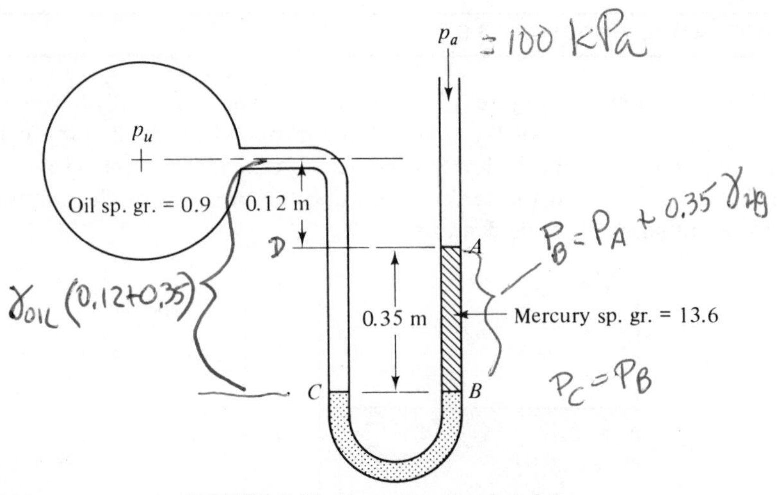

FIGURE 3.12 Illustrative Problem 3.6.

Solution

Starting with the top of the manometer that is open to the atmosphere, we have atmospheric pressure until elevation *A*. If we proceed down the right side of the manometer, we reach level *B*. The pressure at this level, $p_B = p_a + 0.35\,\gamma_{Hg}$. Since *B* and *C* are at the same level, the equal level/equal pressure principle yields $p_C = p_B$. We are now at the left leg of the manometer. If we now proceed up the column to level *D*, the pressure must have *decreased* by an amount $\gamma_{oil}(0.12 + 0.35)$, and this must be p_u, the unknown pressure in the tank. Putting this information together,

$$p_a + \gamma_{Hg}(0.35) - \gamma_{oil}(0.12 + 0.35) = p_u$$

Using $\gamma_{water} = 9.806\ \text{kN/m}^3$ and the data of the problem,

$$100\ \text{kPa} + 13.6 \times 9.806\ \text{kN/m}^3 \times 0.35\ \text{m} - 0.9 \times 9.806\ \text{kN/m}^3 \times 0.47\ \text{m} = p_u$$

$$142.53\ \text{kPa} = p_u$$

This is obviously an absolute pressure. For gage pressure, which is not defined in SI units, we will simply use $p_u - p_a$. Therefore,

$$p_u - p_a = 142.53 - 100 = 42.53\ \text{kPa}$$

The basic utility of the manometer lies in the fact that a linear distance readily permits us to obtain the pressure difference between two points. Therefore, a manometer can and is used to determine pressure differences between the pressures in two containers, or between two points in the same pipe through which a fluid is flowing. Figure 3.13 shows one possible manometer

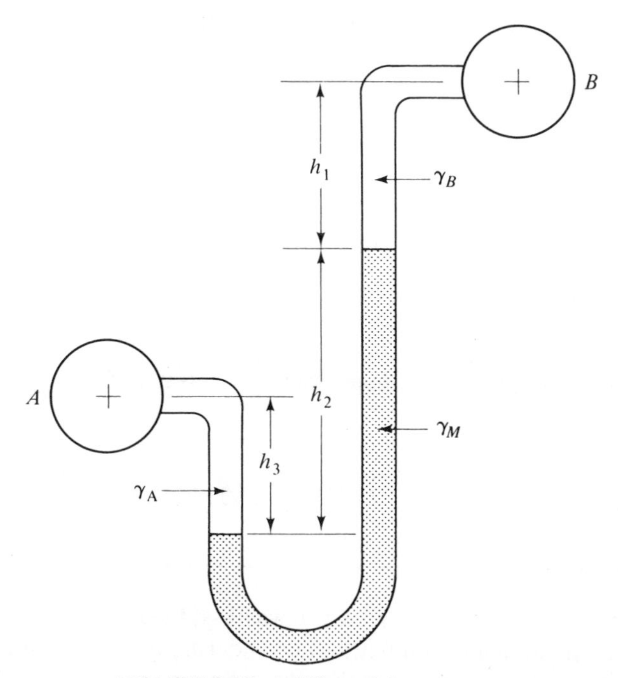

FIGURE 3.13 A differential manometer.

arrangement for measuring the pressure difference between the pressures in two pipes through which different fluids are flowing. This type of manometer is called a *differential manometer* and we can use the principles already developed to obtain the pressure difference between the two pipes. The best way is to start systematically at one end of the system and to apply the equal level/equal pressure principle as required until the other end of the system is reached. Illustrative Problem 3.7 illustrates this procedure.

ILLUSTRATIVE PROBLEM 3.7

If $\gamma_A = 8830$ N/m³, $\gamma_B = 9806$ N/m³, $\gamma_M = 13\,330$ N/m³, $h_1 = 1$ m, $h_2 = 2$ m, and $h_3 = 1.5$ m, determine $p_A - p_B$ for the configuration shown in Figure 3.13.

Solution

Starting at the left, we have p_A. Proceeding down a distance h_3 yields an *increase* in pressure $= \gamma_A h_3$. The same pressure must exist *at the same level* in the right arm. We now proceed up the right arm a distance h_2, giving us a pressure *decrease* of $\gamma_M h_2$. Proceeding further up the right arm yields another pressure *decrease* of $\gamma_B h_1$ and now we are at pipe B whose pressure is p_B. If we now write these relationships in sequence, we have

$$p_A + \gamma_A h_3 - \gamma_M h_2 - \gamma_B h_1 = p_B$$

or

$$p_A - p_B = \gamma_M h_2 + \gamma_B h_1 - \gamma_A h_3$$

Substituting the given data,

$$p_A - p_B = 13\ 330 \frac{\text{N}}{\text{m}^3} \times 2\text{ m} + 9806 \frac{\text{N}}{\text{m}^3} \times 1\text{ m} - 8830 \frac{\text{N}}{\text{m}^3} \times 1.5\text{ m}$$

and

$$p_A - p_B = 23\ 221 \frac{\text{N}}{\text{m}^2} = 23.221\text{ kPa}$$

In the simple U-tube manometer it was assumed that the pressure was acting on the manometer fluid without considering how this came about. Let us consider the case of water (or other fluid) flowing in a pipe and a different fluid (say, mercury or oil) being used in the manometer. This situation is shown in Figure 3.14. The pressure at level A is found from the equilibrium

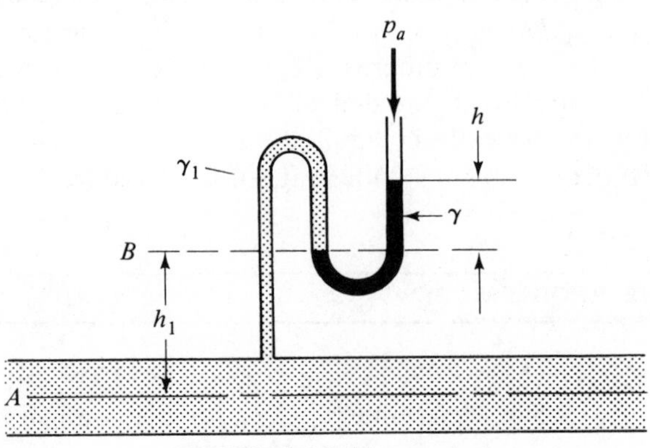

FIGURE 3.14 Manometer.

expression as follows:

$$p_a + \gamma h + \gamma_1 h_1 = p_A \tag{3.11}$$

If gage pressure is desired,

$$p_A = \gamma h + \gamma_1 h_1 \tag{3.12}$$

In equations (3.11) and (3.12) it is assumed that γ_1 is constant and the same in the manometer leg and the pipe. If this is not the case (due, say, to heat transfer), then a correction must be made.

ILLUSTRATIVE PROBLEM 3.8

Water ($\gamma_1 = 62.4\ \text{lb/ft}^3$) flows in a pipe. If a manometer connected as in Figure 3.14 is used with mercury as the fluid ($\gamma = 850\ \text{lb/ft}^3$) and h is 5 in. when the level B is 15 in. above the center line of the pipe, what is the pressure in the pipe?

Solution

Referring to Figure 3.14 and using equation (3.12),

$$p_A = \gamma h + \gamma_1 h_1$$

$$= 850\ \frac{\text{lb}}{\text{ft}^3} \times \frac{5\ \text{in.}}{12\ \text{in./ft}} + \frac{15\ \text{in.}}{12\ \text{in./ft}} \times 62.4\ \frac{\text{lb}}{\text{ft}^3} = 432.2\ \text{psfg}$$

or

psf into psi

$$p_A = 432.2\ \frac{\text{lb}}{\text{ft}^2} \times \frac{1}{(12\ \text{in./ft})^2} = 3\ \text{psig}$$

or

$$p_A = 14.7 + 3 = 17.7\ \text{psia}$$

To achieve greater accuracy and sensitivity in manometers, several different arrangements have been used. Perhaps the simplest of these is the inclined manometer. Consider a relatively large reservior of liquid connected to a small-bore tube that makes an angle θ with the horizontal. The pressure or pressure differential to be measured is connected to the large reservior, while the inclined tube is open-ended. This is shown schematically in Figure 3.15. The unknown pressure p_u is given by

$$p_u = p_A + \gamma h \tag{3.13a}$$

or

$$p_u - p_A = \gamma h \tag{3.13b}$$

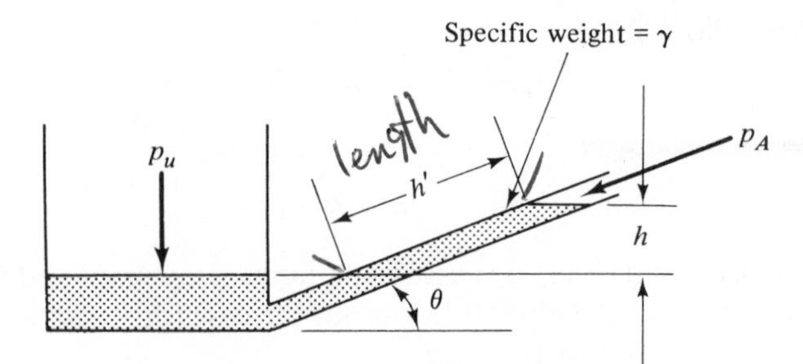

FIGURE 3.15 Inclined manometer.

However, h is $h' \sin \theta$. Thus,

$$p_u - p_A = \gamma h' \sin \theta \tag{3.14}$$

Since θ is fixed, a scale placed along the tube can be calibrated to read directly in units of h of a fluid. Usually, this is done by reading directly in. of water or any other desired unit, such as cm of water, for $p_u - p_A$.

Another method of achieving greater sensitivity and accuracy is to use a manometer with more than one fluid. The arrangement shown in Figure 3.16

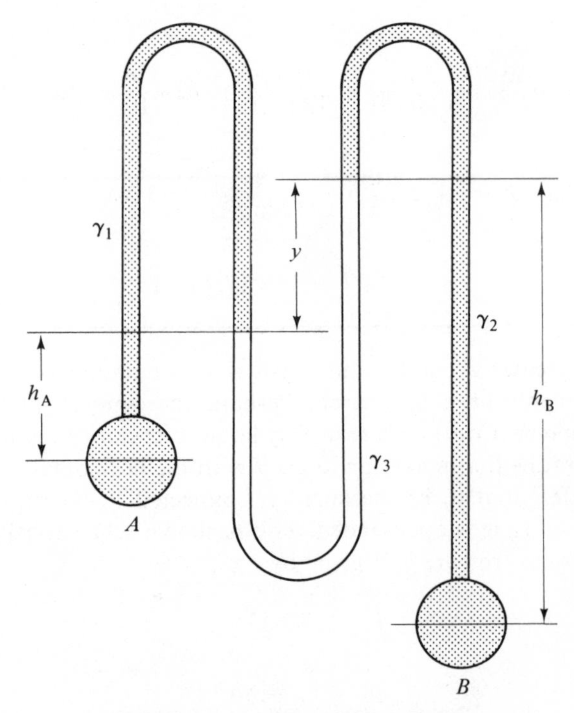

FIGURE 3.16 Three-fluid manometer.

can be used in this manner. Starting at level A,

$$p_A - h_A\gamma_1 - y\gamma_3 + h_B\gamma_2 = p_B \tag{3.15a}$$

or

$$p_A - p_B = (-h_B\gamma_2 + h_A\gamma_1) + y\gamma_3 \tag{3.15b}$$

In the usual case the manometer is connected to different positions on the same pipe and γ_1 can be taken to be equal to γ_2. Also, $h_B - h_A = y$. Thus,

$$p_A - p_B \simeq y(\gamma_3 - \gamma_1) \qquad \text{or} \qquad y(\gamma_3 - \gamma_2) \tag{3.16}$$

For small differences in $p_A - p_B$ it is apparent that the manometer fluid (γ_3) should have a specific weight very nearly equal to the specific weight of the fluid in the pipes. For large pressure differences one can use a heavy fluid such as mercury to increase $\gamma_3 - \gamma_2$ and reduce the manometer reading.

Another way to increase the sensitivity of a manometer is shown in Figure 3.17. The ends of this manometer consist of larger (and equal) areas contain-

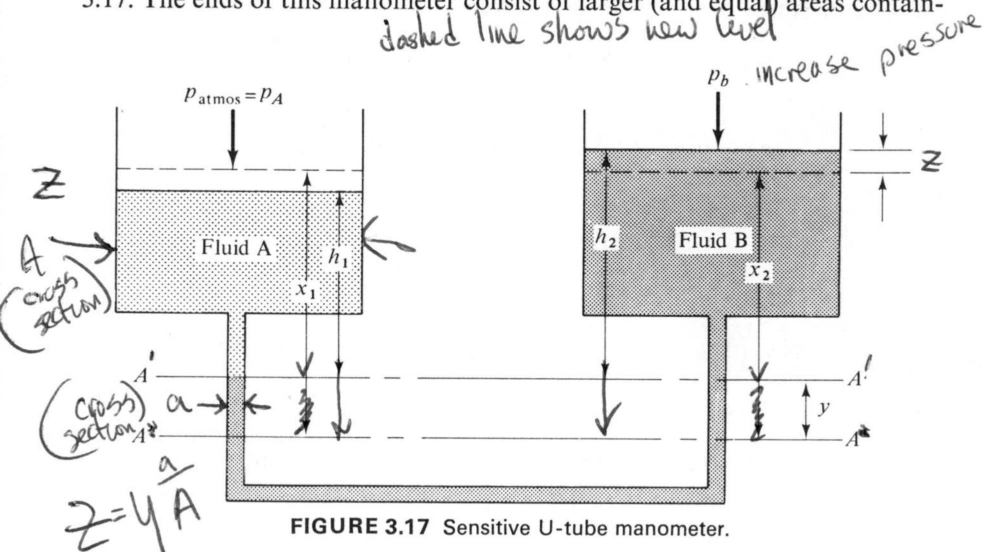

FIGURE 3.17 Sensitive U-tube manometer.

ing two fluids that will not mix. Before any pressure is applied, the common surface of the two liquids is at AA and the heights of the two liquids above this level are h_1 and h_2. Since the fluids must be in equilibrium,

$$h_1\gamma_1 = h_2\gamma_2 \tag{3.17}$$

where γ_1 and γ_2 are the specific weights of the fluids. A pressure is applied to the right arm of the tube, which causes the fluids to be displaced, and the common surface descends an amount y to the new level $A'A'$. The levels of the two liquids above the common surface are now x_1 and x_2, and the applied pressure p_b is given by

$$p_A + x_1\gamma_1 = x_2\gamma_2 + p_b \tag{3.18a}$$

or

$$p_b - p_A = x_1\gamma_1 - x_2\gamma_2 \tag{3.18b}$$

Since mass must be conserved, the mass of fluid displaced in one end must equal the mass of fluid displaced in the other end, or

$$AZ = ay \tag{3.19}$$

where A is the area of cross section of large ends—both taken to be equal—and a the area of cross section of the tube. From geometry,

$$\begin{aligned} X_1 &= h_1 + Z + y \\ X_2 &= h_2 - Z + y \end{aligned} \tag{3.20}$$

Substituting these in equation (3.18) gives

$$p_b - p_A = \gamma_1(h_1 + Z + y) - \gamma_2(h_2 + Z + y)$$

and

$$p_b - p_A = \gamma_1(Z + y) - \gamma_2(y - Z)$$

but

$$Z = \frac{a}{A}y$$

Thus,

$$p_b - p_A = y\left[\gamma_1\left(1 + \frac{a}{A}\right) - \gamma_2\left(1 - \frac{a}{A}\right)\right] = y\left[(\gamma_1 - \gamma_2) + (\gamma_1 + \gamma_2)\frac{a}{A}\right] \tag{3.21}$$

From equation (3.21) we conclude that for greatest sensitivity γ_1 and γ_2 should be as close as possible and the area ratio a/A should be as small as possible.

ILLUSTRATIVE PROBLEM 3.9

A three-fluid manometer such as the one shown in Figure 3.16 is to be used to measure large pressure differences. Assuming that the same liquid is in the legs with $\gamma = 62.4$ lb/ft³ (water) and that the manometer fluid is mercury ($\gamma_3 = 850$ lb/ft³), determine the pressure difference between A and B if y is equal to 1 ft and A and B are at the same level.

Solution

From equation (3.16),

$$p_A - p_B \simeq y(\gamma_3 - \gamma_1) = 1 \text{ ft} \times (850 - 62.4)\frac{\text{lb}}{\text{ft}^3} = 787.6 \text{ psf}$$

or

$$p_A - p_B = \frac{787.6 \text{ lb/ft}^2}{144 \text{ in.}^2/\text{ft}^2} = 5.47 \text{ psi}$$

ILLUSTRATIVE PROBLEM 3.10

If in Illustrative Problem 3.9, A is 6 in. higher than B, determine $p_A - p_B$.

Solution

Refer to Figure 3.16. Starting from A,

$$p_A - h_A\gamma_1 - y\gamma_3 + h_B\gamma_2 = p_B$$

Assuming that $\gamma_1 = \gamma_2 = \gamma$, we obtain

$$p_A - p_B = \gamma(h_A - h_B) + y\gamma_3$$

From the statement of the problem,

$$h_B - (\tfrac{1}{2} + h_A) = y$$

or

$$h_B - h_A = y + \tfrac{1}{2}$$

and thus

$$p_A - p_B = \gamma(-y - \tfrac{1}{2}) + y(\gamma_3) = y(\gamma_3 - \gamma) - \tfrac{1}{2}\gamma$$

This result could almost have been anticipated by observation; that is, elevating one leg by $\frac{1}{2}$ ft is equivalent to decreasing the pressure difference by $\frac{1}{2}$ ft of fluid in the leg. Therefore,

$$p_A - p_B = 1(850 - 62.4) - \tfrac{1}{2}(62.4) = 756.4 \text{ psf}$$

or

$$p_A - p_B = \frac{756.4}{144} = 5.25 \text{ psi}$$

3.4 FORCES ON PLANE-SUBMERGED SURFACES

It was seen in Chapter 2 that the shear stress in a fluid is proportional to the velocity difference between the shearing surfaces and inversely proportional to the distance separating the surfaces. If a fluid is at rest, the shear stress must be zero, which means that no tangential forces can exist in the fluid. The only forces that can exist are forces that are normal to surfaces in contact with the fluid. If a plane is completely immersed in a fluid it will experience a normal force on both sides of the plane, which will tend to compress the plane. However, when a plane is placed in a fluid so that only one side is subjected to the fluid pressure, it will experience a net unbalanced force. Since pressure varies with depth, the pressure on a nonhorizontal plane will in general be variable.

First let us consider the special case of a vertical plane and subdivide this topic into (1) a vertical plane just touching a free surface, (2) a completely submerged vertical plane, and (3) a vertical plane with different liquid levels on each side.

For the vertical plane just touching a free surface, case 1, the pressure on one side will vary linearly from zero to γh due to the column of fluid on this side. This is shown schematically in Figure 3.18a, where the pressure is indi-

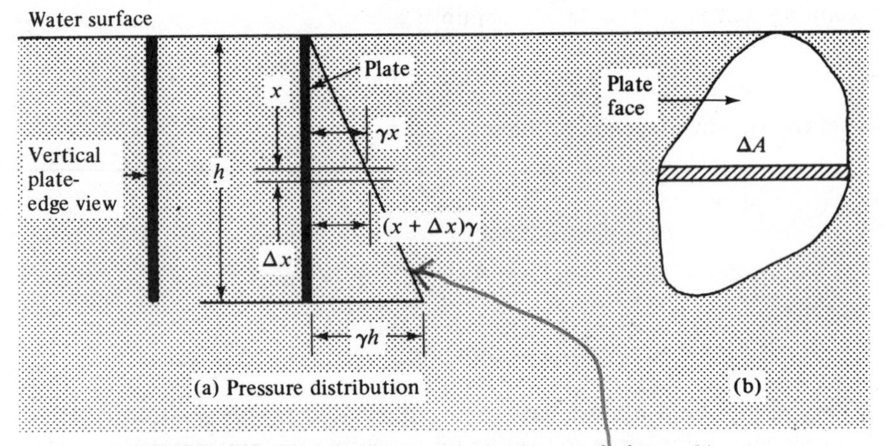

FIGURE 3.18 Pressure on a vertical submerged plane with one edge touching a free surface.

cated to vary linearly, giving rise to a triangular pressure distribution. The force exerted on the area, ΔA, is the average pressure on this area multiplied by the area. The average pressure is

$$p_{\text{avg}} = \gamma x + \frac{\gamma \, \Delta x}{2} \tag{3.22}$$

The force on the area ΔA is the product of the pressure and area,

$$F_x = \gamma x(\Delta A) + \frac{\gamma(\Delta x)(\Delta A)}{2} \tag{3.23}$$

Since Δx is small, the term $(\Delta x)(\Delta A)$ is indeed very small and may be neglected when compared to terms containing only Δx. Therefore, the second term in equation (3.23) can be neglected when compared with the first term. Thus,

$$F_x = \gamma x \, \Delta A \tag{3.24}$$

It is necessary to sum each of the F_x terms over all values of x in order to evaluate the total force on the plane. For the special case where the width of the plane is constant, it will be immediately seen that A, the sum of all the ΔA's, is not a function of x and one need only use the average value of x to obtain the total force on the plane. Thus, since x varies from zero to h,

$$F_x = \frac{\gamma h A}{2} \tag{3.25}$$

Note that $h/2$ represents the *center of gravity* (or *centroid*) of a rectangle and that the average pressure on the area is just $\gamma h/2$ or $\gamma \bar{h}$, where $\bar{h}$ is the location of the center of gravity from the liquid surface.

Let us now consider a plane whose width is not constant but varies in some manner with liquid depth. This situation is shown in Figure 3.18b. The pressure on the small area ΔA can be written as

$$p = \frac{F_x}{A} = \frac{\gamma x \, \Delta A}{A} \tag{3.26}$$

and the pressure over the entire plate as

$$p = \frac{F_x}{A} = \sum \frac{\gamma x \, \Delta A}{A} \tag{3.27a}$$

or

$$p = \sum \frac{\gamma x \, \Delta A}{A} \tag{3.27b}$$

where $\sum$ denotes the sum of all such quantities.

The student will note from elementary physics that $\sum (x \, \Delta A/A)$ is the definition of the *location of the center of gravity* or centroid of any area. Thus,

$$p = \gamma \bar{h} \tag{3.28}$$

where $\bar{h}$ is the location of the center of gravity of the area from the surface.

Thus, from equation (3.28), we see for a vertical submerged plane touching the surface of a fluid that the average pressure on the plane is simply the product of the specific weight of the fluid multiplied by the depth of the center of gravity from the surface of the fluid. The total force is the total area multiplied by this average pressure. Therefore,

$$F_{\text{total}} = \gamma \bar{h} A \tag{3.29}$$

The second case of a vertical plane, case 2, is one that is completely submerged. Figure 3.19 shows any shaped vertical plane whose top edge is located

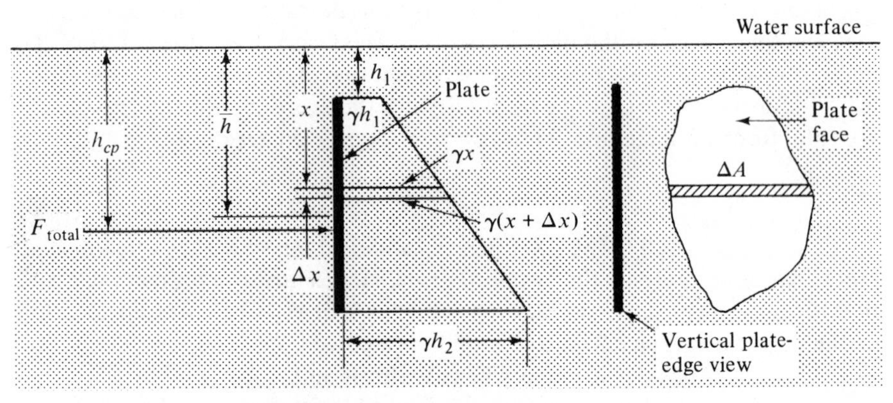

FIGURE 3.19 Submerged vertical surface.

some distance below the free surface of the liquid. Again, the pressure on one side will vary linearly with the depth of fluid, and the plot of pressure on one side of the plane against the distance below the top edge of the plate is a trapezoid whose value of pressure on the top edge is γh_1 and on the bottom edge is γh_2. The force on the area ΔA is the average pressure multiplied by the area. Therefore,

$$F_x = pA \tag{3.30a}$$

or

$$p = \frac{F_x}{A} \tag{3.30b}$$

and

$$p = \frac{F_x}{\Delta A} = \gamma x + \frac{\gamma \, \Delta x}{2} \tag{3.31}$$

The force on ΔA is therefore

$$F_x = \left(\gamma x + \frac{\gamma \, \Delta x}{2}\right) \Delta A \tag{3.32}$$

Carrying out the multiplication indicated by equation (3.32) and neglecting products of small numbers and adding all such items yields

$$F_x = \sum \gamma x \, \Delta A \tag{3.33}$$

Since pressure is force divided by area, the average pressure on the plate is

$$p = \sum \frac{\gamma x \, \Delta A}{A} \tag{3.34}$$

where A is the total plate area and $\sum$ is the sum of all such quantities over the plate. Again $\sum (x \, \Delta A/A)$ is simply the location of the center of gravity or centroid of the plate. Thus, we have the result that for a vertical submerged plate the average pressure on a surface is simply the product of the specific weight of the fluid multiplied by the depth of the center of gravity below the surface of the fluid. Also, as before, the total force is the total area multiplied by the average pressure.

This leads to the result that equation (3.29) is general for any vertical surface, namely $F_{\text{total}} = \gamma \bar{h} A$, where $\bar{h}$ is the location of the center of gravity of the plane below the surface of the fluid.

ILLUSTRATIVE PROBLEM 3.11

As shown in Figure 3.20, a vertical plate is used to dam a channel 2 ft wide. If the plate is $2\frac{1}{2}$ ft high and the water on one side of the plate is at a level $\frac{1}{2}$ ft below the top, what is the total force on the plate?

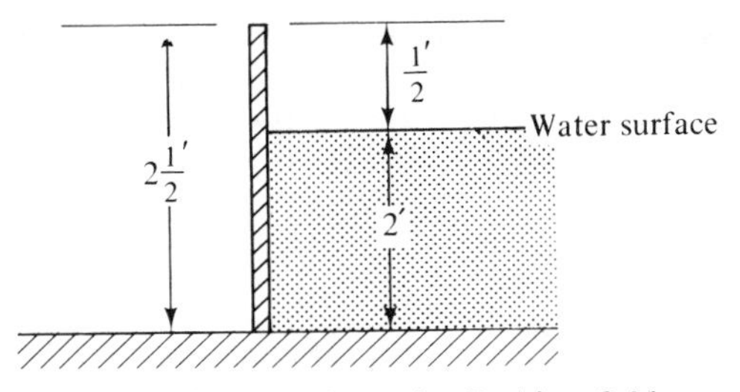

FIGURE 3.20 Illustrative Problem 3.11.

Solution

For this problem, the $2\frac{1}{2}$-ft dimension is not important. Only the portion of the plate that is submerged is important. Its center of gravity is 1 ft below the free liquid surface and the area in contact with the liquid is 2 × 2 ft, or 4 ft². The total force on the plate is therefore

$$F_{\text{total}} = \gamma \bar{h} A = 62.4 \frac{\text{lb}}{\text{ft}^3} \times 4\ \text{ft}^2 \times 1\ \text{ft} = 249.6\ \text{lb}$$

As the last case of a vertical surface subjected to hydrostatic forces, case 3, consider the plane subjected to different liquid levels (but the same liquid) on both sides. It is possible to obtain the net force acting on this plane by considering each side separately and vectorially summing the forces. By considering Figure 3.21 a simple expression for the net force for this case can readily

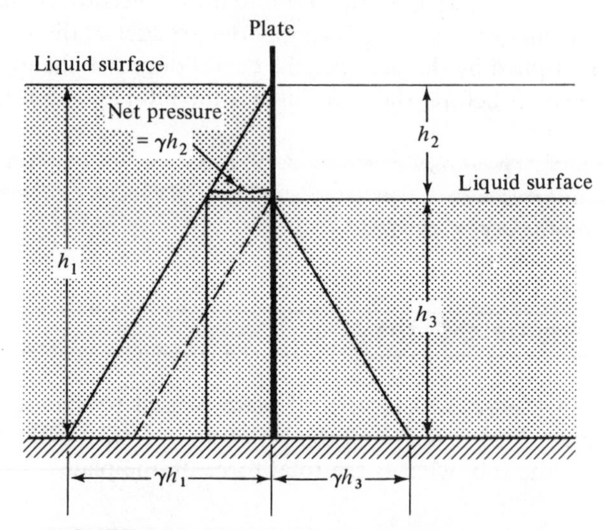

FIGURE 3.21 General submerged vertical surface.

be developed. The pressure distribution for the left side is a triangle extending from zero at the upper free surface to γh_1 at the lower edge of the plate. On the right-hand side the pressure distribution is also a triangle extending from zero at its free surface to γh_3 at the base of the plate. By superposing these two diagrams as shown by the dashed line on the figure, we reach the conclusion that from h_2 down the pressure is constant and equal to γh_2 on the plate. The total force on the plate is therefore given by

$$F_{\text{total}} = \gamma h_2 A_1 + \gamma \frac{h_2}{2} A_2 \tag{3.35}$$

where A_1 is the plate area below the right liquid surface and A_2 the area above the right liquid surface.

ILLUSTRATIVE PROBLEM 3.12

As shown in Figure 3.22, a vertical plate extends from the surface of water on one side to a depth of 10 m. On the other side there is water to within 2 m of the top. If the plate is 5 m wide, what is the net force on the plate?

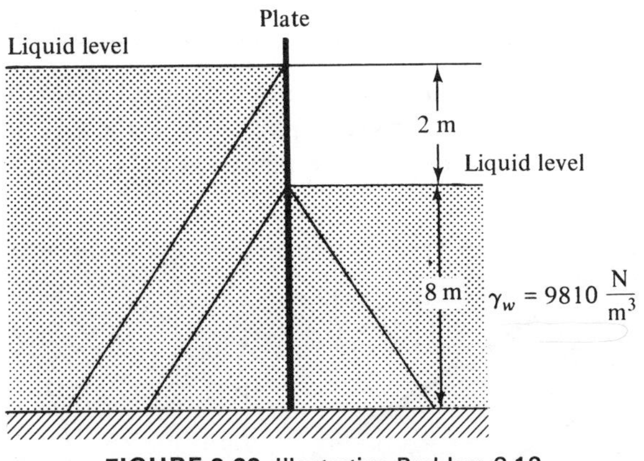

FIGURE 3.22 Illustrative Problem 3.12.

Solution

This problem can be solved by considering each side separately. For the left side,

$$F_{\text{total}} = \gamma \bar{h} A = 9810 \frac{N}{m^3} \times 5 \text{ m} \times 10 \text{ m} \times 5 \text{ m} = 2\,452\,500 \text{ N}$$

On the right side,

$$F_{\text{total}} = \gamma \bar{h} A = 9810 \frac{N}{m^3} \times 4 \text{ m} \times 8 \text{ m} \times 5 \text{ m} = 1\,569\,600 \text{ N}$$

Thus,

$$\text{the net force} = 2\,452\,500 - 1\,569\,600 = 882\,900 \text{ N} = 882.9 \text{ kN}$$

An alternative solution is obtained by considering the superposed pressure destribution as shown in Figure 3.22. For the upper 2 m, we have

$$F_{\text{total}} = 1 \text{ m} \times 2 \text{ m} \times 5 \text{ m} \times 9810 \frac{N}{m^3} = 98\ 100 \text{ N}$$

The lower 8 m can be taken to have a uniform pressure distribution. The force is therefore

$$9810 \frac{N}{m^3} \times 2 \text{ m} \times 8 \times 5 = 784\ 800 \text{ N}$$

The total force is the sum of these two forces, or

$$F_{\text{total}} = 98\ 100 + 784\ 800 = 882\ 900\ \text{N} = 882.9\ \text{kN}$$

Both methods yield the same result and it is left to the student to use the one of his or her choice, or, better yet, to do this type of problem by both methods to obtain an independent check on the solution.

Vertical and horizontal surfaces represent special cases of submerged planes, and, in general, the submerged plane can make any angle with the free surface of the liquid. Consider the plane shown in Figure 3.23, which

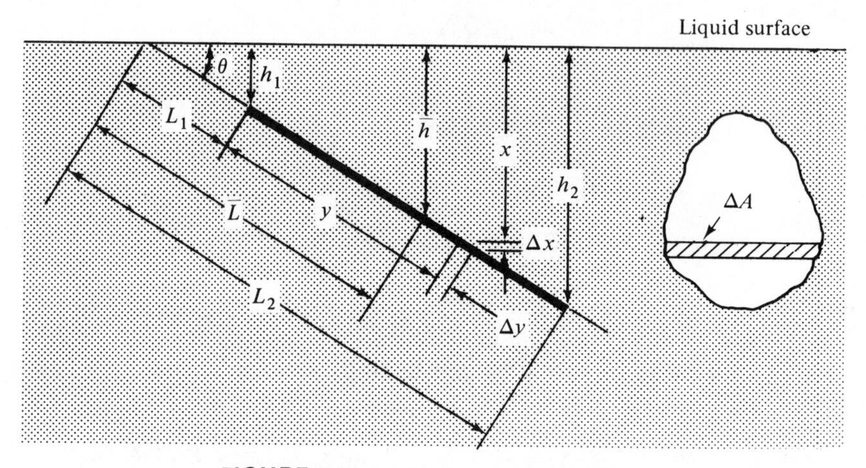

FIGURE 3.23 Inclined submerged plane.

makes an angle θ with the liquid surface. Also recall that it was proved earlier in this chapter that the pressure intensity on any horizontal plane must be the same anywhere on the plane. Thus, the pressure at the level x is the same regardless of the orientation of the surface. However, this pressure acts normal to the surface, which in this case is normal to Δy and over the area ΔA. Using the same procedure as before leads us to the solution that

$$p = \gamma \bar{L} \sin \theta \tag{3.36}$$

and since $L \sin \theta = \bar{h}$,

$$p = \gamma \bar{h} \tag{3.37}$$

Note that in equations (3.36) and (3.37) $\bar{L}$ and $\bar{h}$ are the distances to the center of gravity of the plate shown in Figure 3.23. Thus, we obtain

$$F_{\text{total}} = \gamma \bar{h} A \tag{3.38}$$

It can now be stated as a general result that the total force acting on a submerged plane area is the product of the specific weight of the liquid multiplied by the area and multiplied by the depth of the center of gravity (or centroid) of the area from the liquid surface. This conclusion does not depend on the angle of the plane area.

ILLUSTRATIVE PROBLEM 3.13

A rectangular gate 1 m × 1 m is mounted in the side of a wall having the slope shown in Figure 3.24. Water fills the container until it is 6 m above the top of the gate. What is the total force on the gate?

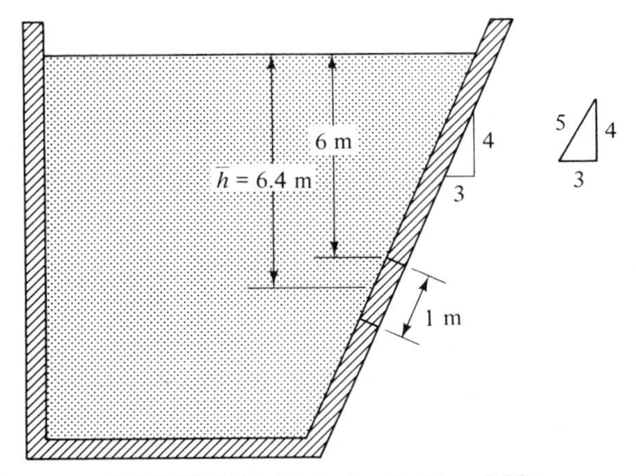

FIGURE 3.24 Illustrative Problem 3.13.

Solution

From the figure we have $\bar{h} = 6.4$ m, therefore,

$$F = \gamma \bar{h} A = 9810 \frac{\text{N}}{\text{m}^3} \times 6.4 \text{ m} \times 1 \text{ m} \times 1 \text{ m} = 62\,784 \text{ N}$$

ILLUSTRATIVE PROBLEM 3.14

As shown in Figure 3.25, a valve (plate) is placed in a channel to control the flow. Assume that the valve consists of a plane that makes an angle of 30° with the free surface. If the channel is 5 × 5 ft in cross section and is flowing full, what is the total force on the valve after the flow is cut off? Assume that $\gamma = 62.4$ lb/ft³.

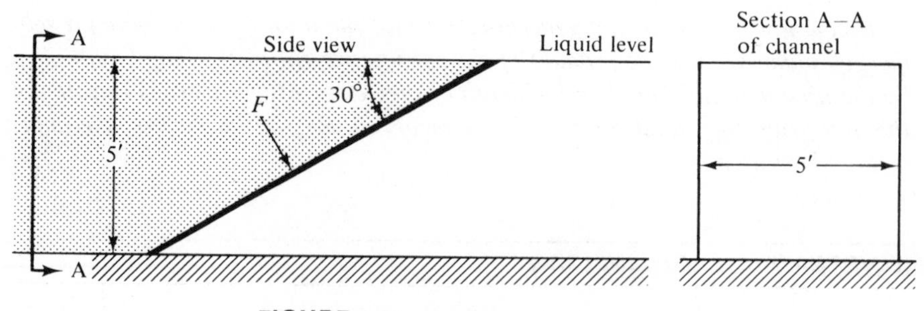

FIGURE 3.25 Illustrative Problem 3.14.

Solution

The area of the valve is $(5 \times 5)/\sin 30 = 50\ \text{ft}^2$. The centroid of the plate is located $2\frac{1}{2}$ ft below the surface. The total force is therefore

$$F_{\text{total}} = \gamma \bar{h} A = 62.4\,\frac{\text{lb}}{\text{ft}^3} \times 2\tfrac{1}{2}\ \text{ft} \times 50\ \text{ft}^2 = 7800\ \text{lb}$$

3.5 LOCATION OF THE CENTER OF PRESSURE

In Section 3.4 the discussion was concerned with the magnitude of the force on a plane area. Since force is a vector quantity, a force is completely described by knowing its magnitude, direction, and point of application. Thus far we have determined the magnitude and direction but not the point of application of these vectors (forces). To determine the point of application of each of the forces, let us once more consider the case of a fully submerged vertical plane such as is shown in Figure 3.19. Since the area is in equilibrium, the sum of the moments about any point must be zero. Therefore, by referring to Figure 3.19 and using the notation h_{cp} for the vertical location of the point of application of the resultant force,

$$(F_{\text{total}})h_{cp} = \sum \gamma x\, \Delta A(x) \tag{3.39a}$$

and

$$h_{cp} = \sum \frac{\gamma x^2 A}{F_{\text{total}}} \tag{3.39b}$$

But

$$F_{\text{total}} = \gamma \bar{h} A$$

Therefore,

$$h_{cp} = \sum \frac{x^2 \Delta A}{\bar{h} A} \tag{3.40}$$

The summation $\sum x^2 \Delta A$ term in equation (3.40) is the second moment of area, or more conventionally it is called the moment of inertia (I) of the area A about the axis at the surface of the liquid. Actually, it is more convenient to express equation (3.40) in terms of the moment of inertia about the center of gravity of the area. To do this we utilize the transfer theorem of mechanics, and inserting the results into equation (3.40), we obtain

$$h_{cp} = \frac{\bar{I}}{\bar{h}A} + \bar{h} \tag{3.41}$$

where $\bar{h}$ is the location of the center of gravity and $\bar{I}$ is the moment of inertia of the plane area about a centroidal axis. The location of the resultant force is known as the center of pressure. Table 3.2 gives the properties of some selected areas.

The student should note that it is always possible to evaluate either $\bar{h}$ or $\bar{I}$ numerically by subdividing the area in question into many small parts and performing the indicated mathematical operations and summation. Even a relatively coarse subdivision quickly leads to a good numerical approximation of the exact solution.

ILLUSTRATIVE PROBLEM 3.15

As shown in Figure 3.26, a sluice gate is 6 ft square. It is hinged at the top and held by two horizontal pins 1 ft from the bottom of the gate. What force must each of these pins resist if the water level is at the top of the gate?

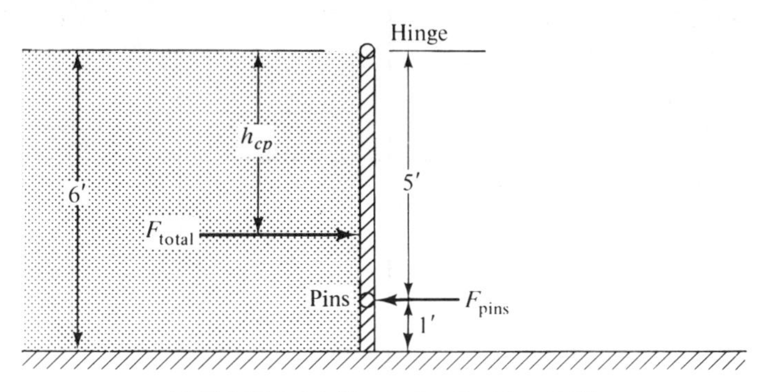

FIGURE 3.26 Illustrative Problem 3.15.

Solution

For the gate to be in equilibrium under the action of the forces shown, it is necessary that the moment of these forces about the hinge be zero. The

TABLE 3.2 Properties of an Area

Section	Area	$\bar{h}$ location of center of gravity	$\bar{I}$ moment of inertia about the center of gravity
Square 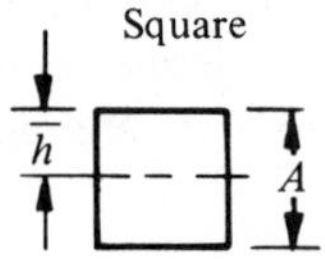	A^2	$\frac{1}{2}A$	$\bar{I} = \frac{1}{12}A^4$
Rectangle 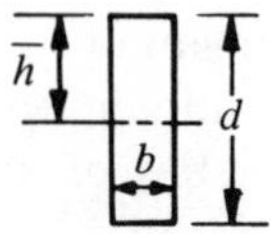	bd	$\frac{1}{2}d$	$\frac{1}{12}bd^3$
Trapezoid 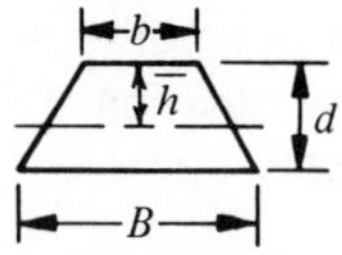	$\frac{1}{2}(B + b)\,d$	$d\left\{\frac{2B + b}{3(B + b)}\right\}$	$\frac{d^3(B^2 + 4Bb + b^2)}{36\,(B + b)}$
Triangle 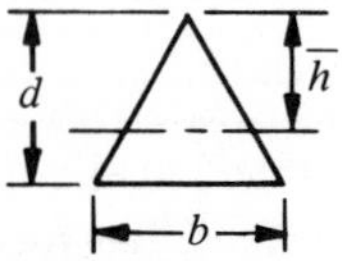	$\frac{1}{2}bd$	$\frac{2}{3}d$	$\frac{bd^3}{36}$
Circle 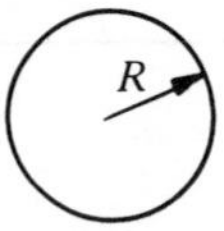	πR^2	R	$\frac{\pi R^4}{4}$
Ellipse 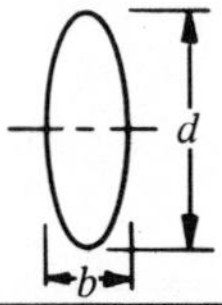	$\frac{\pi bd}{4}$	$\frac{1}{2}d$	$\frac{\pi bd^3}{64}$
Quarter Circle – Also * for Semicircle 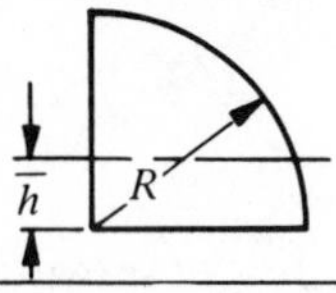	$\frac{\pi R^2}{4}$ * $\frac{\pi R^2}{2}$	$0.4244\,R$ *$0.4244\,R$	$\frac{0.055R^4}{8}$ *$\frac{0.11\,R^4}{8}$

hydrostatic force

$$F_{total} = \tfrac{6}{2} \times 6 \times 6 \times 62.4 = 6740 \text{ lb}$$

The location of h_{cp} is

$$\frac{\bar{I}}{\bar{h}A} + \bar{h} = \frac{bd^3/12}{d/[2(d)(b)]} + \frac{d}{2} = \frac{d}{6} + \frac{d}{2} = 4 \text{ ft}$$

The moment of these forces about the hinge is

$$(F_{total})4 = (F_{pins})5$$

Therefore,

$$F_{pins} = \tfrac{4}{5}(6740) = 5392 \text{ lb}$$

and for one pin

$$F_{pin} = 2696 \text{ lb}$$

ILLUSTRATIVE PROBLEM 3.16

Flashboards are installed on top of a dam and are supported by pipes spaced 4 m apart along the dam. If the water is flowing over the flashboard 1 m deep as shown in Figure 3.27, what is the bending moment produced at the bottom of the flashboard on one pipe?

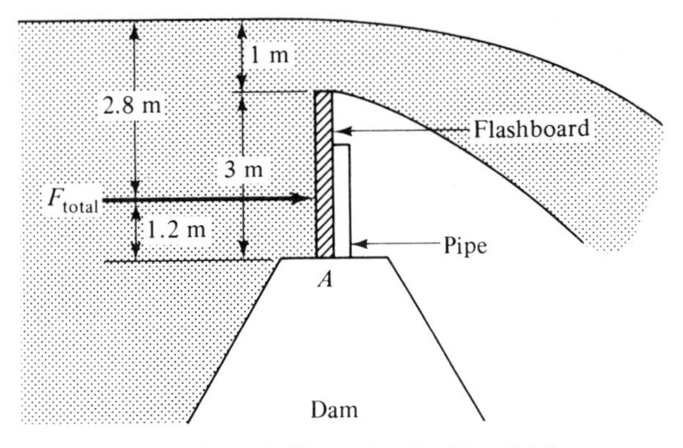

FIGURE 3.27 Illustrative Problem 3.16.

Solution

To solve this problem, let us select a repeating section of flashboard 4 m wide. The total force on the flashboard is

$$F_{\text{total}} = 9810\,\frac{\text{N}}{\text{m}^3} \times 4\text{ m} \times 3\text{ m} \times (1 + 1.5)\text{ m} = 294.3\text{ kN}$$

The location of h_{cp} is

$$h_{cp} = \frac{bd^3/12}{\bar{h}A} + \bar{h}$$

and

$$h_{cp} = \frac{(4 \times 3^3)/12}{(1 + 1.5)(4 \times 3)} + 2.5 = 2.8\text{ m from the water surface}$$

Taking moments about A, $1.2 \times 294.3 = 353.16$ kNm. Since per repeating section there is the equivalent of one pipe, this is the moment at the bottom of a pipe.

For a nonvertical plane, as is shown in Figure 3.23, we can also evaluate the location of the center of pressure in a manner similar to that used for the vertical plane. When this is done using calculus, it can be demonstrated that

$$L_{cp} = \frac{\bar{I}}{\bar{L}A} + \bar{L} \tag{3.42}$$

where L_{cp} denotes the location of the center of pressure along the plane. Thus, the distance from a liquid surface to the center of pressure along the plane is the same regardless of the angle of inclination as long as the angle is not zero. The center of pressure is always below the center of gravity.

ILLUSTRATIVE PROBLEM 3.17

Determine the center of pressure for the gate of Illustrative Problem 3.13.

Solution

Referring to Illustrative Problem 3.13, we have $\bar{h} = 6.4$ m. Therefore, $\bar{L} = 6.4/0.8 = 8$ m,

$$\bar{I} = \frac{bd^3}{12} = \frac{1 \times 1^3}{12} = \frac{1}{12}$$

and

$$A = 1 \times 1 = 1 \text{ m}^2$$

From equation (3.42),

$$L_{cp} = \frac{\bar{I}}{\bar{L}A} + \bar{L} = \frac{1/12}{8 \times 1} + 8 = 8.0104 \text{ m along the plane}$$

Vertically, this corresponds to $8.0104 \times 0.8 = 6.408$ m below the surface.

ILLUSTRATIVE PROBLEM 3.18

Determine the center of pressure for the value of Illustrative Problem 3.14.

Solution

$$L_{cp} = \frac{\bar{I}}{\bar{L}A} + \bar{L}$$

For this problem all dimensions will be considered *along* the plane:

$$\bar{I} = \frac{bd^3}{12} = \frac{5 \times (5/\sin 30)^3}{12} = \frac{5 \times (10)^3}{12} = \frac{5000}{12}$$

$$\bar{L} = \frac{2.5}{\sin 30} = 5 \text{ ft}$$

$$A = 5 \times 10 = 50 \text{ ft}$$

Therefore,

$$L_{cp} \text{ along plane} = \frac{5000/12}{5 \times 50} + 5 = 1.67 + 5 = 6.67 \text{ ft}$$

For the vertical height h_{cp},

$$h_{cp} = 6.67 \sin 30 = 3.34 \text{ ft vertically below the surface.}$$

3.6 BUOYANCY OF SUBMERGED BODIES

If a body is weighed in a vacuum and then is weighed in a fluid, it will be found to have a different weight in both cases. This is due to the buoyant action of the fluid displaced. To evaluate this effect, let us consider the body shown in Figure 3.28. This body will be considered to be floating in a sub-

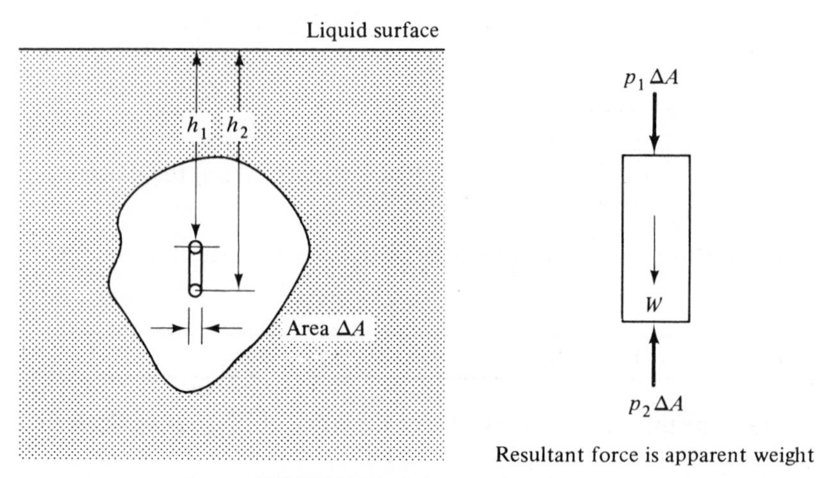

FIGURE 3.28 Submerged body.

merged position, and therefore it will be in equilibrium in this position. Obviously, this is a problem in hydrostatics that can be resolved by applying the principles already developed in this chapter—by considering the small volume shown in Figure 3.28 and summing all such volumes.

From the free-body diagram shown, we need consider only the vertical forces, since all horizontal forces on the body are equal and opposite and give no net horizontal force. The vertical forces are the weight of the body and the forces on each end of the body, with the resultant of these forces equal to the sum of these three vectors. Thus,

$$\text{apparent weight (resultant)} = W + p_1 \,\Delta A - p_2 \,\Delta A \tag{3.43a}$$

$$= \gamma_s \,\Delta V + \gamma h_1 \,\Delta A - \gamma h_2 \,\Delta A \tag{3.43b}$$

where γ_s is the specific weight of the solid and γ the specific weight of the liquid. The volume of the element is given as ΔV and equals $(h_2 - h_1)\,\Delta A$. Therefore,

$$\text{apparent weight} = \gamma_s \,\Delta V - \gamma \,\Delta V = \Delta V(\gamma_s - \gamma) \tag{3.44}$$

Notice that the apparent weight equals the weight of the body in vacuum less the weight of the fluid displaced. If all the small volumes in the body are summed, the total volume of the body will be obtained. Thus,

$$\text{apparent weight of the submerged body} = V(\gamma_s - \gamma) \tag{3.45}$$

Stated in words, *a submerged body is buoyed up by a force equal to the weight of the displaced fluid*, and the buoyant force must also act through the center of gravity of the displaced fluid for the body to be in equilibrium. This is known as *Archimedes' principle*.

ILLUSTRATIVE PROBLEM 3.19

A large stone weighs 100 lb in air, and when the stone is immersed in water it weighs 60 lb. Calculate the volume of the stone in cubic feet and its specific weight.

Solution

Refer to Figure 3.28 and neglect the buoyant effect in air. Thus, the buoyant effect in water is the difference between the weight in air and the weight in water or 100 − 60 = 40 lb. This buoyant effect equals the weight of water displaced or the volume of displaced water multiplied by the specific weight of water:

$$\gamma V = 40$$

$$V = \frac{40}{62.4} = 0.64 \text{ ft}^3$$

The specific weight is simply weight divided by volume:

$$\gamma_s = \frac{W}{V} = \frac{100}{0.64} = 156 \text{ lb/ft}^3$$

ILLUSTRATIVE PROBLEM 3.20

A raft is made of planks that weigh 7000 N/m³. The raft is 4 m × 3 m and the planks are 0.15 m thick. The raft is mounted on six 55-gal drums (Figure 3.29). Neglecting the weight of the drums, what weight will the raft support before it becomes completely submerged?

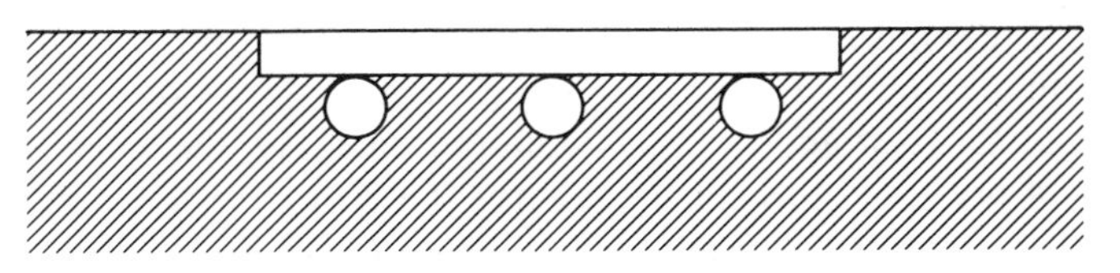

FIGURE 3.29 Illustrative Problem 3.20.

Solution

The volume of planking is 0.15 m $\times$ 4 m $\times$ 3 m $=$ 1.8 m^3. The volume of drums $= 6 \times 55$ gal $\times 3.785 \times 10^{-3} = 1.249$ m^3. The raft weighs 1.8 m^3 $\times$ 7000 N/m^3 $=$ 12 600 N. The total volume of water displaced is 1.8 $+$ 1.249 $=$ 3.049 m^3. The buoyant force on the raft when it is first submerged equals the weight of the displaced water, 3.049 m^3 $\times$ 9810 N/m^3 $=$ 29 911 N. The net buoyant force on the raft is therefore the buoyant force due to the displaced water minus the weight of the raft:

$$F_{\text{net buoyant}} = 29\,911 - 12\,600 = 17\,311 \text{ N}$$

3.7 FORCES ON CURVED SURFACES

Thus far we have considered forces acting on flat plane surfaces. Very often a curved surface is used in engineering work, and it is necessary to determine the forces acting on such surfaces. In general the problem of a resultant force on an irregular surface reduces to the solution of a nonparallel, noncoplanar force system. However, for most cases of interest it is found that the surface in question is symmetrical and that the horizontal component of the total force and the vertical component of the total force can be directly combined to yield a single force. It is therefore convenient to evaluate the horizontal and vertical forces on a curved submerged surface to obtain the resultant total force.

The student will again note that a force is a vector quantity. Since vectors cannot be added algebraically, the force on each unit of area must be broken into two mutually perpendicular components, and these components must be added. Since each of these unit areas makes a different angle with the coordinates, the vectors also make differing angles with the coordinates, and this process is sometimes virtually impossible to do analytically.

The principles used in solving problems involving curved surfaces are best illustrated by discussing a specific problem. Consider the quarter-circle shown in Figure 3.30. For the surface AE to be in equilibrium, it is necessary that the reactant equal the sum of the forces acting on the surface. The sum of the horizontal forces must equal the force on the projected area AB. This follows since the fluid cannot sustain shear forces. If this were not true, a vertical plane placed at AB would not be in equilibrium. Also, the location of F_H is the center of pressure of the projected area AB. For vertical equilibrium, the surface AE must support the weight of fluid above it. This weight is best considered to consist of the force on the horizontal projected area BE plus

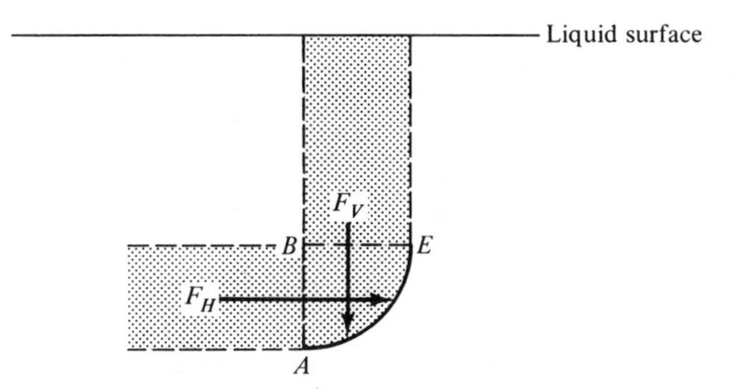

FIGURE 3.30 Submerged vertical surface.

the weight of fluid contained by the curved volume ABE. Each of these vertical forces will act through their respective centers of gravity, which can be obtained from Table 3.2 for the various geometric shapes.

ILLUSTRATIVE PROBLEM 3.21

On the top of a dam a radial gate is mounted as shown in Figure 3.31 with its top edge even with the water level. The gate extends down to become tangent to the top of the dam and has a radius of 12 ft and a length of 12 ft. What is the total water pressure against the gate?

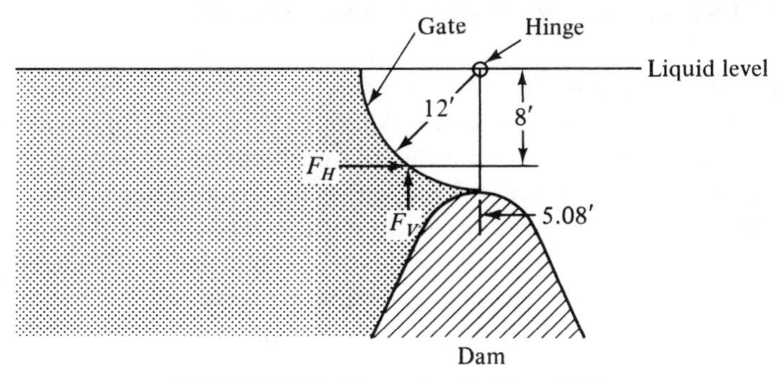

FIGURE 3.31 Illustrative Problem 3.21.

Solution

In this case the hydrostatic pressure acts on the outside of the curved surface, and there is a vertical buoyant force on the surface. This is equivalent to a negative (vertically upward) force of the same magnitude and location as if the gate were filled with fluid. Proceeding to calculate the vertical force, we simply have the weight of the displaced fluid in the quarter-circle multiplied by the specific weight of the fluid;

$$F_V = \tfrac{1}{4}\pi R^2 L\gamma$$

where L is the length of the gate,

$$\therefore F_V = \tfrac{1}{4}\pi(12)^2 \times 12 \times 62.4 = 84{,}600 \text{ lb}$$

The location of the center of gravity $\bar{h}$ is given from Table 3.2 as $0.4244R = 0.4244 \times 12 = 5.08$ ft. For the horizontal force, we simply apply the results determined previously in this chapter. Therefore,

$$F_H = 12 \times 12 \times 6 \times 62.4 = 53{,}900 \text{ lb}$$

$$h_{cp} = \frac{\bar{I}}{\bar{h}A} + \bar{h} = \frac{[12(12)^3]/12}{6 \times 12 \times 12} + 6 = 8 \text{ ft}$$

The total force is the resultant of F_V and F_H and equals $\sqrt{(84{,}600)^2 + (53{,}900)^2} = 100{,}000$ lb. This forces acts through the point 8 ft below the water level and 5.08 ft to the left of the vertical, as shown in the figure.

3.8 STRESSES IN CYLINDERS AND SPHERES

When a fluid is contained, it exerts forces on the walls of the container that stress the material of the container. As an extension of the principles discussed in this chapter, let us consider the problem of evaluating the tensile stresses in thin cylinders and spheres subjected to an internal pressure. To evaluate these stresses we shall assume that the ratio of the thickness to the diameter of the cylinder or sphere is small and that the stresses developed in these pressure vessels are uniform over the cross section of the vessel. By *thin* we shall mean that the ratio of thickness to diameter is 0.1 or less.

With these assumptions in mind, let us consider the cylinder shown in Figure 3.32. Due to the internal pressure there is a stress in the circumferential direction (S_2) and a stress in the longitudinal direction (S_1), as shown in Figure 3.32a. If the pipe is cut perpendicular to the longitudinal axis (and far from the ends) the resultant free-body diagram will be as shown in Figure 3.32b. The total resisting force in the cylinder will be the stress S_1 multiplied by the area over which the stress acts. Thus, the resisting force is $S_1(2\pi R)(t)$ since the material area is $2\pi Rt$. The applied load is due to the pressure in the tube acting over the tube area. The applied load is therefore $p\pi R^2$. For equi-

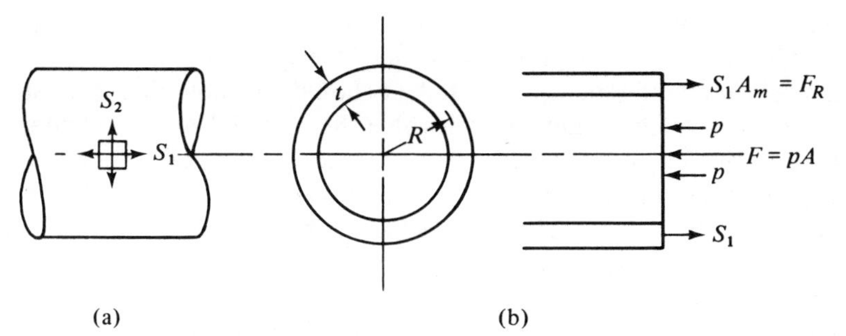

FIGURE 3.32 Thin cylinder under internal pressure.

librium these forces must be equal. Therefore,

$$S_1 2\pi R t = p \pi R^2$$

or

$$S_1 = \frac{pR}{2t} \tag{3.46}$$

Let us now consider a diametrical cut through the cylinder as show in Figure 3.33. The resisting force, F_R equals $S_2 A_m$, where A_m is the metal area. Since there are two metal areas, $A_m = 2tL$ and $F_R = S_2 \times 2tL$. The applied force F consists of the pressure p acting on the projected area, which is $p \times 2RL$. For equilibrium we equate the applied and resisting force to obtain

$$2tLS_2 = p2RL$$

Solving for S_2, yields

$$S_2 = \frac{pR}{t} \tag{3.47}$$

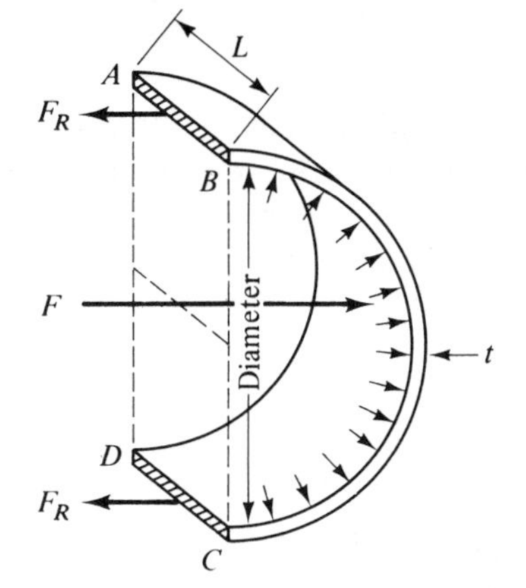

FIGURE 3.33 Thin cylinder under internal pressure.

From equations (3.46) and (3.47) it can be seen that S_2 is twice S_1. S_2 is the tensile stress in the longitudinal seam of the cylinder and S_1 is the transverse stress (circumferential). The student should also note that these stresses were arrived at without having to involve the material of the cylinder. The strength of the material will determine the thickness of the cylinder required for a given internal pressure, and the interested student is referred to any standard text on strength of materials for further discussion of this subject.

Let us now consider a thin sphere subjected to an internal pressure. In this sphere S_1 will be equal to S_2 by symmetry. As can be seen from Figure 3.34, the stress S_2 acts over an area equal to $2\pi Rt$. The sum of the force com-

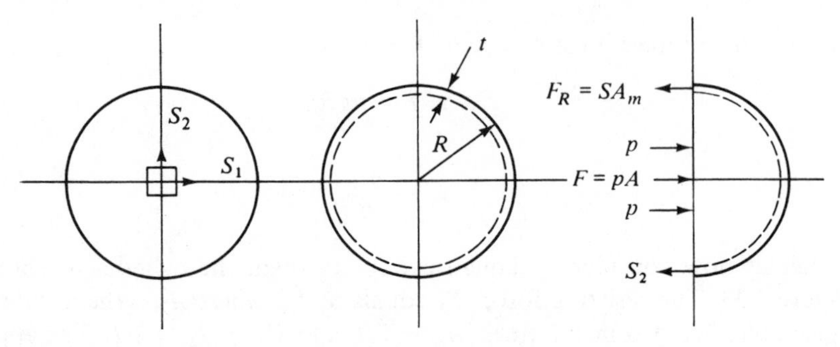

FIGURE 3.34 Thin sphere under internal pressure.

ponents in the horizontal direction is once again just the product of the pressure multiplied by the projected area, that is, $p\pi R^2$. Therefore,

$$S_2 2\pi Rt = p\pi R^2$$

or

$$S_2 = S_1 = \frac{pR}{2t} \tag{3.48}$$

Equation (3.48) yields the result that S_2 in a thin sphere is half of S_2 in a thin cylinder and equal to S_1 when both are subjected to the same internal pressure. The student is cautioned to consult a text on strength of materials or applicable codes before attempting to design cylinders and spheres for either internal or external pressure.

ILLUSTRATIVE PROBLEM 3.22

A 48-in.-diameter steel pipe 0.25 in. thick carries water under an internal pressure of 100 psi.

1. Compute the circumferential stress in the steel.
2. If the pressure is raised to 250 psi and the allowable unit stress in the steel is 15,000 psi, what thickness of steel is required?

Solution

1. The circumferential stress is S_2. Thus, from equation (3.47),

$$S_2 = \frac{pR}{t} = \frac{100 \times 24}{0.25} = 9600 \text{ psi}$$

2.

$$t = \frac{pR}{S_2} = \frac{250 \times 24}{15{,}000} = 0.4 \text{ in.}$$

3.9 CLOSURE

Fluid statics is simply an extension of the principles of statics that the student has studied for solid bodies in his physics and mechanics courses. However, this is not to be taken to minimize this division of fluid mechanics, since the applications of these principles occur frequently and in all fields of engineering. Some of these applications have been explicitly pointed out and studied in this chapter, but there are many others that are encountered, and even though we are, of necessity, limited to the number of items studied, mastery of these items will enable the student to readily analyze the novel situations that he may meet in his career.

In later chapters we utilize many of the principles developed in our study of fluid statics. However, of necessity we require further information when dealing with fluids that are in motion relative to a given gravitational field. These items will be developed in each of the chapters that follow, as required, and we may consider this chapter on fluid statics as the first, necessary step toward the more complicated study of fluid dynamics.

PROBLEMS

3.1 If a column of water is 20 ft high, what is the pressure at its base due to the water only, in psf, psi, and kPa? Use $\gamma = 62.4$ lb/ft^3.

3.2 A column of water is 8 m high. What is the pressure at its base due to the water only in psi, psf, and kPa? Use $\gamma = 9400$ N/m^3.

3.3 If the atmospheric pressure is 14.7 psia, what are the absolute pressures in Problem 3.1?

3.4 If the atmospheric pressure is 100 kPa, what are the absolute pressures in Problem 3.2?

3.5 A column of mercury ($\gamma = 133.5$ kN/m^3) is 25 in. high. What is the pressure at its base due to the mercury only?

3.6 A skin diver descends to a depth of 60 ft in fresh water. What is the total pressure on his body if $\gamma = 62.4$ lb/ft^3?

3.7 The pressure at a level of 100 ft in the Great Salt Lake is 46 psig. What is the specific weight of the lake water if it is constant?

3.8 If the specific weight of air at sea level is taken to be 0.075 lb/ft^3, how high would the atmosphere extend if γ of air is assumed to remain constant?

3.9 If the specific weight of a liquid varies linearly with gage pressure, determine the pressure at a level of 1000 ft. Assume that the specific weight under atmospheric pressure is 40 lb/ft^3 and that the variation of specific weight with pressure can be given as a change in specific weight of 10 lb/ft^3 for every 100-psi change in pressure.

3.10 If the atmospheric pressure at some unknown elevation is half that at sea level, determine the elevation. Assume the temperature of the air to be constant and equal to 70°F.

3.11 Using equation (3.8), evaluate the height at which the pressure of the atmosphere will become zero. Would you expect this to be the case in the actual atmosphere?

3.12 A barometer reads 755 mm Hg at sea level and 20°C. If it is now taken up a hill that is 1000 m high, what is the expected barometer reading? Assume that the temperature remains constant.

3.13 Table 3.1 for the ICAO Standard Atmosphere gives a pressure of 12.243 psia at 5000 ft. Assuming air to have the constant temperature given at zero altitude (59°F), what pressure would equation (3.8) have given for this altitude? Use $p_0 = 14.696$ psia.

3.14 An airplane uses a barometer as an altimeter. While approaching an airport, the pilot is told that the local barometer stands at 29.8 in. Hg. Using this value to set zero, the pilot reads 1500 ft altitude. However, the instrument does not compensate for temperature and is only correct at 70°F. If the air is at 0°F, will the plane clear a 1400-ft mountain in the flight path?

3.15 If a barometer is read to be 750 mm Hg, what is the absolute pressure if a pressure gage reads 70 kPa (gage)? Use the specific gravity of mercury to be 13.6 and take $\gamma_w = 62.4$ lb/ft^3.

3.16 A pressure gage reads 90 kPa above atmospheric pressure. If a barometer reads 760 mm Hg (sg = 13.6, $\gamma_w = 9810$ N/m^3), what is the absolute pressure?

3.17 Table A2.3 gives a specific weight for mercury at 32°F of 848.714 lb/ft^3 and at 100°F it gives a specific weight of 842.925 lb/ft^3. What maximum percentage error will a barometer have if no temperature compensation is made over this temperature range?

3.18 A U-tube mercury manometer that is open at one end is connected to a pressure source as shown in Figure P3.18. What is the unknown pressure in psfa? Assume that the unknown fluid is water, atmospheric pressure is 14.7 psia, and that the mercury has a specific gravity of 13.6.

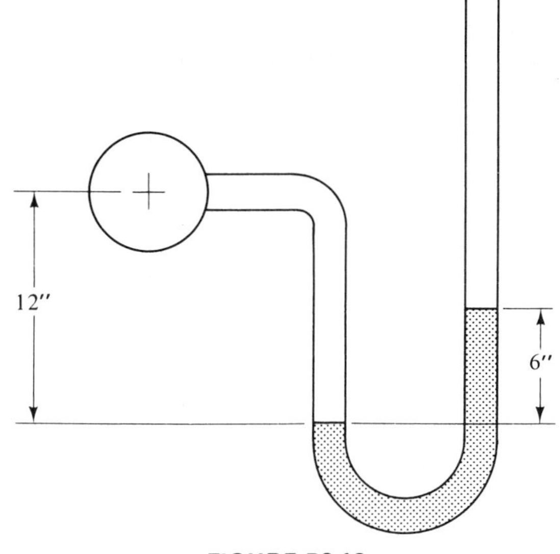

FIGURE P3.18

3.19 In Figure P3.19, two sources of pressure, M and N, are connected by a water–mercury differential gage. What is the difference in pressure between M and N in psi?

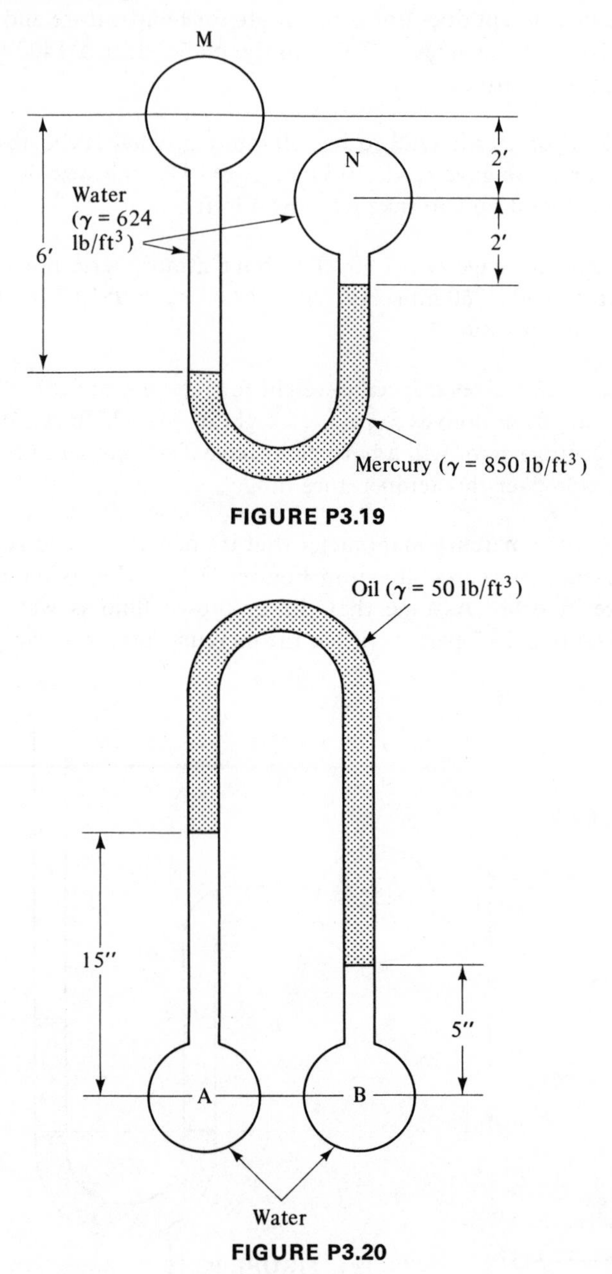

FIGURE P3.19

FIGURE P3.20

3.20 In Figure P3.20, determine the difference in pressure between A and B if the specific weight of water is 62.4 lb/ft^3.

3.21 If the liquid in pipe B in Problem 3.20 is carbon tetrachloride, whose specific weight is 99 lb/ft^3, determine the pressure difference between A and B.

3.22 For the arrangement shown in Figure P3.22, determine $p_A - p_B$.

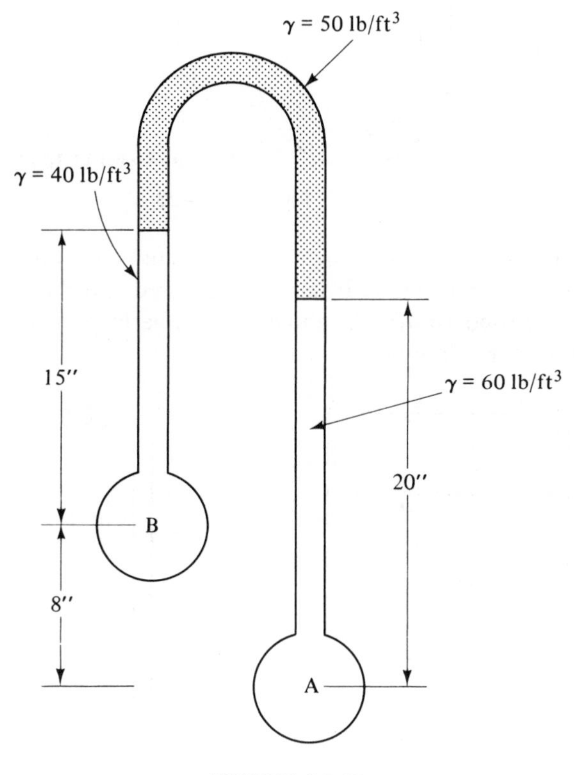

FIGURE P3.22

3.23 Two sources of pressure A and B are connected by a water–mercury differential gage as shown in Figure P3.23. What is the difference in pressure between A and B in Pa?

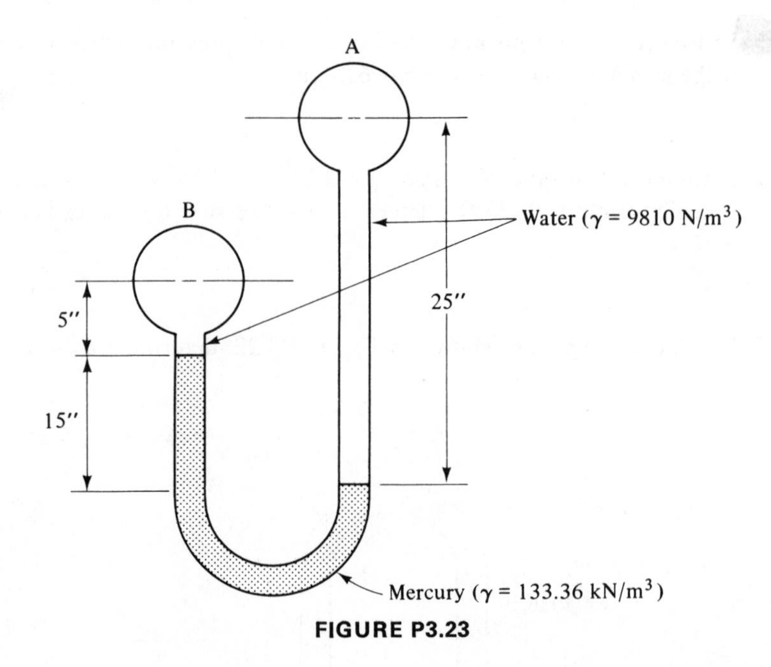

FIGURE P3.23

3.24 In Figure P3.24, a manometer is connected to a tank that is initially open to the atmosphere. If the tank is closed and the pressure in the air space is raised to 70 kPa above atmospheric pressure, what will the manometer reading be?

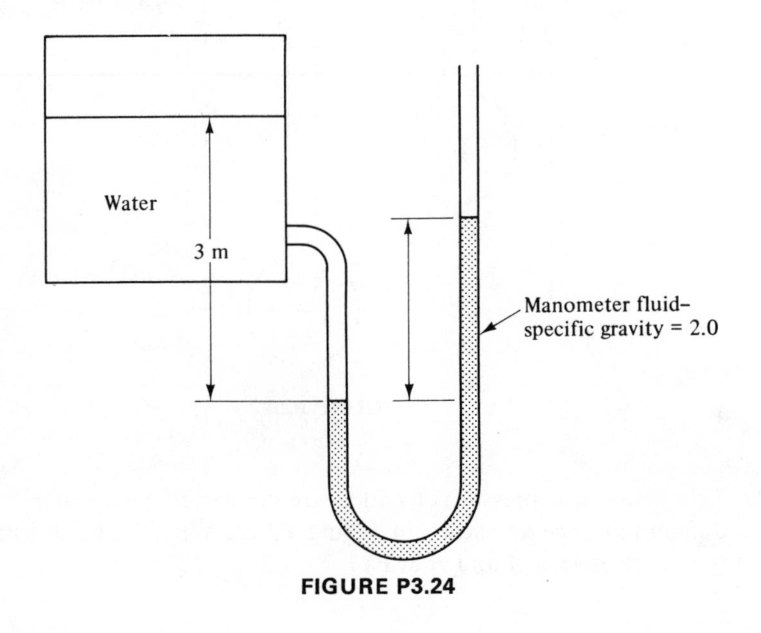

FIGURE P3.24

3.25 What is p_u for the manometer shown in Figure P3.25?

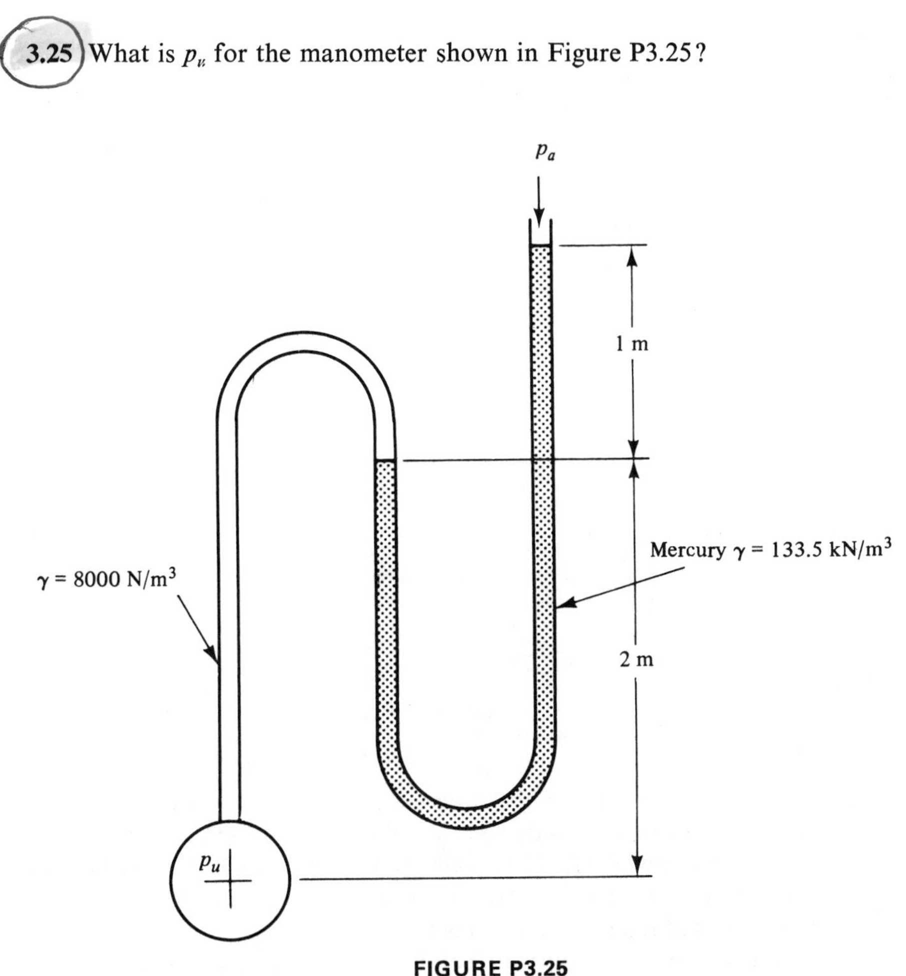

FIGURE P3.25

3.26 In a U-tube manometer, one end is closed, trapping atmospheric air in the column. The other end is connected to a pressure supply of 5 psig. If the level of mercury in the closed end is 2 in. higher than that in the open end, what is the pressure of the trapped air?

3.27 Calculate the pressure of the trapped air and the reading of the pressure gage for the manometer shown in Figure P3.27. Atmospheric pressure is 100 kPa.

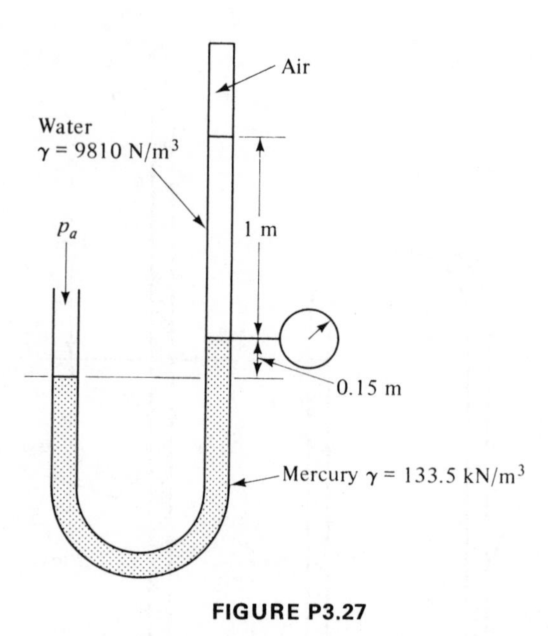

FIGURE P3.27

3.28 A chimney depends on the difference in static pressure between the column of air on the inside and the column of air on the outside to provide the required draft. The specific weight of air at 70°F is 0.075 lb/ft³. Assume that the stack is 50 ft high and that the gas inside the stack can be considered to be air and is at 300°F. Determine the difference in pressure between the outside air column and the chimney column if the specific weight of each is constant.

3.29 To measure the draft in a chimney (see Problem 3.28) an inclined manometer is connected so that the inclined leg (at 10° to the horizontal) is connected to the hot-gas side and the large reservoir is connected to the atmosphere. If the draft gage uses water as the manometer fluid, determine how far along the tube the water will travel. Assume 0.1 psig to be the difference in pressure being measured.

3.30 What pressure will the gage read in Figure P3.30?

3.31 In Figure P3.31, what will the gage read if barometric pressure is 100 kPa?

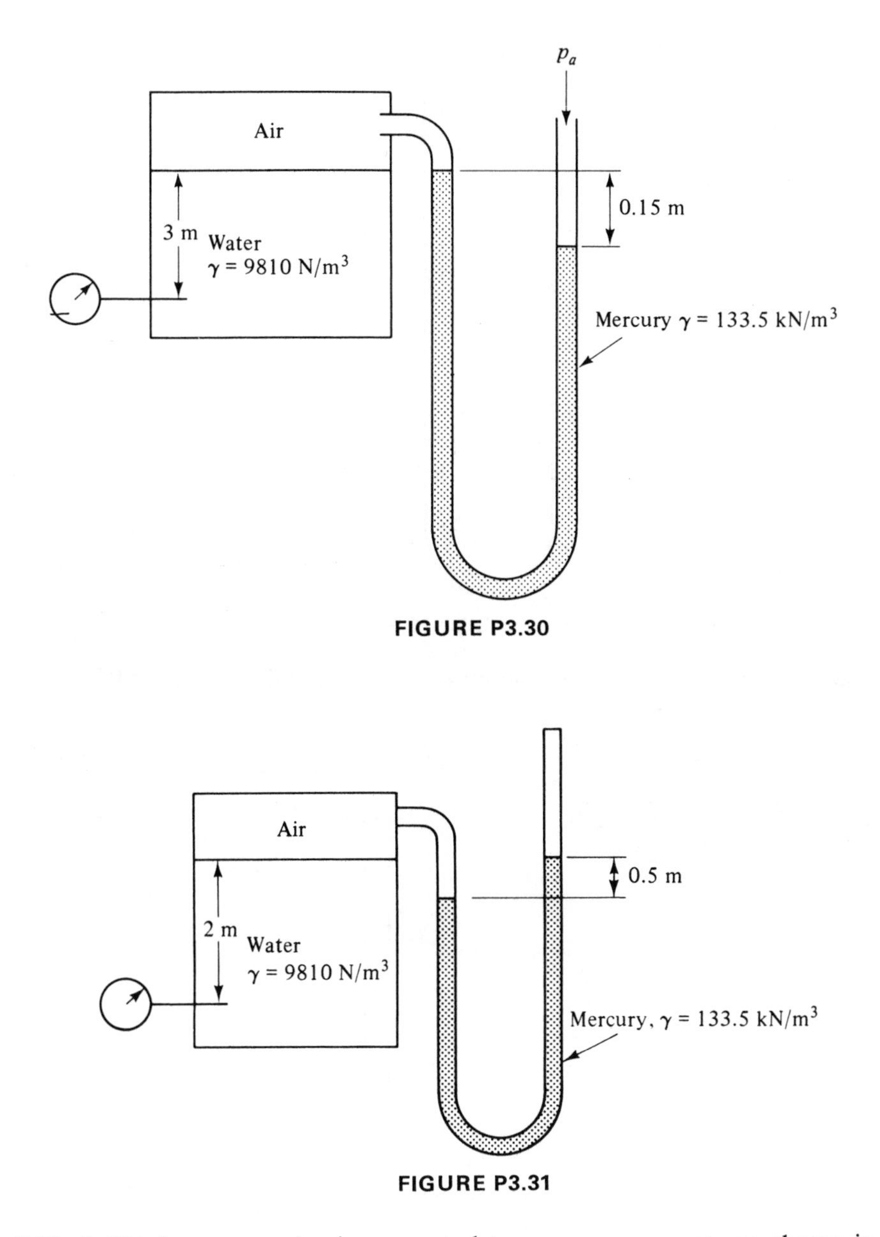

FIGURE P3.30

FIGURE P3.31

3.32 A U-tube manometer is connected to a pressure source as shown in Figure P3.32. What is the unknown pressure if p_a is 100 kPa? Assume the fluid to be air of negligible specific weight.

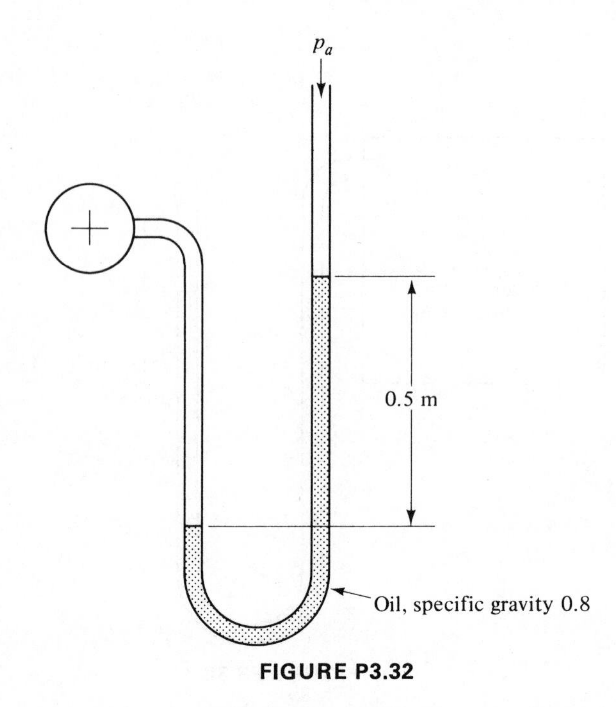

FIGURE P3.32

3.33 Determine the pressure difference between A and B in Figure P3.33.

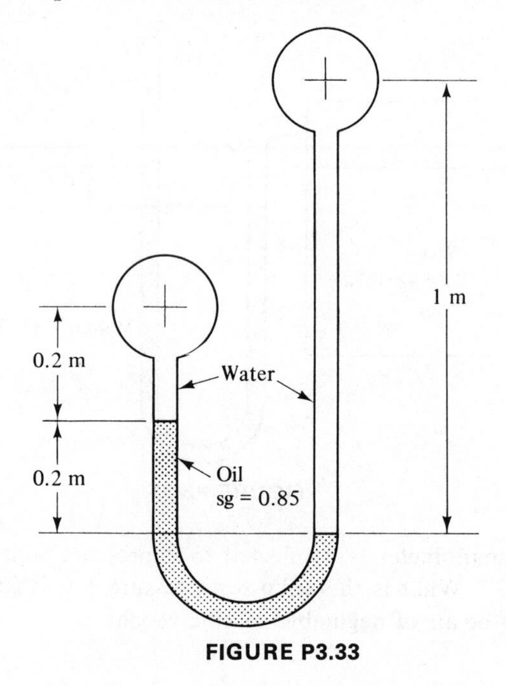

FIGURE P3.33

3.34 A manometer is connected to a pipe in which water is flowing as shown in Figure 3.14. Assume that the pipe is 20 in. in diameter and that the specific weight of the water in the pipe is 50 lb/ft³. Because of heat transfer, the specific weight of the water in the manometer leg is 55 lb/ft³. If B is 10 in. above the top of the pipe, h is 5 in., and $\gamma = 850$ lb/ft³, determine the pressure at the center of the pipe.

3.35 A sensitive U-tube manometer is to be used to measure a small pressure difference. If the ends are each 2 in. in diameter and the inside diameter (i.d.) of the tubing is $\frac{1}{4}$-in., what is the movement of the common liquid surface in the tube for a pressure differential of 0.1 psi if the fluids have specific weights of 50 and 60 lb/ft³, respectively?

3.36 In Figure P3.36, what is the pressure in the tank above the water surface?

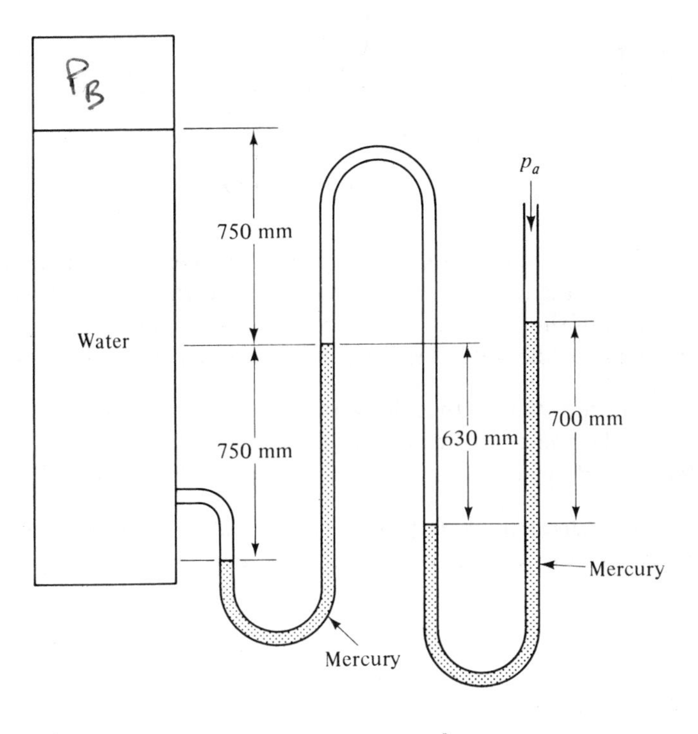

Mercury γ = 133.5 kN/m³
Water γ = 9810 N/m³

FIGURE P3.36

3.37 In Figure P3.37, what is the pressure in the line A?

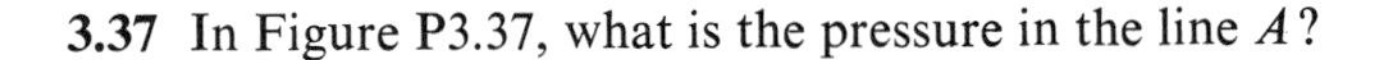

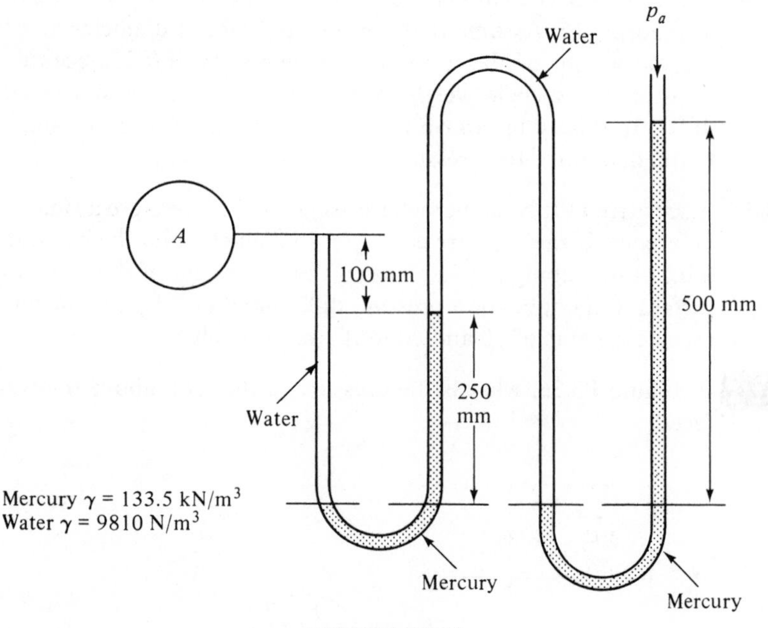

FIGURE P3.37

3.38 A cylindrical tank having flat ends has its long axis vertical. If the top is open to the atmosphere and there is a height of 3 m of water above the base of the tank ($\gamma_w = 9810\ \text{N/m}^3$), what is the force on the bottom plate due to the water. Assume the tank to have a diameter of 2 m.

3.39 A cylindrical tank with its long axis horizontal is filled with a fluid to its center. Derive an expression for the force on the end of the cylinder in terms of R, the cylinder radius, and the specific weight of the fluid.

3.40 If a cylindrical tank is placed horizontally and is 10 ft in diameter, determine the force on the end if the water level is at the center of the tank.

3.41 A cylindrical tank is placed horizontally and is filled with oil to its center. If the tank has a 3-m diameter and the oil has a specific gravity of 0.85, what is the force on the end of the cylinder?

3.42 A vertical square plate is used as a dam in a channel, as shown in Figure P3.42. If the plate is 8 ft square and the liquid level on the upstream side is 6 ft high, determine the total force on the plate.

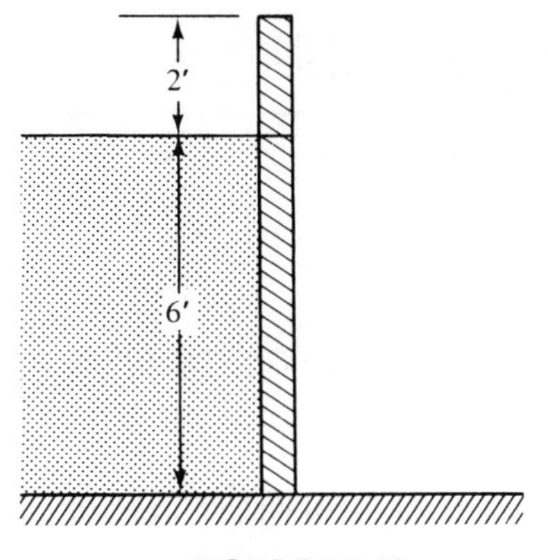

FIGURE P3.42

3.43 Determine the location of the resultant force in Problem 3.42.

3.44 What is the moment of the force on the plate of Figure P3.42 about its base?

3.45 What is the total force on the rectangular plate shown in Figure P3.45, and where is its center of pressure located? The plate is 5 m wide.

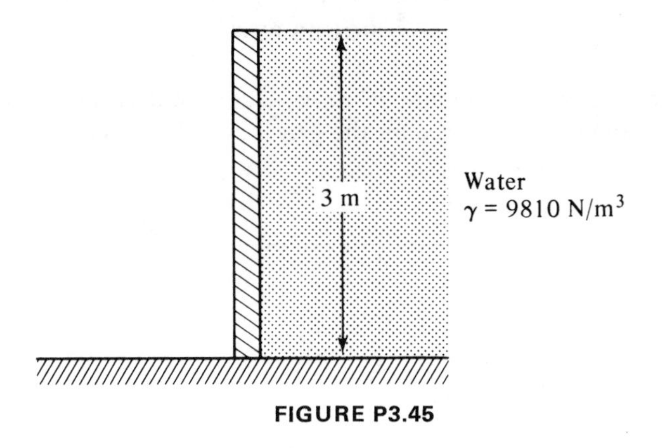

FIGURE P3.45

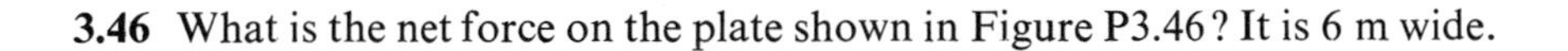

3.46 What is the net force on the plate shown in Figure P3.46? It is 6 m wide.

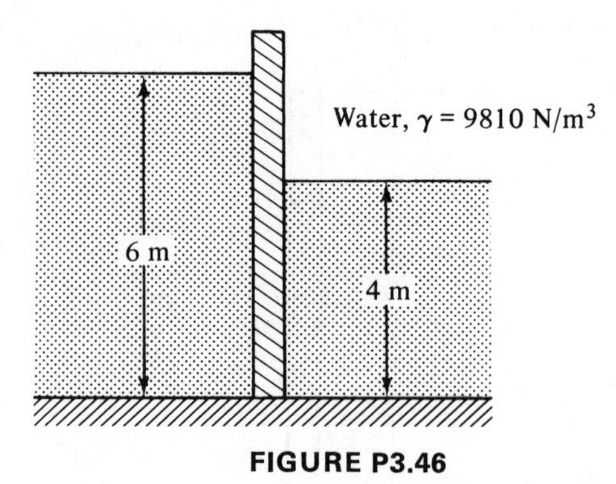

FIGURE P3.46

3.47 Calculate the force on the wall shown in Figure P3.47 if the wall is 6 m long.

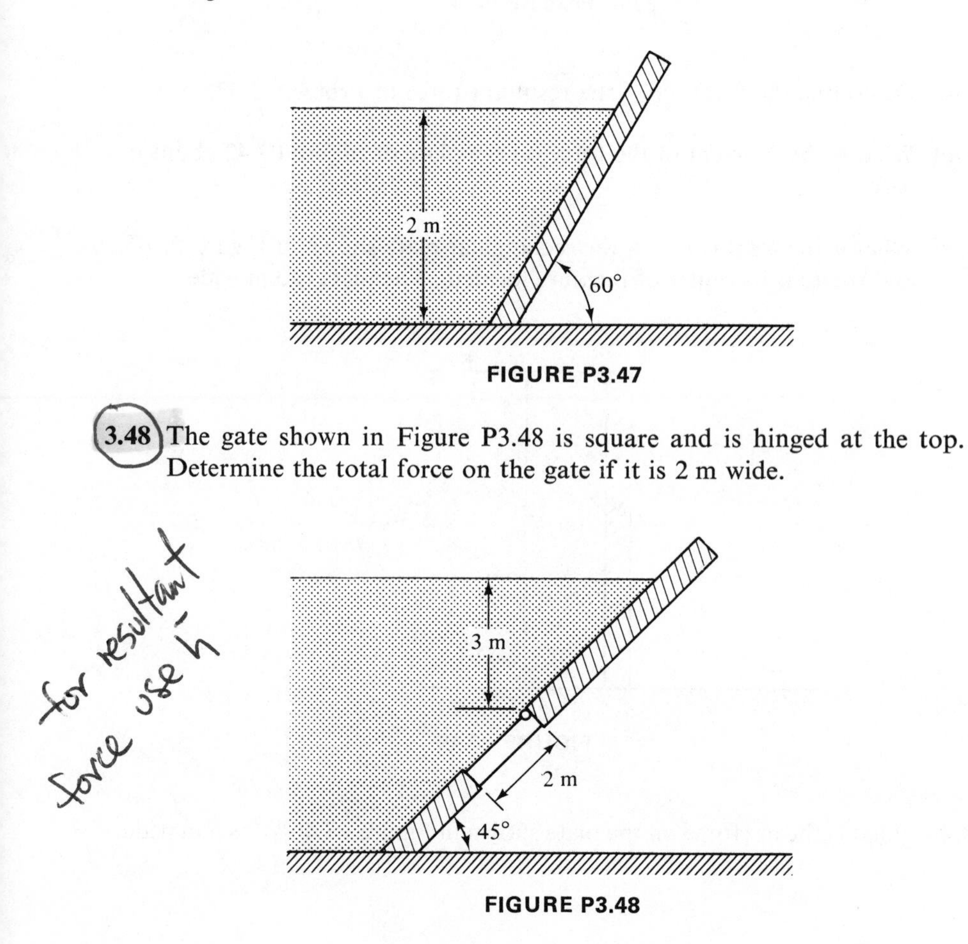

FIGURE P3.47

3.48 The gate shown in Figure P3.48 is square and is hinged at the top. Determine the total force on the gate if it is 2 m wide.

FIGURE P3.48

3.49 Determine the moment of the hydrostatic force about the hinge in Problem 3.48.

3.50 If the gate of Problem 3.48 is circular, determine the total force on the gate and the moment about the pin.

3.51 If the gate shown in Figure P3.51 is circular, determine the moment of the hydrostatic force about the hinge pin.

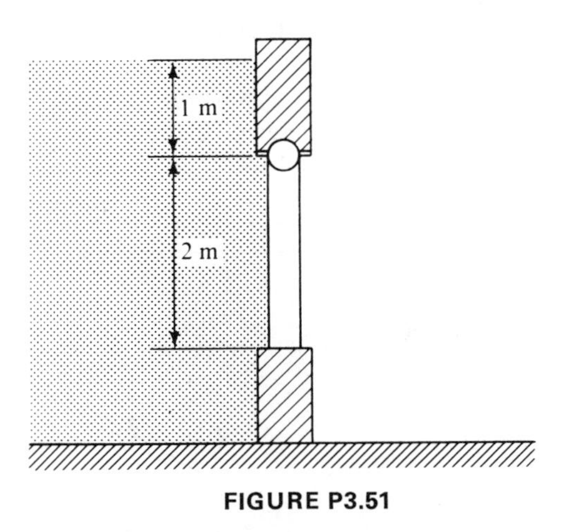

FIGURE P3.51

3.52 What is the net moment about the base of the plate in Problem 3.46?

3.53 A trash rack 5 ft wide protects the intake to a power plant as shown in Figure P3.53. In the event that the intake is blocked by ice and debris, what would be the total force acting on the rack and where would this force be applied?

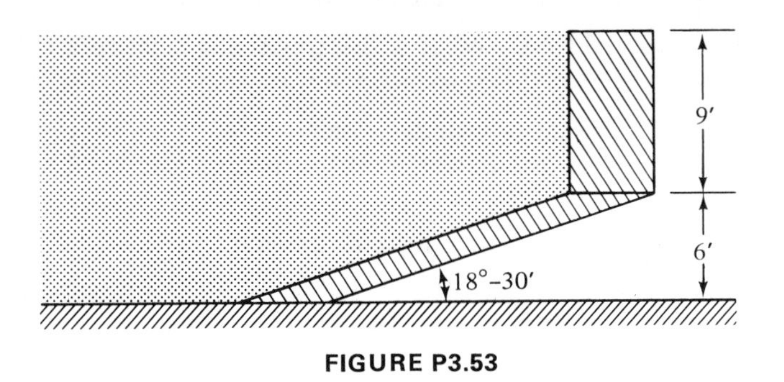

FIGURE P3.53

3.54 A vertical plate is used to dam a channel. If the plate has the shape shown in Figure P3.54, determine the total force on the plate. Assume that $\theta_1 = \theta_2$.

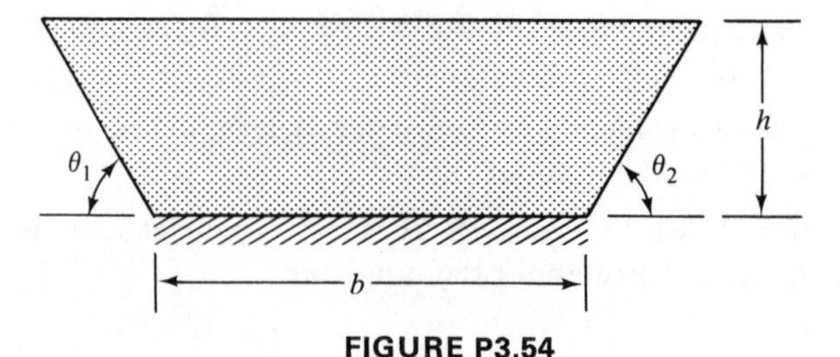

FIGURE P3.54

3.55 As shown in Figure P3.55, an aquarium installed a viewing window having a diameter of 1.5 m. Determine the magnitude and location of the hydrostatic force on the window.

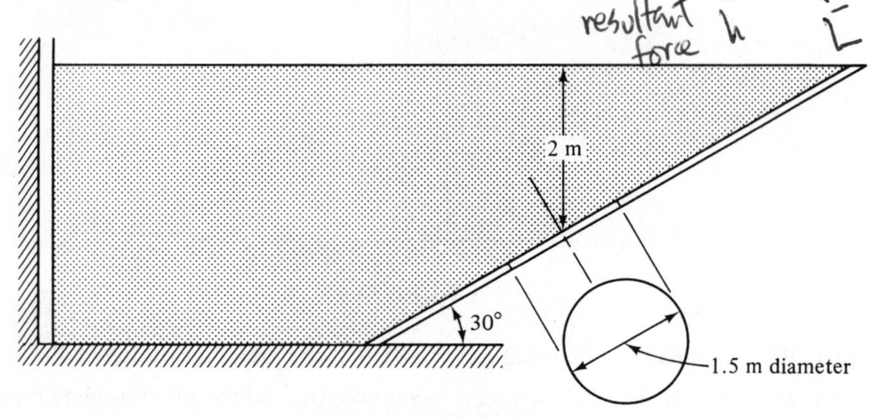

FIGURE P3.55

3.56 If the specific weight of air is 0.075 lb/ft³, solve Illustrative Problem 3.19 accounting for the buoyant effect of the air.

3.57 Steel weighs 490 lb/ft³. What is the apparent weight of a cube of steel 4 in. on a side when it is suspended in water having $\gamma_w = 62.0$ lb/ft³?

3.58 A piece of material is found to weigh 7.848 N in air and 5.00 N in water. What is the specific weight of the material? $\gamma_{\text{air}} = 11.8$ N/m³ and $\gamma_{\text{water}} = 9810$ N/m³.

3.59 A cylindrical log is 1 ft in diameter and 20 ft long. What weight of iron must be tied to one end of the log to keep it floating in an upright position in seawater with 18 ft of the log submerged? ($\gamma_{\text{log}} = 43.7$ lb/ft³, $\gamma_{\text{iron}} = 493$ lb/ft³, and $\gamma_{\text{sea water}} = 64.3$ lb/ft³.)

3.60 A ship is placed in a dry dock and the dry dock is filled until the ship just floats. By measuring the volume of water in the dock with and without the ship in it, it is found that 10,000 ft³ of seawater is displaced by the ship. If γ is 64.3 lb/ft³, determine the weight of the ship.

3.61 A quarter-circle is to be used as a gate, with the center of the quarter-circle at the surface of the liquid. If the radius is 10 in. and the gate has a length of 10 in., determine the horizontal and vertical forces acting on the gate. Also determine the location of the point of application of these forces.

3.62 Determine the force F to keep the gate closed in Figure P3.62. Assume that γ_w is the specific weight of the water and that the gate is W ft wide.

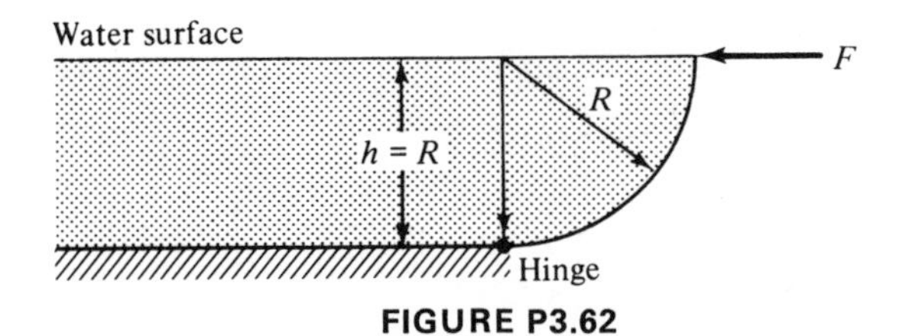

FIGURE P3.62

3.63 Determine the thickness of a cylinder subjected to internal pressure if the pressure is 100 psi and the diameter is 4 ft. Assume an allowable stress of 20,000 psi for the material of the cylinder.

3.64 Solve Problem 3.63 if the container is a sphere.

3.65 A pipe 2 ft in diameter has a wall thickness of $\frac{1}{4}$ in. If the allowable stress of the material is 15,000 psi, what internal pressure can it sustain?

REFERENCES

1. *Fundamentals of Hydro- and Aero-Mechanics* by L. PRANDTL and O. G. TIETJENS, McGraw-Hill Book Company, New York, 1934.
2. *Basic Fluid Mechanics* by J. L. ROBINSON, McGraw-Hill Book Company, New York, 1963.
3. *Fluid Mechanics For Engineers* by P. S. BARNA, Butterworth & Co. (Publishers) Ltd., London, 1957.
4. *Fluid Mechanics* by V. L. STREETER, 3rd ed., McGraw-Hill Book Company, New York, 1962.
5. *Elementary Fluid Mechanics* by J. K. VENNARD, John Wiley & Sons, Inc., New York, 1961.
6. *Elementary Theoretical Fluid Mechanics* by K. BRENKERT, Jr., John Wiley & Sons, Inc., New York, 1960.
7. *Engineering Applications of Fluid Mechanics* by J. C. HUNSAKER and B. G. RIGHTMIRE, McGraw-Hill Book Company, New York, 1947.

8. *Electronics in Engineering* by W. R. HILL, McGraw-Hill Book Company, New York, 1961.
9. "Pressure and Its Measurement" by R. P. BENEDICT, *Electro-Technology* (New York), **80,** October 1967, pp. 69–90.
10. *Applied Fluid Mechanics* by R. L. MOTT, 2nd ed., Charles E. Merrill Publishing Co., Columbus, Ohio, 1979.
11. *Fluid Mechanics* by A. G. HANSEN, John Wiley & Sons, Inc., New York, 1965.
12. *Fundamentals of Fluid Mechanics* by J. A. SULLIVAN, Reston Publishing Co., Reston, Va., 1978.

Energetics of Steady Flow

4.1 INTRODUCTION

In Chapter 3, we studied fluids at rest, and by applying the requirement that static systems must be in equilibrium, we were able to analyze many applications that occur in engineering. The next logical extension of the study of fluid mechanics is to those situations in which the fluid is flowing relative to its boundaries or to a fixed datum. A system has already been defined as a grouping of matter taken in any convenient or arbitrary manner. However, when dealing with fluids in motion, it is more convenient to utilize the concept of an arbitrary volume in space, known as a *control volume*, that can be bounded by either a real or imaginary surface, known as the *control surface*. By correctly noting all of the forces acting on the fluid within the control volume, the energies crossing the control surface, and the mass crossing the control surface it is possible to derive mathematical expressions that will evaluate the flow of the fluid relative to the control volume. In this chapter, we are concerned with fluids flowing steadily through the control volume, and for this type of system a monitoring station anywhere in the control volume will indicate no change in the fluid properties or energy quantities crossing the control surface with time. These quantities can and will vary from position to position in the control volume.

A word of caution is in order at this point—the concepts and equations that will be developed in this chapter are applicable to fluids flowing steadily. They should not be extended to unsteady (time-varying) cases unless the effects of time-varying terms, which are not considered in this chapter, are properly accounted for. Also, the flow will be assumed to be one-dimensional; that is, the flow parameters are constant in planes perpendicular to the mean flow direction.

4.2 CONSERVATION OF MASS—THE CONTINUITY EQUATION

As noted in Section 4.1, both energy and mass can enter and leave a control volume and cross the control surface of a system. Since we are considering steady-flow systems, we can express the fact that the principle of conservation of mass for these systems requires that the mass of the fluid in the control volume at any time be constant. *In turn, this requires that the net mass flowing into the control volume must equal the net mass flowing out of the control volume at any instant of time.* To express these concepts in terms of a given system, let us consider the system schematically in Figure 4.1.

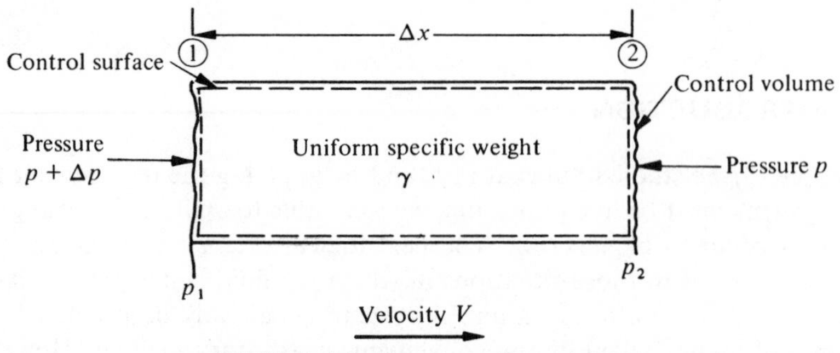

FIGURE 4.1 Elementary flow system.

Let us assume that at a certain time, fluid starts to enter the control volume by crossing the control surface at section 1 and that after a small interval of time the flowing fluid fills the pipe for a short distance x. If it is further assumed that in this short section of uniform pipe no heat is added or work is exchanged and that the specific weight stays constant, we can evaluate the amount of fluid that flowed in between sections 1 and 2. The weight contained between these sections is equal to the volume contained between the sections multiplied by the specific weight of the fluid. The volume is Ax and the specific weight is γ; therefore, the contained weight is γAx. The distance between stations, x, is simply Vt, where V is the velocity of the fluid and t is the flow time required to fill the pipe between sections 1 and 2. Substituting this for x,

$$W = \gamma AVt \tag{4.1a}$$

or

$$\dot{W} = \gamma AV \tag{4.1b}$$

where $\dot{W}$ is the weight rate of flow per unit time, W/t. Also, we can write equation (4.1) in terms of mass flow rate per unit time:

$$\dot{m} = \rho AV \tag{4.1c}$$

In order for the mass within the control volume to remain constant, the amount of mass entering the system must equal the amount of mass leaving the system. Therefore, using the subscripts 1 and 2 to denote any two stations,

$$\dot{m}_1 = \dot{m}_2 \tag{4.2a}$$

Equation (4.2a) can also be expressed in the following equivalent forms:

$$\rho_1 A_1 V_1 = \rho_2 A_2 V_2 \tag{4.2b}$$

$$\gamma_1 A_1 V_1 = \gamma_2 A_2 V_2 \tag{4.2c}$$

$$\frac{A_1 V_1}{v_1} = \frac{A_2 V_2}{v_2} \tag{4.2d}$$

where the specific volume $v = 1/\rho$. For an incompressible fluid, $\gamma_1 = \gamma_2$, $\rho_1 = \rho_2$. Therefore, $A_1 V_1 = A_2 V_2$.

Equation (4.2) is known as the *continuity equation*, and as written for a pipe or duct we have made the assumption that the flow is normal to the pipe cross section and that the velocity V is either constant across the section or is the average value over the cross section of the pipe. The average value in this case is obtained by summing the products of velocity multiplied by the corresponding areas and dividing by the sum of all of the areas.

ILLUSTRATIVE PROBLEM 4.1

At the entrance to a steady flow device it is found that the pressure is 100 psia and that the specific weight of the fluid is 62.4 lb/ft^3. If 10,000 ft^3/min of fluid enters the system and the exit area is 2 ft^2, determine the mass flow rate and the exit velocity.

Solution

Refer to Figure 4.1. The weight flow rate is $\dot{W}$, which equals $\gamma A V =$ 62.4(10,000) = 624,000 lb/min. Since this is the same at entrance and exit, the exit velocity is

$$V = \frac{\gamma A V}{\gamma A} = \frac{10{,}000}{2} = 5000 \text{ ft/min}$$

In each form of equation (4.2) we find the product of AV. Since this combined term appears very frequently in fluid mechanics applications, it is usually given a separate designation,

$$Q = AV \tag{4.3}$$

where Q is the *volume flow rate*. In English units,

$$Q = \text{ft}^2 \times \frac{\text{ft}}{\text{sec}} = \frac{\text{ft}^3}{\text{sec}}$$

and in SI units,

$$Q = \text{m}^2 \times \frac{\text{m}}{\text{s}} = \frac{\text{m}^3}{\text{s}}$$

We can obtain the mass and weight flow rates in both systems as

$$\begin{aligned} \dot{m} &= \rho Q \\ \dot{W} &= \gamma Q \end{aligned} \tag{4.4}$$

where $\dot{m}$ is kg/s or slugs/sec and $\dot{W}$ is N/s or lb/sec.

ILLUSTRATIVE PROBLEM 4.2

Water flows in a pipe at the rate of 100 gallons per minute (gpm). What is Q in cubic feet per second (cfs)?

Solution

A gallon is a volume equal to 231 in.^{3}. Thus, this problem is one in the conversion of units.

$$Q = 100 \frac{\text{gal}}{\text{min}} \times \frac{231 \text{ in.}^3}{\text{gal}} \times \frac{1}{(12 \text{ in./ft})^3} \times \frac{1}{60 \text{ sec/min}}$$

$$= 0.2228 \frac{\text{ft}^3}{\text{sec}} = 0.2228 \text{ cfs}$$

ILLUSTRATIVE PROBLEM 4.3

How many m^3/s is flowing in Illustrative Problem 4.2?

Solution

Using 1 in. = 0.0254 m,

$$Q = 100\,\frac{\text{gal}}{\text{min}} \times 231\,\frac{\text{in.}^3}{\text{gal}} \times \left(0.0254\,\frac{\text{m}}{\text{in.}}\right)^3 \times \frac{1}{60\text{ sec/min}}$$

$$= 0.006\,309\ \text{m}^3/\text{s}$$

ILLUSTRATIVE PROBLEM 4.4

Water is flowing from one pipe to another as shown in Figure 4.2. Determine the velocity in each section of pipe.

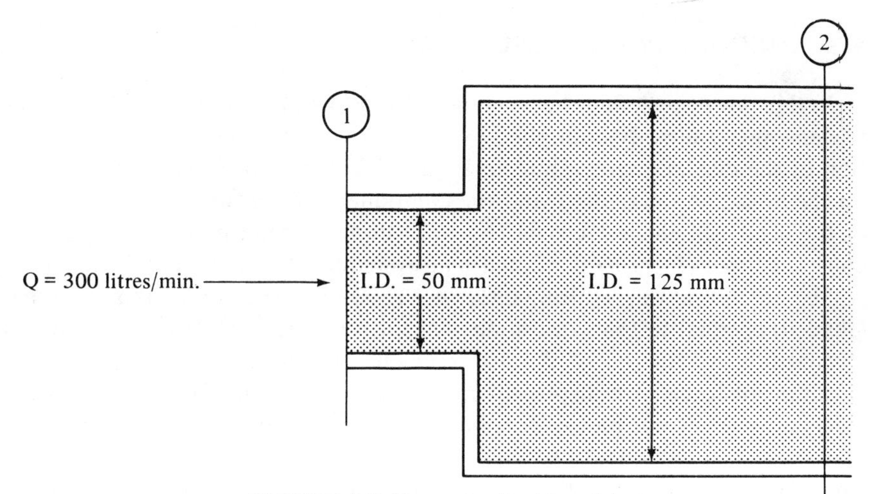

FIGURE 4.2 Illustrative Problem 4.4.

Solution

One litre is 1000 cm^3 = $1000 \times (10^{-2})^3 = 10^{-3}\ m^3$. Therefore, Q is

$$Q = 300\,\frac{l}{\text{min}} \times 10^{-3}\,\frac{\text{m}^3}{l} \times \frac{1}{60\text{ sec/min}} = 0.005\ \text{m}^3/\text{s}$$

The flow areas of each pipe are $\pi d^2/4$, where d is the inside diameter. Therefore,

$$A_1 = \frac{\pi d_1^2}{4} = \frac{\pi(0.050)^2}{4} = 1.963 \times 10^{-3}\ \text{m}^2$$

$$A_2 = \frac{\pi d_2^2}{4} = \frac{\pi(0.125)^2}{4} = 1.227 \times 10^{-2}\ \text{m}^2$$

Since water is incompressible, $\gamma_1 = \gamma_2$ or $\rho_1 = \rho_2$. This gives us the condition that $Q_1 = Q_2 = Q$. Therefore,

$$Q = A_1V_1 = A_2V_2$$

Thus,

$$V_1 = \frac{Q}{A_1} = \frac{0.005\ \text{m}^3/\text{s}}{1.963 \times 10^{-3}\ \text{m}^2} = 2.55\ \text{m/s}$$

$$V_2 = \frac{Q}{A_2} = \frac{0.005\ \text{m}^3/\text{s}}{1.227 \times 10^{-2}\ \text{m}^2} = 0.407\ \text{m/s}$$

4.3 ENERGY, WORK, AND HEAT

In this book we define the *work* done by a force as the product of the displacement of the body multiplied by the component of the force in the direction of the displacement. Thus, in Figure 4.3a the displacement of the body on the horizontal plane is x and the component of the force in the direction of the displacement is $F \cos \theta$. The work done is therefore $(F \cos \theta)(x)$. The constant force $(F \cos \theta)$ is plotted as a function of x in Figure 4.3b, and it will be noted that the resulting figure is a rectangle. The area of this rectangle (shaded) is equal to the work done, since it is $(F \cos \theta)(x)$. If the force varies so that it is a function of the displacement, it is necessary to consider the variation of force with displacement in order to find the work done. Figure 4.3c shows a general plot of force as a function of displacement. If the displacement is subdivided into many small parts, Δx, and for each of these small parts F is assumed to be very nearly constant, it is apparent that the sum of the small areas $(F)(\Delta x)$ will represent the total work done when the body is displaced from x_1 to x_2. Thus, the area under the curve of F as a function of x represents the total work done if F is the force component in the direction of x.

At this point let us define the term *energy* in terms of work. Energy can be described (in an elementary manner) as the capacity to do work. At first it may appear that this definition of energy is too restrictive. Yet in all instances the observed effects on a system can (ideally) be converted to mechanical

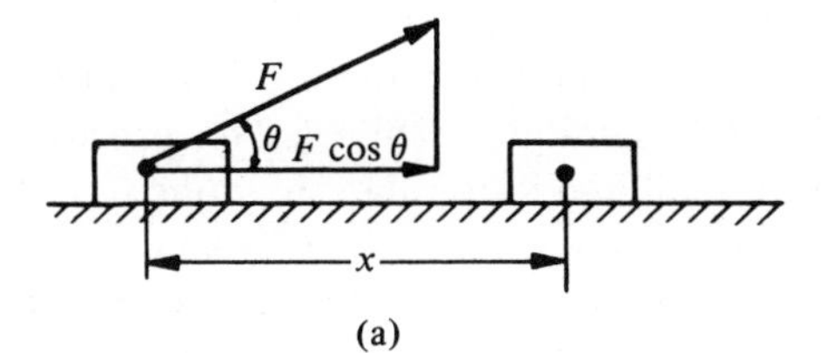

(a)

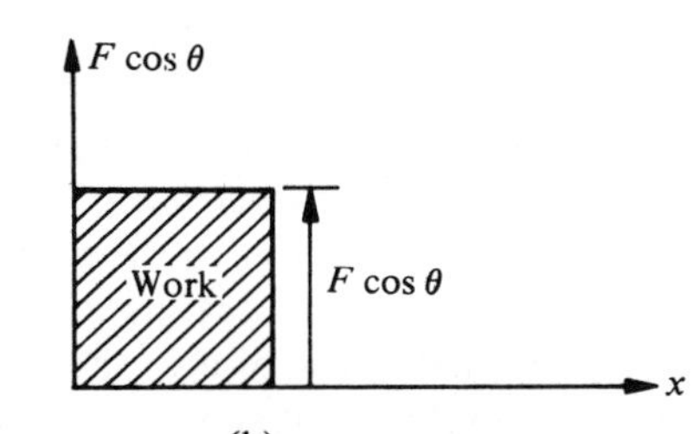

(b)

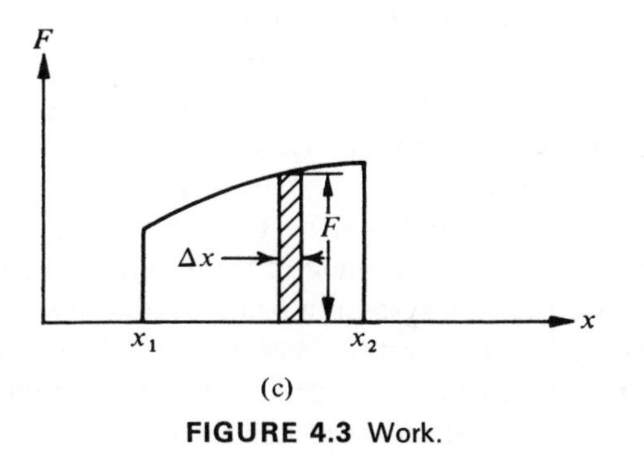

(c)

FIGURE 4.3 Work.

work. Since work has been defined as the product of a force multiplied by a displacement, it is not stored in a system. It represents energy that must be crossing the boundaries (real or imaginary) of the system and can properly be denoted to be energy in transition.

To distinguish between the transfer of energy as work to or from a system, we shall adopt the convention that *work done by a system on its surroundings* is *positive* and *work done by the surroundings on the system represents negative work*. For the student it is best to think of useful *work out of a system as being a desirable quantity and therefore as positive*. For example, consider a spring that is compressed by a force. If the spring is the system, the work is negative by convention. If the external force system acting on the spring is the system, the work is positive by convention.

The work that a system can perform on its surroundings is not an intrinsic property of the system. Work is a function of the path taken and is a transitory effect that is not stored in the system and that is not a property of the system. There is one process, however, that does permit the evaluation of the work done since the path is uniquely defined. This process is called the quasi-static process, and we shall find it useful in subsequent work. The quasi-static process is a process carried out infinitely slowly so that it is in equilibrium at all times. Before considering several processes, let us first discuss another form of energy that can cross the boundaries of a system—*heat*.

When a *heat* interaction occurs in a system, it is observed that two distinct events occur. The first is that an interchange of energy has occurred between the system and its surroundings. The second is that this would not have taken place if there were no temperature difference between the system and its surroundings. We may therefore define *heat* as the energy in transition crossing the boundaries of a system due to a temperature difference between the system and its surroundings. In this definition of heat the transfer of mass across the boundaries of the system is excluded. It should be noted that this definition of heat indicates a similarity between heat and work. Both are energies in transition and both are not properties of the system in question. Just as was the case for work, we can have quasi-static heat transfer to or from a system. In this case the difference in temperature between the system and its surroundings can be only an infinitesimal amount at any time. Once again it is necessary to adopt a convention for the energy interchanged by a system with its surroundings as heat. We shall use the convention that heat to the system from its surroundings is positive and that heat out of the system is negative. To learn these conventions it is convenient to think of the conventional situation where heat is transferred to a system in order to obtain useful work from the system. This sets the convention that heat into a system is positive and work out of the system is also positive. *Positive* in this sense means either desirable or conventional from the viewpoint of conventional power cycles.

Since work and heat are both forms of energy in transition, it follows that the units of work should be capable of being expressed as heat units and vice versa. In the SI system there is no need for a conversion factor, since the joule (N·m) is the basic energy unit. In the English system, this conversion factor is defined as 778.16 ft lb/Btu and is conventionally given the symbol J. In this text we shall use the symbol J to be 778 ft lb/Btu since this is sufficiently accurate.

4.3a Potential Energy

Let us consider the following problem. A body of mass m is in a locality where the local gravitational field is constant and equal to g. A force is applied to the body, and it is raised a distance Z from its initial position. The force will be assumed to be only infinitesimally greater that the mass, so that the process is carried out as a quasi-static process. In the absence of electrical, magnetic, and other extraneous effects, determine the work done on the body. The solution to this problem is obtained by noting that the equilibrium of the body requires that a force must be applied to it equal to its weight. The weight of the body is simply w. In moving through a distance Z, the work done by this force will therefore equal

$$\text{work} = wZ \tag{4.5}$$

The work done on the body can be returned to the external environment by simply reversing this quasi-static process. We therefore conclude that this system has had work done on it equal to wZ and that, in turn, the system has stored in it an amount of energy in excess of the amount it had in its initial position. The energy added to the system in this case is called *potential energy*. Thus,

$$\text{potential energy (P.E.)} = wZ \tag{4.6}$$

or

$$\text{potential energy (P.E.)} = mgZ \tag{4.7}$$

since $w = mg$.

ILLUSTRATIVE PROBLEM 4.5

In Figure 4.4, a pump "lifts" water from a well that is 10 m deep. What is the change in the potential energy of the water?

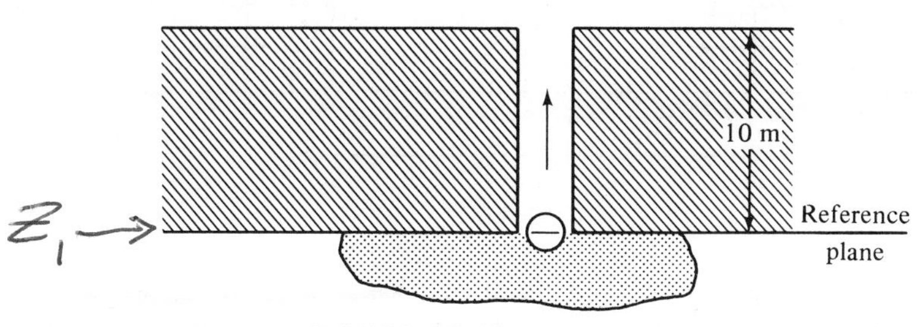

FIGURE 4.4 Illustrative Problem 4.5.

Solution

It is convenient to select a reference plane that is at the lowest point in the system. This helps to avoid negative numbers. Initially, the water is at the reference plane. Consequently, Z_1 is zero and the initial potential energy is zero. Finally, $Z_2 = 10$ m, therefore, for 1 N,

$$(\text{P.E.})_2 = 1 \text{ N} \times 10 \text{ m per Newton since 1 N was assumed}$$

and

$$(\text{P.E.})_2 = 10 \text{ N}\cdot\text{m/N}$$

The change in P.E. is therefore

$$(\text{P.E.})_2 - (\text{P.E.})_1 = (10 - 0) = 10 \text{ N}\cdot\text{m/N}$$

A feature of importance of potential energy is that a system can be said to possess potential energy only with respect to an arbitrary initial or datum plane.

4.3b Kinetic Energy

Let us consider another situation in which a body of mass m is at rest on a frictionless plane. If a force F is applied to the mass, it will be accelerated in the direction of the force. After moving through a distance S, the velocity of the body will have increased from 0 to V. The only effect of the work done on the body will be an increase in its velocity. Since the force F in Figure 4.5 is constant, the acceleration of the block will be constant.

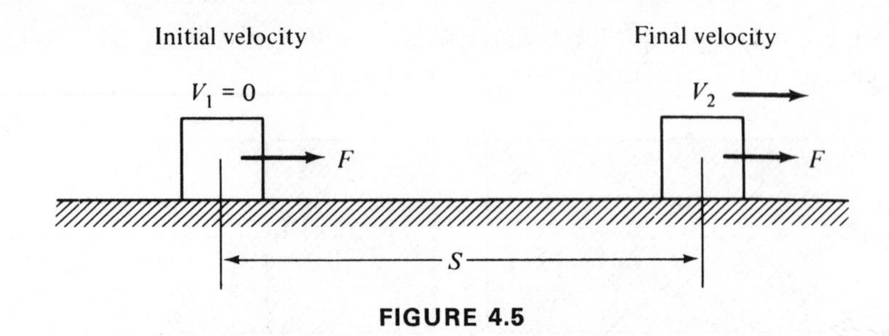

FIGURE 4.5

From elementary physics we will recall that the velocity V and displacement for a body having constant acceleration is given for zero initial velocity as

$$V^2 = 2aS \tag{4.8}$$

But Newton's second law gives us

$$a = \frac{F}{m} \tag{4.9}$$

If equation (4.9) is inserted into equation (4.8), we have

$$V^2 = 2\frac{F}{m}S \tag{4.10}$$

Transposing terms yields

$$\frac{mV^2}{2} = FS \tag{4.11}$$

Since FS is the work done by the constant force F acting on the body for a displacement of S, the body is said to possess *kinetic energy* (K.E.). By virtue of its velocity, it has the ability to do work. Since $m = w/g$, we can rewrite equation (4.11) to yield the familiar form for kinetic energy as

$$\text{K.E.} = \frac{wV^2}{2g} \tag{4.12}$$

having units of N·m or ft lb.

ILLUSTRATIVE PROBLEM 4.6

A body having a mass of 10 kg falls freely from rest (Figure 4.6). After falling 10 m, what will its kinetic energy be and what will its velocity be?

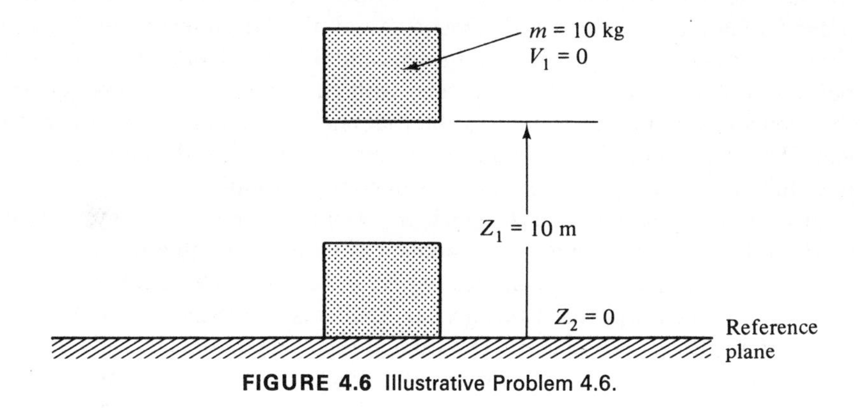

FIGURE 4.6 Illustrative Problem 4.6.

Solution

Since there are no energy losses in the system, we conclude that the sum of the initial potential energy plus kinetic energy must equal the sum of the final potential energy plus kinetic energy. Thus,

$$(\text{P.E.})_1 + (\text{K.E.})_1 = (\text{P.E.})_2 + (\text{K.E.})_2$$

Notice that the selected reference plane and the condition that the body falls from rest gives us $(P.E.)_1 = (K.E.)_2$, that is potential energy lost equals the kinetic energy gained.

$$10(10 \times 9.81) = \frac{10(9.81)V^2}{2 \times 9.81}$$

Proceeding,

$$V^2 = 2 \times 9.81 \times 10$$

and

$$V = 14.0 \text{ m/s}$$

4.3c Internal Energy

To this point we have considered the energy in a system that arises from the work done on the system. However, a body possesses energy by virtue of the molecular motion of the molecules of the body. In addition to the energy possessed by the body due to the motion of its molecules, it also possesses energy due to the internal attractive and repulsive forces between molecules. These forces give rise to the storage of energy internally as potential energy. The energy that a body possesses from all such sources is called the internal energy of the body and is designated by the symbol U. Per unit mass w/g the specific internal energy is denoted by the symbol u, where $wu/g = U$. From a practical standpoint, the measurement of the absolute internal energy of a system in a given state presents an insurmountable problem and is not essential to our study. We shall be concerned with changes in internal energy, and the arbitrary datum for the zero of internal energy will not enter these problems. Internal energy is a property of a system and is not dependent on the path taken to place the system in a given configuration.

Just as it is possible to distinguish the various forms of energy, such as work and heat in a mechanical system, it is equally possible to distinguish the various forms of energy associated with electrical, chemical, and other systems. For the purposes of this text, these forms of energy other than work and heat will not be considered. The student is cautioned that should a system include any forms of energy other than mechanical, these items must be included. Thus, the energy that is dissipated in a resistor as heat when a current flows through it must be taken into account when all the energies of an electrical system are being considered.

4.3d Flow Work

When a fluid is caused to flow in a system, it is necessary that somewhere in the system work must have been supplied. At this time let us evaluate the net work required to push the fluid into and out of the system. Consider the system shown in Figure 4.7, where a fluid is flowing steadily across the system boundaries as shown. At the inlet section ①, the pressure is p_1, the area is A_1, the mass flow rate is $\dot{m}$ and the fluid density is ρ_1 (or its reciprocal specific volume, $1/v_1$); at the outlet section ②, the pressure is ρ_2, the area is A_2, the mass flow rate is still $\dot{m}$ and the fluid density is ρ_2 (or its reciprocal

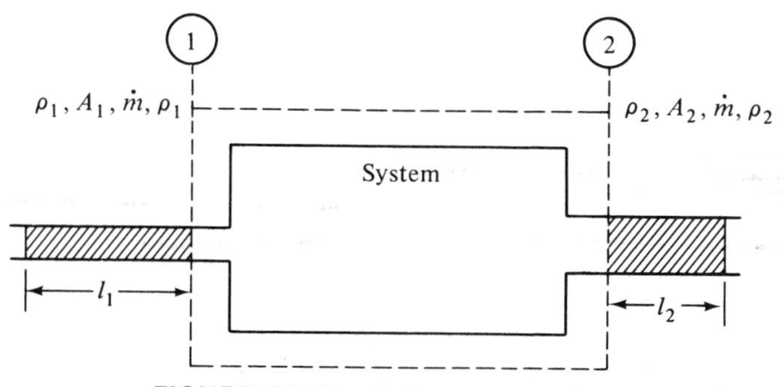

FIGURE 4.7 Steady-flow system—flow work.

specific volume $1/v_2$). Let us now consider a plug of fluid of length l_1 entering the system such that the amount of fluid contained in the plug is numerically $\dot{m}$. The force acting on the inlet cross-sectional area A_1 is p_1A_1. In order to push the plug into the system it is necessary for this force to move the plug a distance equal to l_1. In so doing, the work done will be $p_1A_1l_1$. However, A_1l_1 is the volume of the plug containing a mass m. Using this, we find the work W to be

$$W = p_1A_1l_1 = p_1V_1 \tag{4.13}$$

However, the weight is

$$w = \gamma V \tag{4.14}$$

Therefore, combining equations (4.13) and (4.14), we obtain

$$W = w\frac{p_1}{\gamma_1} \tag{4.15a}$$

If we now consider the outlet section, using the same reasoning,

$$W = w\frac{p_2}{\gamma_2} \tag{4.15b}$$

Each of the p terms in equations (4.15a) and (4.15b) is known as flow work

(FW). Thus, the net flow work in the situation depicted in Figure 4.7 is

$$\text{net flow work} = w\left(\frac{p_2}{\gamma_2} - \frac{p_1}{\gamma_1}\right) \tag{4.16}$$

The units of equation (4.16) are N·m or ft lb.

4.4 STEADY-FLOW ENERGY EQUATION

The *First Law of Thermodynamics* is the statement of the conservation of energy; it can be expressed by stating that energy can be neither created nor destroyed but only converted from one form to another. Since we shall not be concerned with nuclear reactions, it will not be necessary to invoke the interconvertiblity of energy and mass in this study.

In the derivations made in earlier sections, certain assumptions were made, and these are repeated here for emphasis. The term *steady*, when applied to a flow situation, means that the condition at any section of the system is independent of time. Even though the velocity, specific weight, and temperature of the fluid can vary in an arbitrary manner across the stream, they are not permitted to vary with time. The weight entering the system per unit time must equal the weight leaving the system in the same period of time; otherwise, the system would either store or be depleted of fluid.

To summarize, we have identified six energy terms that apply to various situations. Assigning the symbol u for the internal energy per unit weight, q to be the heat interchange per unit weight, and J to be the conversion factor for converting Btu to ft lb, we have

Item	*Value*	
	ft lb/lb	***N·m/N***
Potential energy	Z	Z
Kinetic energy	$V^2/2g$	$V^2/2g$
Internal energy	Ju	u
Flow work	p/γ	p/γ
Work	W	W
Heat	Jq	q

Figure 4.8 shows such a system, where it is assumed that each form of energy can both enter and leave the system. At the entrance, $\dot{w}$ lb of fluid/sec enters and the same amount leaves at the exit. At the entrance the fluid has a pressure of p_1, a specific weight of γ_1, an internal energy of u_1, and a velocity

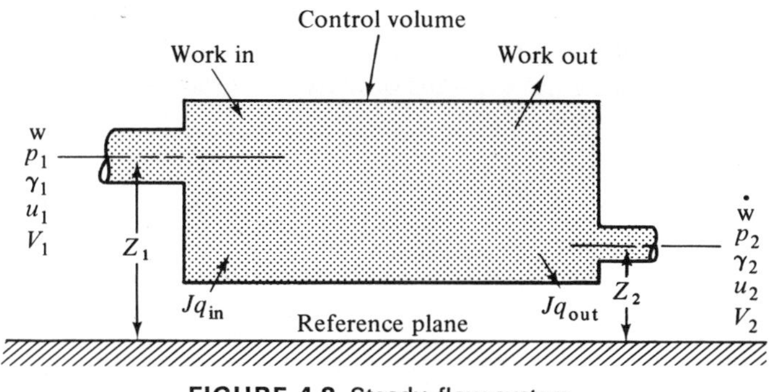

FIGURE 4.8 Steady-flow system.

of V_1. At the exit we have similar quantities expressed as p_2, γ_2, u_2, and V_2. The fluid enters and leaves at different elevations, and both work and heat cross the boundary in both directions. We express the conservation of energy for this steady-flow system by stating that all of the energy entering the system must equal all of the energy leaving the system. Setting up this energy balance:

	Energy In		*Energy Out*	
	ft lb/lb	*J/N*	*ft lb/lb*	*J/N*
Potential energy	Z_1	Z_1	Z_2	Z_2
Kinetic energy	$V_1^2/2g$	$V_1^2/2g$	$V_2^2/2g$	$V_2^2/2g$
Internal energy	Ju_1	u_1	Ju_2	u_2
Flow work	p_1/γ_1	p_1/γ_1	p_2/γ_2	p_2/γ_2
Work	W_{in}	W_{in}	W_{out}	W_{out}
Heat	Jq_{in}	q_{in}	Jq_{out}	q_{out}

Equating all of these terms;

$$Z_1 + \frac{V_1^2}{2g} + Ju_1 + \frac{p_1}{\gamma_1} + W_{in} + Jq_{in} = Z_2 + \frac{V_2^2}{2g} + Ju_2 + \frac{p_2}{\gamma_2} W_{out} + Jq_{out} \tag{4.17}$$

In SI units,

$$Z_1 + \frac{V_1^2}{2g} + u_1 + \frac{p_1}{\gamma_1} + W_{in} + q_{in} = Z_2 + \frac{V_2^2}{2g} + u_2 + W_{out} + q_{out}$$

Equation (4.17) is quite general and is a consequence of the principle of conservation of energy for this system.

If at this point we note that the heat and work terms can each be combined to form individual terms of net heat and net work and if we take proper cognizance of the mathematical signs of these net terms, we can rewrite equation (4.17) as

$$Z_1 + \frac{V_1^2}{2g} + Ju_1 + \frac{p_1}{\gamma_1} + Jq = Z_2 + \frac{V_2^2}{2g} + Ju_2 + \frac{p_2}{\gamma_2} + W \tag{4.18}$$

or in SI units,

$$Z_1 + \frac{V_1^2}{2g} + u_1 + \frac{p_1}{\gamma_1} + q = Z_2 + \frac{V_2^2}{2g} + u_2 + \frac{p_2}{\gamma_2} + W \tag{4.19}$$

Using the net values of heat and work as is done in equation (4.18) is equivalent to adopting the convention that net heat in is positive and net work out is positive. This corresponds to the conventional power plant cycle where heat is released (heat input) to obtain useful work (work output). In equation (4.18) or (4.19) all terms are written as foot pounds per pound or joules/newton and q and W represent net values. If we now further note that the grouping of terms $u + p/\gamma$ appears on both sides of equations (4.18) and (4.19) and that this combined term is a property, we obtain

$$h = u + \frac{p}{\gamma} \tag{4.20}$$

where h is given the name enthalpy. The student should note the necessity to use consistent units in equation (4.20). Regrouping terms in equations (4.18) and (4.19) and using equation (4.20), we obtain

$$q - \frac{W}{J} = (h_2 - h_1) + \frac{Z_2 - Z_1}{J} + \frac{V_2^2 - V_1^2}{2gJ} \quad \text{in Btu/lb} \tag{4.21}$$

or

$$q - W = (h_2 - h_1) + (Z_2 - Z_1) + \frac{V_2^2 - V_1^2}{2g} \quad \text{in J/N} \tag{4.22}$$

The student should notice that equations (4.17) through (4.22) are essentially the same and are called the *steady-flow energy equations.* To intelligently apply these equations, it is important that each of the terms be completely understood. Although these equations are not difficult, the student may encounter difficulty at this point. Most of this difficulty arises from a lack of understanding of these energy equations and the basis for each of the terms in them.

4.5 THE BERNOULLI EQUATION

In fluid mechanics, use is frequently made of an equation known as the Bernoulli equation. In Chapter 5 this equation will be studied in detail; however, it is in order to undertake a brief discussion of it at this point. Let us consider a system in which the flow is steady, there is no change in internal energy, no work is done on or by the system, no energy as heat crosses the boundaries of the system, and the fluid is incompressible. We shall further assume that all processes are ideal in the sense that they are frictionless. For this system the energy equation reduces to

$$Z_1 + \frac{V_1^2}{2g} + \frac{p_1}{\gamma_1} = Z_2 + \frac{V_2^2}{2g} + \frac{p_2}{\gamma_2} \tag{4.23}$$

Equation (4.23) is the usual form of the Bernoulli equation. It should be noted that in arriving at equation (4.23) we have essentially taken each energy term for w lbs and then canceled out the w, as in equation (4.23a):

$$\cancel{w}Z_1 + \frac{\cancel{w}V_1^2}{2g} + \cancel{w}\frac{p_1}{\gamma_1} = \cancel{w}Z_2 + \cancel{w}\frac{V_2^2}{2g} + \frac{\cancel{w}p_2}{\gamma_2} \tag{4.23a}$$

Each term of equation (4.23) can represent a height since the dimension of foot pounds per pound corresponds numerically (and dimensionally) to a height. Hence, the term *head* is frequently used to denote each of the terms in equation (4.23). In SI units, each term is N·m/N. Again, this unit is dimensionally a height or "head." It is more fundamental to interpret each term as the energy per unit weight of fluid. From the derivation of this equation we note that its use is restricted to those situations where the flow is steady, there is no friction, no shaft work is done on or by the fluid, the flow is incompressible, there is no change in internal energy during the process, and there is no heat transfer to or from the system. Although these restrictions severely limit the use of equation (4.23), several modifications have been made to it to try to rationally approach the generality of the steady-flow energy equation. We shall study the Bernoulli equation extensively in Chapter 5.

ILLUSTRATIVE PROBLEM 4.7

Water ($\gamma = 9810$ N/m^3) flows in a pipe. At a section where the inside diameter is 150 mm, the velocity is 3 m/s and the pressure is 350 kPa. At a section located 10 m from the first section, the inside diameter reduces to

75 mm. Calculate the pressure at the second section if

(a) The pipe is horizontal.
(b) The pipe is vertical and flow is downward.

Solution

Our first task is to obtain the velocity in the pipe at the second section. Since γ does not change, $Q_1 = Q_2$. Therefore,

$$A_2 V_2 = A_1 V_1$$

$$\frac{\pi}{4}(75)^2 \times V_2 = \frac{\pi}{4}(150)^2 \times 3$$

and

$$V_2 = 12 \text{ m/s}$$

Considering the pipe to be horizontal, $Z_1 = Z_2$, and writing equation (4.23),

$$Z_1 + \frac{V_1^2}{2g} + \frac{p_1}{\gamma_1} = Z_2 + \frac{V_2^2}{2g} + \frac{p_2}{\gamma_2}$$

$$\frac{3^2}{2 \times 9.81} + \frac{350 \times 10^3}{9810} = \frac{12^2}{2 \times 9.81} + \frac{p_2}{9810}$$

and

$$p_2 = 9810 \times 28.80 = 282.5 \text{ kPa}$$

For the case of flow being vertically downward, we have (see Figure 4.9)

$$Z_1 + \frac{V_1^2}{2g} + \frac{p_1}{\gamma_1} = Z_2 + \frac{V_2^2}{2g} + \frac{p_2}{\gamma_2}$$

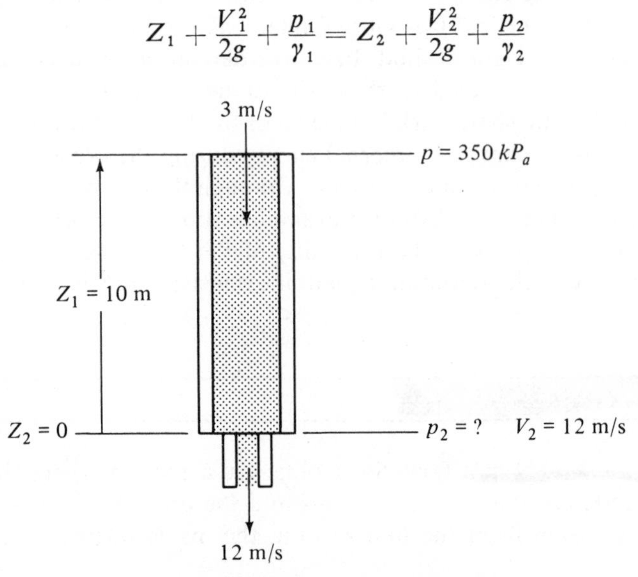

FIGURE 4.9 Illustrative Problem 4.7.

but $Z_2 = 0$ by the selection of the reference plane as shown. Thus,

$$10 + \frac{3^2}{2 \times 9.81} + \frac{350 \times 10^3}{9810} = \frac{12^2}{2 \times 9.81} + \frac{p_2}{9810}$$

from which $p_2 = 380.6$ MPa.

4.6 SOME APPLICATIONS OF THE ENERGY EQUATION

Extensive use is made of the energy equation in the study of thermodynamics. It is our present purpose to illustrate the application of this equation to certain cases of interest. As noted earlier, Chapter 5 will be concerned with fluid dynamic applications.

As our first application, let us consider a pump whose main purpose is to do work on the fluid to cause an appreciable change in potential energy (i.e., pump it to an elevation appreciably different from the initial elevation).

ILLUSTRATIVE PROBLEM 4.8

Calculate the work required for a pump to pump water from a well to ground level 125 m above the bottom of the well (see Figure 4.10). At the inlet to the pump the pressure is 96.5 kPa and at the system outlet it is 103.4 kPa. Assume no heat flow, constant pipe diameter, and constant internal energy. Use $\gamma = 9810$ N/m³ and assume it to be constant.

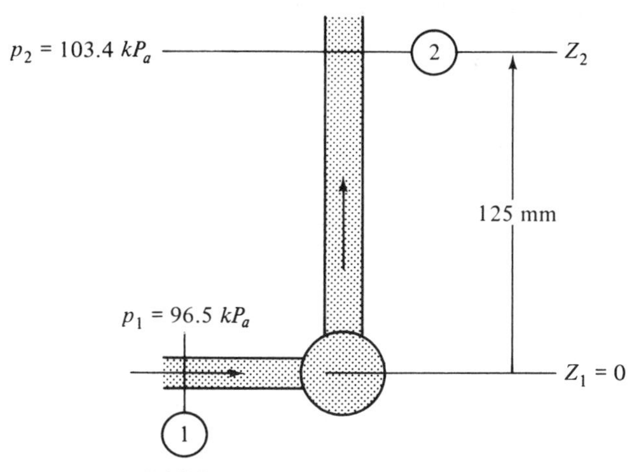

FIGURE 4.10 Illustrative Problem 4.8.

Solution

Writing the full energy equation,

$$Z_1 + \frac{V_1^2}{2g} + \frac{p_1}{\gamma_1} + u_1 + q = Z_2 + \frac{V_2^2}{2g} + \frac{p_2}{\gamma_2} + u_2 + W$$

and noting that $V_1^2/2g = V_2^2/2g$, $u_1 = u_2$, and $q = 0$, we have

$$0 + \frac{p_1}{\gamma_1} = Z_2 + \frac{p_2}{\gamma_2} + W$$

Solving for W,

$$W = \frac{p_1}{\gamma_1} - \frac{p_2}{\gamma_2} - Z_2$$

Inserting the values from the problem,

$$W = \frac{96.5 \times 1000}{9810} - \frac{103.4 \times 1000}{9810} - 125$$

$$= -125.7 \text{ N}\cdot\text{m/N}$$

Note that the negative sign indicates *work into* the system, in accordance with our convention.

ILLUSTRATIVE PROBLEM 4.9

Steam is expanded in a nozzle from an initial enthalpy of 1220 Btu/lb to a final enthalpy of 1100 Btu/lb (see Figure 4.11). If the initial velocity of the

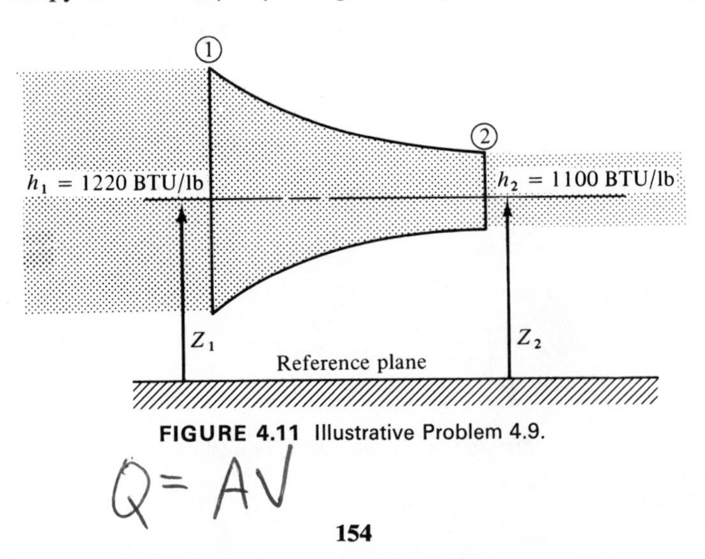

FIGURE 4.11 Illustrative Problem 4.9.

Q= AV

steam is negligible, what is the final velocity? If the initial velocity is 1000 ft/sec, what is the final velocity?

Solution

A nozzle is essentially a short device in which no work is added while the fluid expands. The energy equation in English units is

$$Z_1 + \frac{V_1^2}{2g} + u_1 + \frac{p_1}{\gamma_1} + Jq = Z_2 + \frac{V_2^2}{2g} + u_2 + \frac{p_2}{\gamma_2} + W$$

For this device differences in elevation are negligible; no work is done on or by the fluid; friction will be taken to be negligible and due to the speed of the fluid flowing, and the short length of the nozzle, heat transfer to or from the surroundings will also be taken to be negligible. Under these circumstances,

$$Ju_1 + \frac{p_1}{\gamma_1} + \frac{V_1^2}{2g} = Ju_2 + \frac{p_2}{\gamma_2} + \frac{V_2^2}{2g}$$

or

$$J(h_1 - h_2) = \frac{V_2^2}{2g} - \frac{V_1^2}{2g}$$

1. For negligible entering velocity,

$$V_2 = \sqrt{2gJ(h_1 - h_2)}$$

Substituting the data of the problem yields

$$V_2 = \sqrt{2 \times 32.17 \times 778(1220 - 1100)} = 2450 \text{ ft/sec}$$

2. If the initial velocity is appreciable,

$$J(h_1 - h_2) + \frac{V_1^2}{2g} = \frac{V_2^2}{2g}$$

Again inserting numerical values, we obtain

$$1220 - 1100 + \frac{(1000)^2}{2 \times 32.17 \times 778} = \frac{V_2^2}{2 \times 32.17 \times 778}$$

$$120 + 20 = \frac{V_2^2}{2 \times 32.17 \times 778}$$

and

$$2647 = V_2$$

Note that in this part of the problem the entering velocity was nearly 40% of the final velocity, yet neglecting the entering velocity makes only a 10% error in the answer. It is quite common to neglect the entering velocity in many of these problems.

ILLUSTRATIVE PROBLEM 4.10

A fluid flows steadily in a pipe. At some section of the pipe there is an obstruction that causes an appreciable local pressure loss (see Figure 4.12). Derive the energy equation for this process after flow has become uniform in the downstream section of the pipe. This process is known as a throttling process and is characteristic of valves and orifices placed in pipelines.

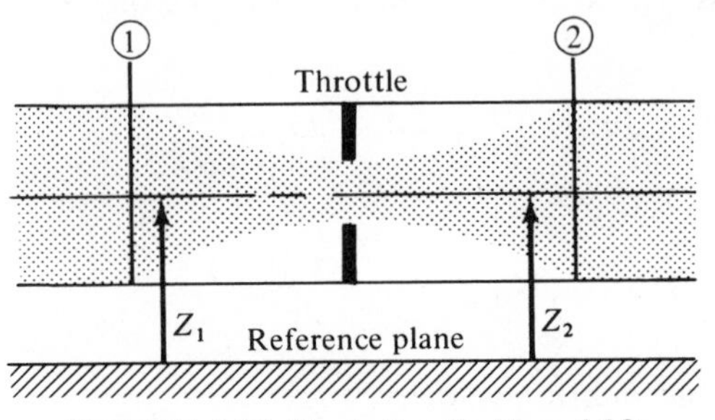

FIGURE 4.12 Illustrative Problem 4.10.

Solution

For the conditions of this problem, we can take differences in elevation to be negligible, differences in kinetic energy terms will be relatively small, and no work or heat crosses the system boundary. The complete energy equation is

$$Z_1 + \frac{V_1^2}{2g} + Ju_1 + \frac{p_1}{\gamma_1} + Jq = Z_2 + \frac{V_2^2}{2g} + Ju_2 + \frac{p_2}{\gamma_2} + W$$

and with the assumptions made for this process,

$$u_1 + \frac{p_1}{\gamma_1 J} = u_2 + \frac{p_2}{\gamma_2 J} \qquad \text{(in appropriate units)}$$

or

$$h_1 = h_2$$

In words, a throttling process is carried out at constant enthalpy. One assumption should be verified; that is, the kinetic energy differences at inlet and outlet are indeed negligible. This can easily be done by using the constant enthalpy condition and using the final properties of the fluid and the cross-sectional area of the duct to determine an approximate final velocity.

The internal energy and enthalpy of a fluid are properties and their values depend upon the fluid and the state of the fluid. For a gas that is considered "ideal" or "perfect,"[1] we can write the following relations for enthalpy and internal energy change:

$$u_2 - u_1 = c_v(T_2 - T_1) \tag{4.24}$$

and

$$h_2 - h_1 = c_p(T_2 - T_1) \tag{4.25}$$

where c_v and c_p are the specific heats at constant volume and constant pressure, respectively. For the English system, typical values for c_v and c_p for air are 0.17 and 0.24 Btu/lb R and in SI units, the corresponding values are 0.719 and 1.006 kJ/kg·K.

ILLUSTRATIVE PROBLEM 4.11

A gas flows steadily in a pipe so that at one section of pipe its pressure and temperature are 100 psia and 950°F. At a second section of the pipe the pressure is 76 psia and the temperature is 580°F. The specific volume of the inlet gas is 4.0 ft³/lb, and at the second section it is 3.86 ft³/lb. Assume that the specific heat at constant volume is 0.32 Btu/lb °F. If no shaft work is done and if the velocities are small, determine the magnitude and direction of the heat transfer.

Solution

First we draw a sketch (Figure 4.13) and writing an energy equation:

$$Z_1 + \frac{V_1^2}{2g} + Ju_1 + \frac{p_1}{\gamma_1} + Jq = Z_2 + \frac{V_2^2}{2g} + Ju_2 + \frac{p_2}{\gamma_2} + W$$

[1]For a further discussion, see *Thermodynamics and Heat Power* by I. Granet, 2nd ed., Reston Publishing Co., Reston, Va., 1980.

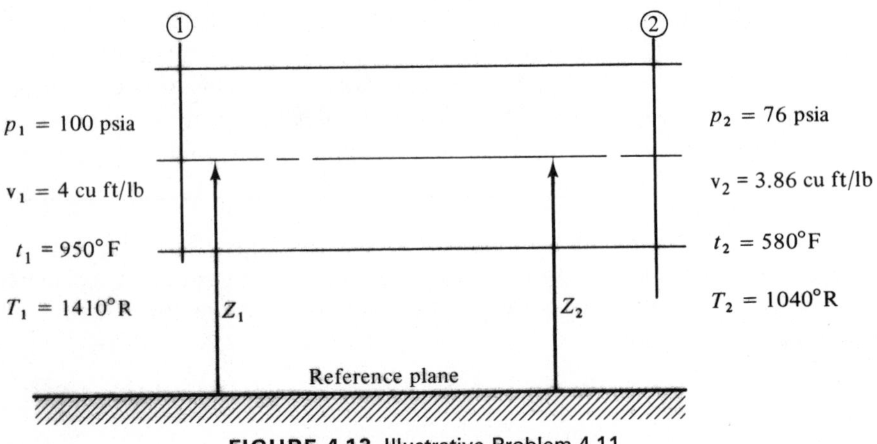

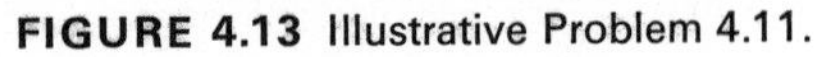
FIGURE 4.13 Illustrative Problem 4.11.

Based upon the statement of the problem,

$$Z_1 - Z_2 \longrightarrow 0$$

$$\frac{V_1^2 - V_2^2}{2g} \longrightarrow 0$$

$$W \longrightarrow 0$$

which reduces the energy equation to

$$u_1 + \frac{p_1}{\gamma_1} + Jq = u_2 + \frac{p_2}{\gamma_2}$$

But equation (4.24) permits us to express the internal energy change in terms of the temperature change as

$$u_2 - u_1 = c_v(T_2 - T_1) \qquad \text{for constant } c_v$$

Therefore,

$$q = c_v(T_2 - T_1) + \frac{p_2 v_2}{J} - \frac{p_1 v_1}{J} \qquad \text{in units of Btu/lb}$$

Inserting numerical quantities yields

$$q = 0.32(1040 - 1410) + \frac{76 \times (144)(3.86)}{778} - \frac{100 \times 144 \times 4}{778}$$

$$= -138.2 \text{ Btu/lb}$$

Thus, 138.2 Btu/lb is transferred *from* the fluid.

4.7 CLOSURE

The rational study of the fluid mechanics of steady flow develops from Newton's laws combined with the principle of conservation of energy. The items that we will need have been developed in this chapter. In Chapter 5, extensive use will be made of the Bernoulli equation, which we developed from the energy equation to indicate its limitations.

The principles utilized and the equations developed in this chapter are not difficult for the student. However, it has been the author's experience that unless these equations are applied with due care to their limitations, with the proper units, and above all with an understanding of their development, they will invariably present a problem. Also, since subsequent developments in this book require the use of the material in this chapter, it is very important that it be mastered before proceeding further into the study of fluid mechanics.

PROBLEMS

4.1 One gallon is a volume measure of 231 $in.^3$. Determine the velocity in a pipe 2 in. in diameter when water flows at the rate of 20 gal/min in the pipe. The specific weight of water is 62.4 lb/ft^3.

4.2 Solve Problem 4.1 if oil having a specific gravity of 0.85 is flowing.

4.3 What weight of water is flowing in Problem 4.1?

4.4 If 30 gpm is flowing in a pipe, how many m^3/s is flowing?

4.5 A chemical process uses 5000 litres/min. How many m^3/s does this correspond to?

4.6 An American manufacturer can process 100 gpm. In talking to his European factory he finds that he must tell them this rate in litres per second. How many litres per second will he tell them that he can process?

4.7 Twenty litres of fluid per second flows in a pipe whose inside diameter is 7.5 cm. Determine the velocity and weight rate of flow if the fluid is water having a specific weight of 9810 N/m^3.

4.8 What is the mass flow rate (kg/s) in Problem 4.7?

4.9 A reservoir contains 1000 gal of water. If a pump can pump at the rate of 4 litres/min, how long will it take for the tank to empty?

4.10 A pipe carrying water at the rate of 100 litres/min is reduced in diameter from 75 mm to 50 mm. What is the velocity of the water in each section of the pipe?

4.11 Water flows at the rate of 50 gpm in a pipe that has an internal diameter of 4 in. The diameter of the pipe is reduced to 2 in. Determine the velocity in each section of the pipe.

4.12 When a fluid of constant density flows in a pipe, show that the velocity is inversely proportional to the square of the diameter.

4.13 A 25-mm-diameter pipe supplies two branch pipes each 100 mm in diameter. If 500 gpm flows in the main pipe and each branch carries equal volumes of flow, determine the velocity in each branch pipe. Assume that γ remains constant.

4.14 A body of mass of 10 slugs is placed 10 ft above an arbitrary plane. If the acceleration of gravity is 10 ft/sec^2, how much work (foot pounds) was done in lifting the body above the plane?

4.15 How much work does it take to lift 10 kg a distance 3 m?

4.16 A body weighing 100 lb is lifted 10 ft. What velocity will it have after freely falling the 10 ft?

4.17 If the body in Problem 4.15 falls freely, what will its velocity be after it travels the 3 m?

4.18 A body weighing 10 lb is lifted 100 ft. What is its potential energy? What velocity will it possess after falling the 100 ft?

4.19 A body has a mass of 5 kg. If its velocity is 10 m/s, what is its kinetic energy?

4.20 A body weighing 10 lb is moving with a velocity of 50 ft/sec. From what height would it have to fall to achieve this velocity? What is its kinetic energy?

4.21 Water flows over the top of a dam and falls freely until it reaches the bottom some 600 ft below. What is the velocity of the water just before it hits the bottom? What is its kinetic energy per pound at this point?

4.22 Show that the kinetic energy of a fluid flowing in a pipe varies inversely as the fourth power of the pipe diameter, all other things being equal.

4.23 The flow in a city water system is being tested by allowing it to flow from a hydrant with an outlet 4 in. in diameter. The outlet is 2.0 ft above the ground, and the issuing stream of water hits the level ground at a distance 9 ft from the outlet. How much water is flowing?

4.24 A water pump is placed at the bottom of a well. If it is to pump 1.0 ft^3/sec to the surface, 100 ft away, through a 4-in.-i.d. pipe, determine the power required in foot pounds per second and the hosepower.

4.25 A hose is 4 in. in diameter and has water flowing in it. If 10 ft^3/sec is flowing, determine the velocity in the hose. If the pressure in the pipe is 20 psig, determine the maximum height the water can reach.

4.26 A water turbine operates from a water supply that is 100 ft above the turbine inlet. It discharges to the atmosphere through a 6-in.-diameter pipe with a velocity of 25 ft/sec. If the reservoir is infinite in size, determine the work out of the turbine.

4.27 A pressure of 5 bars (1 bar = 10^5 Pa) is presented to the piston of a pump, causing it to travel 100 mm. If the cross-sectional area of the piston is 1000 mm^2 and the pressure is constant, how much work was done by the steam on the piston?

4.28 In a constant-pressure process, steam at 500 psia is presented to the piston of a pump and causes the piston to travel 4 in. If the cross-sectional area of the piston is 5 $in.^2$, how much work was done by the steam on the piston?

4.29 Air is compressed steadily until its final volume is half its initial volume. The initial pressure is 35 psia and the final pressure is 100 psia. If the initial specific volume is 1 ft^3/lb, determine the difference in the p/γ terms at entrance and exit in foot pounds per pound.

4.30 If a fluid flows past a section of pipe with a pressure of 100 kPa and a specific volume of 10^{-3} m^3/kg, determine its flow work.

4.31 At the entrance to a steady flow device, the pressure is 350 kPa and the specific volume is 0.04 m^3/kg. At the outlet the pressure is 1 MPa and the specific volume is 0.02 m^3/kg. Determine the flow work change in this device.

4.32 A flow of 2000 kg/min of water is compressed from 100 kPa to 1 MPa. The water density can be taken to be 1000 kg/m^3 and its temperature does not change. The inlet to the pump is 100 mm in diameter and the outlet is 150 mm in diameter. The inlet is 50 m below the outlet. Determine the work of the pump.

4.33 Gas enters a turbine at 600°C and leaves at 350°C. If the specific heats of the gas are $c_p = 0.8452$ kJ/kg·K and $c_v = 0.6561$ kJ/kg·K, determine the work out of the turbine. State all assumptions.

4.34 Solve Illustrative Problem 4.7(b) if the flow is upward.

4.35 Air expands through a nozzle from 1000 psia and 500°F to 700 psia and 50°F. If the specific heat of air at constant pressure is 0.24 Btu/lb °F, what is the final velocity? Assume that the initial velocity is zero.

4.36 Steam expands in a nozzle from an initial enthalpy of 1300 Btu/lb to a final enthalpy of 980 Btu/lb. Determine the final velocity if the entering velocity and heat losses are negligible.

4.37 Solve Problem 4.36 if the initial velocity is 1100 ft/sec.

4.38 Air expands in a turbine from 7 MPa and 250°C to 3.5 MPa and 50°C. If $c_p = 1.0062$ kJ/kg·K, determine the work out of the turbine. State all assumptions. (Note that c_p is given per unit mass.)

4.39 A steam turbine expands steam from an initial condition of 50 bars and 500°C to a final condition of 1 bar. If the respective enthalpies at these conditions are 3433.8 kJ/kg and 2675.5 kJ/kg, determine the work out of the turbine. State all your assumptions.

REFERENCES

1. *Thermodynamics and Heat Power* by I. GRANET, 2nd ed., Reston Publishing Co., Reston, Va., 1980.
2. *Thermodynamics* by E. FERMI, Dover Publications, Inc., New York, 1956.
3. *Heat and Thermodynamics* by M. W. ZEMANSKY, 4th ed., McGraw-Hill Book Company, New York, 1957.
4. *Concepts of Thermodynamics* by E. F. OBERT, McGraw-Hill Book Company, New York, 1960.
5. *Principles of Engineering Thermodynamics* by P. J. KIEFER, G. F. KINNEY, and M. C. STUART, 2nd ed., John Wiley & Sons, Inc., New York, 1954.
6. *Engineering Thermodynamics* by D. B. SPALDING and E. H. COLE, McGraw-Hill Book Company, New York, 1959.
7. *Chemical Engineering Thermodynamics* by B. F. DODGE, McGraw-Hill Book Company, New York, 1944.
8. *Fluid Mechanics* by R. C. BINDER, 4th ed., Prentice-Hall, Inc., Englewood Cliffs, N.J., 1962.
9. *Elementary Theoretical Fluid Mechanics* by K. BRENKERT Jr., John Wiley & Sons, Inc., New York, 1960.
10. *Engineering Thermodynamics* by W. C. REYNOLDS and H. C. PERKINS, McGraw-Hill Book Company, New York, 1970.

Fluid Dynamic Applications

5.1 INTRODUCTION

In Chapter 4, the energy equation and the Bernoulli equation were derived and their limitations were discussed. It is interesting to note that the Bernoulli equation was first proposed by Daniel Bernoulli in 1738 and since that time has been used extensively to solve many problems in fluid mechanics. The application of Bernoulli's equation often requires that it be modified to account for deviations from the ideal conditions to which it is strictly applicable and usually the equation (or its modifications) yield results that are within the accuracy required for engineering applications. However, there are several cases where the application of this equation will yield incorrect results since it is being used beyond the limits that even its modified forms can be hoped to apply. Several examples of this situation are the high-speed flow of compressible fluids, fluid flow with heat transfer causing large density changes, and fluid flow with large pressure changes due to frictional effects. For these cases it is necessary to utilize the continuity equations, the energy equations, the momentum equations, and Newtons' laws to obtain analytical solutions.

For complex flow problems it may also be necessary to resort to experiments to obtain sufficient information to be able to correlate and solve some of these complex fluid mechanics problems.

This chapter is devoted to the study of those one-dimensional steady flow situations to which the Bernoulli equation and momentum relations can be applied to yield reasonable solutions. Many practical cases will be studied and analyzed in this chapter, and their limitations will be noted for each case. Usually, a flow situation will be simplified and idealized, and due to this a word of caution must be emphasized at this point. It is quite easy for the student to extend these idealizations to cases where they do not apply. It is necessary to be aware of the assumptions being made, and care must be exercised not to attempt to extend these simplified analyses to complex flow situations for which they are not intended.

5.2 GENERAL CONSIDERATIONS

When the energy equation was considered in Chapter 4, it was very often convenient to select a control volume whose control surface coincided with the physical surface of a system. However, this is not necessary, and frequently we shall wish to utilize a control volume whose control surface is selected for ease of analysis. Any fluid stream consists of molecules of the fluid having a random motion due to collisions with other molecules and in addition having a directed motion along the flow path. A *streamline* is defined as a line with the tangent to it at any point being the direction of its velocity at that point. Streamlines cannot cross since this would require two molecules having different velocities to be at the same point at the same time, and we also note that in steady flow, streamlines do not change with time. Since the streamline coincides with the macroscopic path of the flow, it must be parallel to the surrounding flow and no flow can cross a streamline. If the flow is accelerating, the streamlines are not parallel to each other and will appear closer together, and if the flow is decelerating, the streamlines will appear farther apart. For the case of uniform steady flow the streamlines are parallel. These concepts are apparent when the ideal airfoil shown in Figure 5.1 is

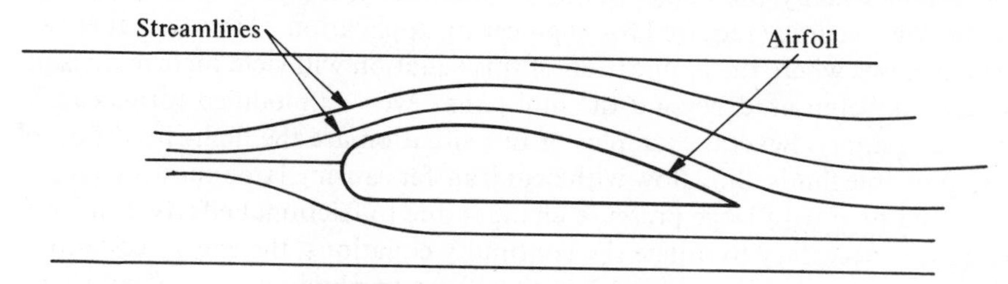

FIGURE 5.1 Flow over an airfoil.

considered. In the undisturbed stream before and after the airfoil, the streamlines are uniformly spaced and parallel. Above the airfoil they are close together, indicating a high-velocity region, and beneath the airfoil they are farther apart, indicating a low-velocity region.

As noted, in steady flow no fluid crosses a streamline, and the flow follows the streamlines. Every streamline is a continuous line that may be considered as starting at an infinite distance upstream and extending to a infinite distance downstream. The streamlines through the points of a closed curve will result in a closed surface being generated that is known as a stream tube. Since the surface of the stream tube (Figure 5.2) consists of streamlines, no

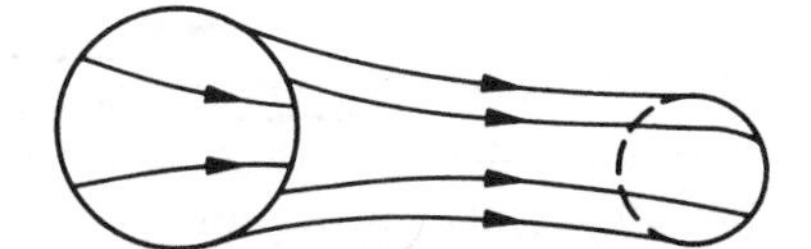

FIGURE 5.2 Stream tube for steady flow.

flow can enter or leave the stream tube through its lateral surface. When a control volume is selected in which the flow is confined by solid walls, streamlines can be considered to start in a reservoir where the fluid is at rest and to extend to the discharge of the system.

The Bernoulli equation derived in Chapter 4 was

$$\frac{p_1}{\gamma_1} + \frac{V_1^2}{2g} + Z_1 = \frac{p_2}{\gamma_2} + \frac{V_2^2}{2g} + Z_2 \tag{5.1}$$

and since the flow was assumed to be incompressible, γ_1 can be taken equal to γ_2 or both can simply be written as γ. Equation (5.1) can also be written as

$$\frac{p_1}{\gamma} + \frac{V_1^2}{2g} + Z_1 = \text{constant} \tag{5.2}$$

When written in this form we state that the sum of flow energy, the kinetic energy, and the potential energy is a constant *along a streamline*. For another streamline the constant can have a different value, but in many problems all of the streamlines have practically the same total energy. For example, a reservoir that has all the streamlines starting in it will have the same constant for all the streamlines. Thus, for these cases the sum of the energies on the left side of equation (5.1) or (5.2) can be equated to the sum of the corresponding terms between positions in an ideal system regardless of which streamline is being considered.

All real fluid-flow situations are irreversible due to viscous effects giving rise to shear stresses in the fluid. Theoretically, the Bernoulli equation as we have derived it must be modified in order to apply it to those cases in which nonideal effects (friction, turbulence, etc.) are relatively large when compared to the constant of equation (5.2). Let us assume that point 1 is upstream and that point 2 is downstream along a streamline. Assuming that there is no addition of energy into or out of the streamline as either work or heat, we can state that the energy at point 1 equals the energy at point 2 plus all of the flow losses between these two points. Mathematically,

$$\frac{p_1}{\gamma_1} + \frac{V_1^2}{2g} + Z_1 = \frac{p_2}{\gamma_2} + \frac{V_2^2}{2g} + Z_2 + \text{losses}_{1\to 2} \tag{5.3}$$

The loss term on the right-hand side of equation (5.3) can be considered to be a term required to ensure energy conservation. If a pump adds energy to the streamline between points 1 and 2 or a turbine extracts energy between these points, the Bernoulli equation is usually further modified to account for these energy additions or depletions. Thus, for a pump,

$$\frac{p_1}{\gamma_1} + \frac{V_1^2}{2g} + Z_1 + E_p = \frac{p_2}{\gamma_2} + \frac{V_2^2}{2g} + Z_2 + \text{losses}_{1\to 2} \tag{5.4}$$

For a turbine,

$$\frac{p_1}{\gamma_1} + \frac{V_1^2}{2g} + Z_1 = \frac{p_2}{\gamma_2} + \frac{V_2^2}{2g} + Z_2 + \text{losses}_{1\to 2} + E_T \tag{5.5}$$

For both a pump and a turbine, we have

$$\frac{p_1}{\gamma_1} + \frac{V_1^2}{2g} + Z_1 + E_p = \frac{p_2}{\gamma_2} + \frac{V_2^2}{2g} + Z_2 + \text{losses}_{1\to 2} + E_T \tag{5.6}$$

The student will note that the modifications to the Bernoulli equation are an attempt to meet the requirement of energy conservation and that the resultant equations bear a resemblance to the energy equation derived in Chapter 4. It should be emphasized that for flows in which there are large density changes due to temperature changes or large additions of depletions of energy from a system causing large density changes, the Bernoulli equation and the energy equation are not equivalent and the apparent resemblance of equations (5.4) through (5.6) to the energy equation should not lead the student to believe that they are applicable to such situations.

In equations (5.3) through (5.6) the term p/γ is referred to as the *pressure head*, the term $V^2/2g$ is referred to as the *velocity head*, and the term Z is referred to as the *potential head*. Using the usual engineering units it will be found that each of these terms has the units of feet, and their sum is called the *total head* or *total energy*. These terms are more readily understood by referring to Figure 5.3, on which each term is shown for sections along an irregular tube.

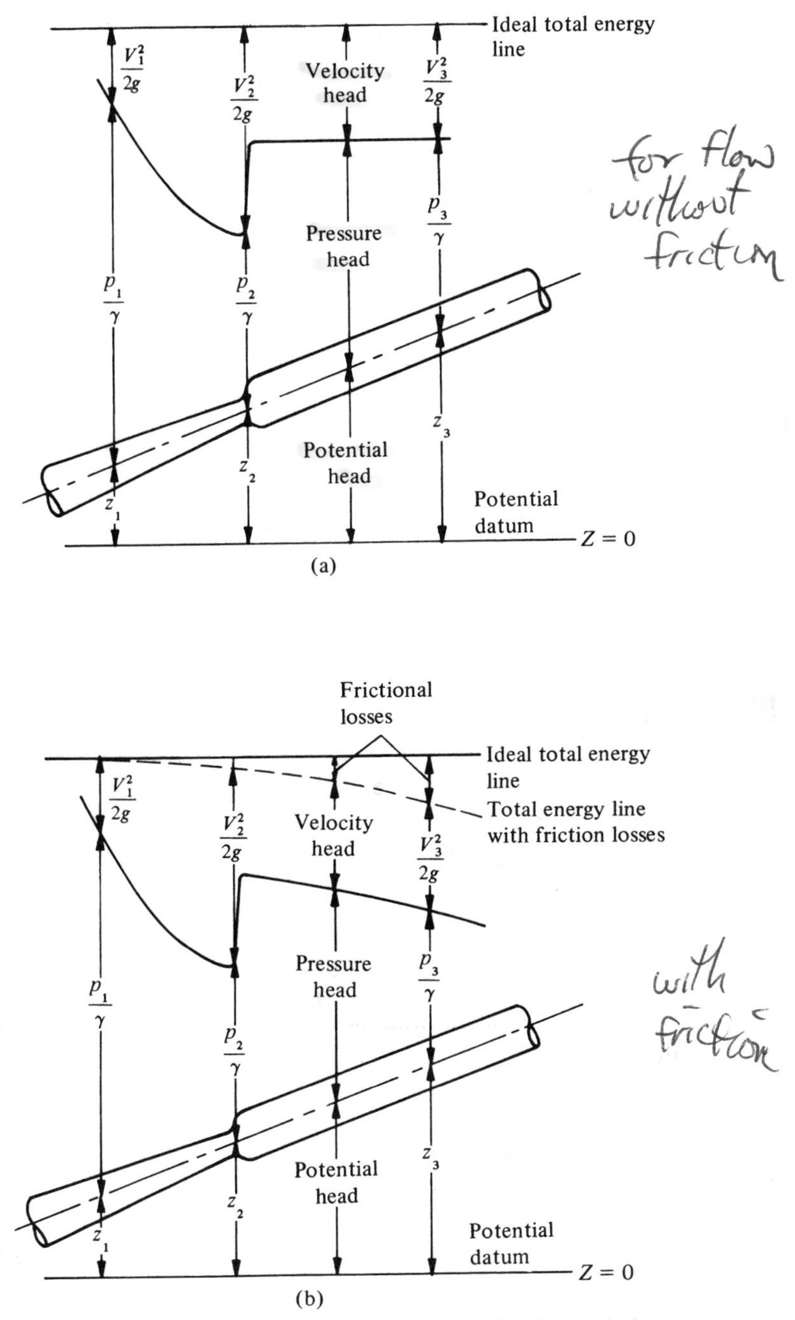

FIGURE 5.3 Head variation in a tube. (Adapted with permission from *Basic Fluid Mechanics* by J. L. Robinson, McGraw-Hill Book Company, New York, 1963, p. 49.)

In Figure 5.3a the total head or total energy is a constant for flow without friction and is represented by the total head or total energy line as a horizontal line at a constant distance from the datum plane. With friction this is not true and the total head line in Figure 5.3b is shown as a dashed curved line dropping off to the right. It will be noticed that when the velocity in the tube increases, the sum of the potential and pressure heads must decrease, and that for a decreased velocity, the sum of the potential and pressure heads increases. For steady flow, $V_3^2/2g$ is a constant, requiring p_3/γ to be constant for the case of no friction (Figure 5.3a); in the case of friction (Figure 5.3b), p_3/γ must decrease since $V_3^2/2g$ is constant.

ILLUSTRATIVE PROBLEM 5.1

Water is to be delivered from an open tank through a pipeline to a lower elevation. Flow rate is to be 100 gal/min of 60°F water ($\gamma = 62.4$ lb/ft³), and the inside diameter of the pipeline is 2 in. At the pipe outlet the desired pressure is to be 4 psi above atmospheric pressure. If the friction drop in the pipeline is estimated to be 38 ft, determine the vertical distance required between the level in the supply tank and the point of water discharge.

Solution

Referring to Figure 5.4, we shall select two stations, one at the water–air free surface in the tank and the other at the outlet of the pipe. Also, we shall select the potential head datum at the pipe outlet as shown. For the conditions of this problem we note that all streamlines have the same total energy at the water level in the tank and that therefore we can properly apply

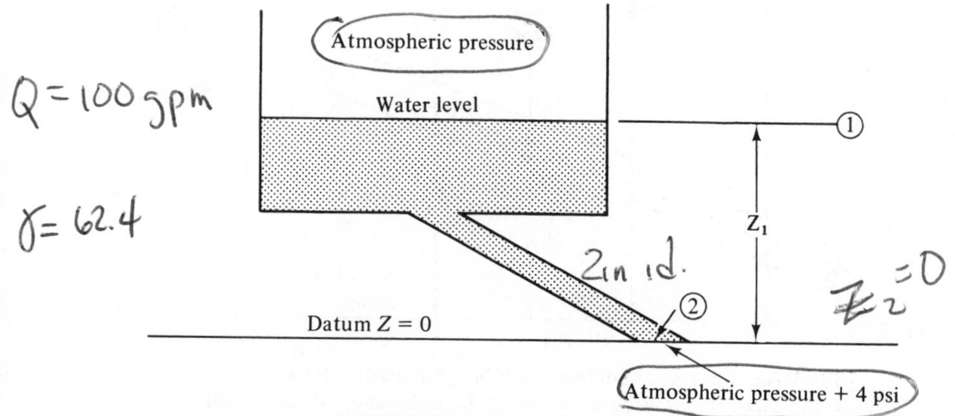

FIGURE 5.4 Illustrative Problem 5.1.

the Bernoulli equation at all positions in the flow regardless of the streamline involved. Using equation (5.3), we obtain

$$\frac{p_1}{\gamma} + \frac{V_1^2}{2g} + Z_1 = \frac{p_2}{\gamma} + \frac{V_2^2}{2g} + Z_2 + \text{losses}_{1\to 2}$$

The selection of the datum as shown automatically makes $Z_2 = 0$, and it will be assumed that V_1 is very small compared to V_2 since the tank is presumably large compared to the pipe. Thus, for this problem we have

$$\frac{p_1}{\gamma} + Z_1 = \frac{p_2}{\gamma} + \frac{V_2^2}{2g} + \text{losses}_{1\to 2}$$

However, p_2 exceeds p_1 by 4 psi. Since a column of 1 ft of water of $\gamma = 62.4$ exerts a static pressure at its base equal to 62.4/144 psi (or 0.433 psi), 4 psi = 4/0.433 = 9.24 ft. Therefore, $(p_2/\gamma) - (p_1/\gamma) = 9.24$ ft and

$$Z_1 = \left(\frac{p_2}{\gamma} - \frac{p_1}{\gamma}\right) + \frac{V_2^2}{2g} + \text{losses}_{1\to 2} = 9.24 + \frac{V_2^2}{2g} + 38$$

By definition, a gallon is a volume measure equal to 231 in.³. Therefore,

$$100\,\frac{\text{gal}}{\text{min}} = \frac{100\text{ gal}}{60\text{ sec}} = \frac{100 \times 231}{60 \times 1728}\,\frac{\text{ft}^3}{\text{sec}}$$

Since volume flow equals AV,

$$\frac{100 \times 231}{60 \times 1728} = \left[\frac{(\pi/4)(2)^2}{144}\right] V_2 \qquad \text{and} \qquad V_2 = 10.21 \text{ ft/sec}$$

Using this value of V_2, we obtain

$$Z_1 = 9.24 + \frac{(10.21)^2}{2 \times g} + 38 = 48.86 \text{ ft} \qquad \text{say 49 ft}$$

ILLUSTRATIVE PROBLEM 5.2

A pipe is connected to a large reservoir as shown in Figure 5.5 and discharges to the atmosphere. If 10 litres/s flows in the pipe, determine the energy losses in the system.

Solution

Taking both the surface of the liquid in the tank and the outlet of the pipe as atmospheric pressure, we have $p_1 = p_2 = p_a$. If the tank is considered to

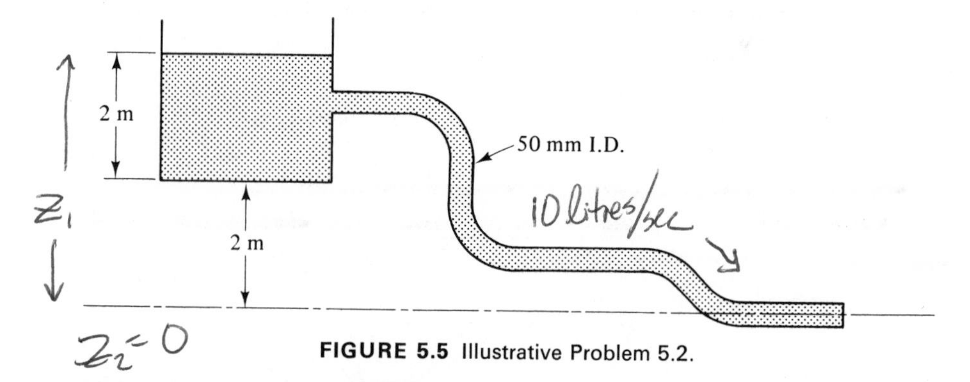

FIGURE 5.5 Illustrative Problem 5.2.

be large, the velocity of the fluid at its surface, V_1, may be taken as zero. The velocity in the pipe, which is also the velocity leaving the pipe, is found from the continuity equation as

$$Q = AV = \frac{10 \text{ litres}}{\text{s}} \times \frac{1000 \text{ cm}^3}{\text{litre}} \times 10^{-6} \frac{\text{m}^3}{\text{cm}^3} = 10 \times 10^{-3} \text{ m}^3/\text{s}$$

Therefore,

$$\frac{\pi}{4}(0.05)^2 V_2 = 10 \times 10^{-3}$$

$$V_2 = 5.09 \text{ m/s}$$

Writing the Bernoulli equation, we obtain

$$\frac{p_1}{\gamma} + \frac{V_1^2}{2g} + Z_1 = \frac{p_2}{\gamma} + \frac{V_2^2}{2g} + Z_2 + \text{losses}_{1 \to 2}$$

Since $p_1 = p_2$ and $V_1 = 0$,

$$4 = \frac{V_2^2}{2g} + 0 + \text{losses}_{1 \to 2}$$

The loss in energy is therefore

$$4 - \frac{(5.09)^2}{2 \times 9.81} = 2.68 \text{ m} \quad \text{or} \quad 2.68 \text{ N} \cdot \text{m/N}$$

5.3 TORRICELLI'S THEOREM

Let us consider a very large reservoir that has a series of well-rounded openings placed in its side as shown in Figure 5.6. Assuming atmospheric pressure at the water level in the tank and that the jet efflux from each orifice

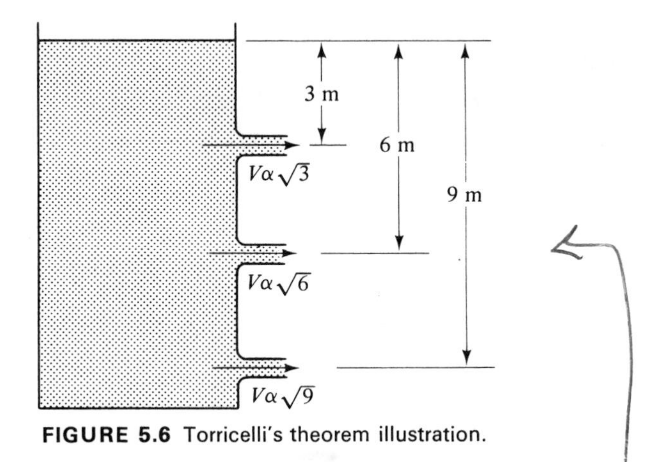

FIGURE 5.6 Torricelli's theorem illustration.

is to the atmosphere, we find that the velocity of a jet is proportional to the square root of the distance from the center line of the jet to the water level in the tank.

We can analyze the flow situation shown in Figure 5.6 by considering a single well-rounded opening in a large tank as indicated in Figure 5.7. The center line of the opening is taken to be a distance h below the level of the water in the reservoir. As before, we will consider atmospheric pressure to exist at the surface of the water in the reservoir and that the jet efflux is also to atmospheric pressure. For convenience, we select the center line of the opening as the datum for potential energy. Also, since the total energy content of all the streamlines in the reservoir is the same, we can select any two arbi-

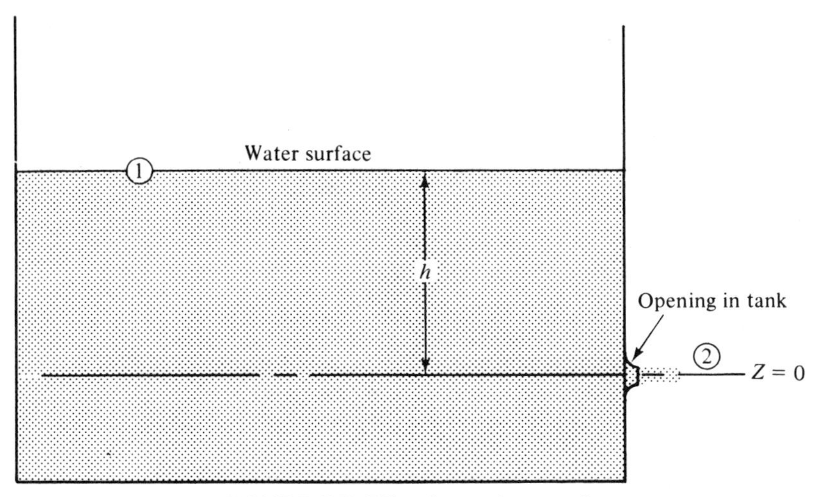

FIGURE 5.7 Efflux from a large tank.

trary and convenient sections in the system to which we shall apply the Bernoulli equation even though the equation is not being applied to the same streamline. Using the conditions of the problem, noting that the density is constant, and neglecting losses, we have

$$\frac{V_1^2}{2g} + Z_1 + \frac{p_1}{\gamma} = \frac{V_2^2}{2g} + Z_2 + \frac{p_2}{\gamma}$$

Since $Z_2 - Z_1 = -h$, and the pressure in the jet must be atmospheric pressure, making $p_1/\gamma = p_2/\gamma$,

$$h = \frac{V_2^2}{2g}$$

or

$$V_2 = \sqrt{2gh} \tag{5.7}$$

Equation (5.7) states simply that the velocity of the water from the jet would equal the velocity of a body in free fall from the surface of the reservoir to the center line of the jet; this result is known as *Torricelli's theorem*.

In deriving equation (5.7) the assumption was made that there were no losses in the flow as it issues from the tank. Actually, even for well-rounded openings, the streamlines will start to converge toward the opening while still in the tank and will continue to converge after leaving the tank. In effect, there is a contraction of the stream after leaving the plane of the opening that gives rise to a section of the stream having an area less than the geometric area of the opening in the tank. For a sharp-edged orifice the streamlines are shown as in Figure 5.8, and the area of minimum cross section is denoted as the *vena contracta*. By making the edges of the opening rounded or by using

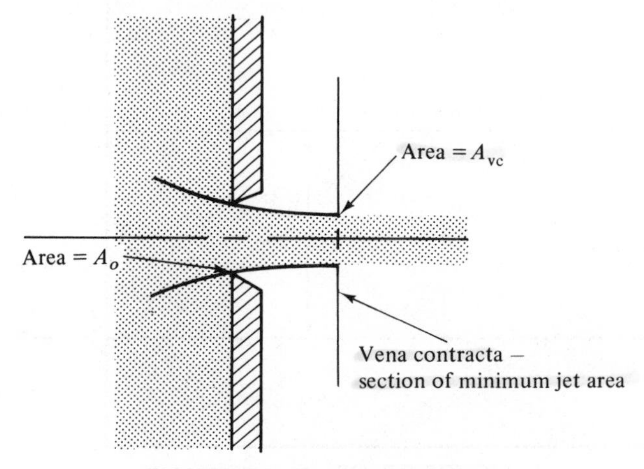

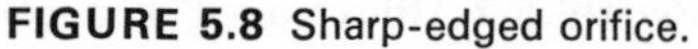

FIGURE 5.8 Sharp-edged orifice.

a tubular insert into the tank, it is possible to cause a rearrangement of the streamlines to yield a larger diameter at the vena contracta than for the case of a sharp-edged orifice. It is usual in practice to modify equation (5.7) to account for these effects by defining a coefficient of contraction, which is the ratio of the area of the vena contracta to the area of the orifice (or opening). Thus,

$$C_C = \frac{A_{vc}}{A_o} \tag{5.8}$$

and we also define the ratio of the actual velocity to the velocity given by equation (5.7) as the coefficient of velocity. This gives

$$V_2 = C_V\sqrt{2gh} \tag{5.9}$$

where C_V is the coefficient of velocity. Since the volume discharging through the orifice or opening is the product of the actual velocity in the jet multiplied by the jet area, we arrive at a new coefficient, called the *coefficient of discharge*, which is the ratio of the volume actually flowing in the jet to the volume calculated using the area of the opening and the ideal velocity given by equation (5.7). This also gives us the following relation between these three coefficients:

$$C_D = C_C C_V \tag{5.10}$$

For sharp-edged orifices some typical values of these coefficients are $C_C = 0.61$, $C_V = 0.98$, and $C_D = 0.60$. For well-rounded openings, C_C can go to 1, which makes C_D go toward unity, since C_V usually lies in the range 0.95 to 0.98.

ILLUSTRATIVE PROBLEM 5.3

A large tank has a well-rounded circular opening in its side located 10 ft below the water level in the tank. If the opening is 1 in. in diameter, determine the velocity of the efflux, the area of the vena contracta, and the volume rate of efflux if $C_C = 0.85$ and $C_V = 0.98$.

Solution

From equation (5.7),

$$V_2 = \sqrt{2gh} = \sqrt{2 \times 32.7 \times 10} = 25.4 \text{ ft/sec}$$

The actual velocity is $C_V V_2 = 0.98(25.4) = 24.9$ ft/sec. The area of the opening is $(\pi/4)[(1)^2/144] = 0.00546$ ft². The area of the vena contracta is $C_c A_o =$

$0.85(0.00546) = 0.00464 \text{ ft}^2$. We can obtain the volume rate of efflux as the product of the actual velocity multiplied by the area of the vena contracta or as C_D multiplied by the product of the area of the opening and the ideal jet velocity. Thus,

$$Q = 24.9 \frac{\text{ft}}{\text{sec}} \times 0.00464 \text{ ft}^2 = 0.115 \text{ ft}^3/\text{sec}$$

and as a check,

$$Q = (A_o V_2) C_C C_v = (25.4)(0.00546)(0.85)(0.98) = 0.116 \text{ ft}^3/\text{sec}$$

In terms of gallons per minute, we have

$$Q = 0.116 \frac{\text{ft}^3}{\text{sec}} \times 1728 \frac{\text{in.}^3}{\text{ft}^3} \times \frac{1}{231 \text{ in.}^3/\text{gal}} \times 60 \frac{\text{sec}}{\text{min}} = 52 \text{ gal/min}$$

ILLUSTRATIVE PROBLEM 5.4

Solve Illustrative Problem 5.3 if the opening is located 3 m below the surface of the water level and if the opening has a 25-mm diameter.

Solution

Proceeding as in Illustrative Problem 5.3, we obtain

$$V_2 = \sqrt{2gh} = \sqrt{2 \times 9.81 \times 3} = 7.67 \text{ m/s}$$

The actual velocity is $C_V V_2 = 0.98 \times 7.67 = 7.52$ m/s. Since the area of the vena contracta is

$$C_c A_o = \frac{\pi}{4}(0.025)^2 \times 0.85 = 4.172 \times 10^{-4} \text{ m}^2$$

the actual discharge will equal the product of the actual velocity and the area of the vena contracta,

$$Q = AV = 4.172 \times 10^{-4} \text{ m}^2 \times 7.52 \frac{\text{m}}{\text{s}} = 3.14 \times 10^{-3} \text{ m}^3/\text{s}$$

Orifices are frequently used in pipelines to meter the quantity of fluid flowing. When used in this manner the orifice usually consists of a thin plate inserted into a pipe and clamped between flanges. The hole in the orifice plate is concentric with the pipe and static pressure taps are provided upstream and downstream of the orifice, as shown in Figure 5.9. The orifice meter causes a

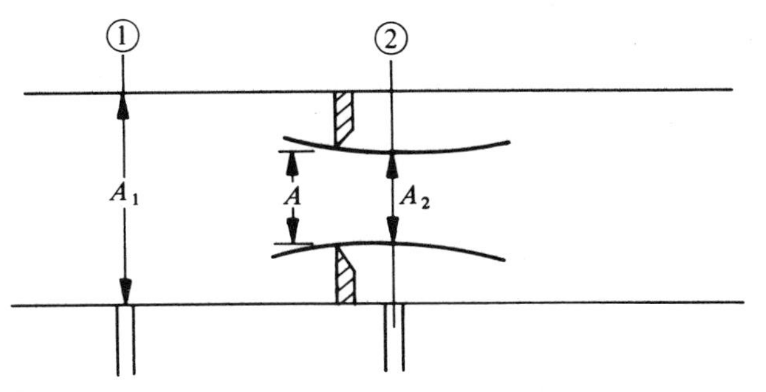

FIGURE 5.9 Orifice meter.

constriction in the flow and creates a jet smaller than itself. Applying the Bernoulli equation to this situation between sections 1 and 2 yields

$$V_2 = \left[\frac{C_v}{\sqrt{1 - C_C^2(A/A_1)^2}}\right]\left[\sqrt{\frac{2g(p_1 - p_2)}{\gamma}}\right] \tag{5.11}$$

and

$$Q = AV = \left[\frac{AC_C C_V}{\sqrt{1 - C_C^2(A/A_1)^2}}\right]\left[\sqrt{\frac{2g(p_1 - p_2)}{\gamma}}\right] \tag{5.12}$$

From a practical standpoint it is almost impossible to determine C_C and C_V separately for the orifice meter, and it is equally difficult to locate the pressure tap at the vena contracta. Therefore, the simplification is frequently made to combine these coefficients and to rewrite equation (5.12) as

$$Q = AC\sqrt{\frac{2g(p_1 - p_2)}{\gamma}} \tag{5.13}$$

where C is defined as

$$\frac{C_D}{\sqrt{1 - (d_0/d_1)^4}}$$

and therefore includes the correction for the contraction of the jet. Figure 5.10 shows the coefficient C for a square-edged orifice as a function of the Reynolds number $d_1 V_1 \gamma/\mu g$. It will be noted that the coefficient becomes constant above certain values of the Reynolds number. At this point, let us note that the term Reynolds number was introduced in our discussion. As we shall find in Chapter 6, this parameter, involving diameter, velocity, specific weight, and viscosity, determines the character of the flow. For our present purposes it is simply a useful parameter. The advantages of the orifice meter are its relatively small size, the ease of installation in a pipe, and the fact that "standard" installations can be used without the need for calibration. However, the orifice meter acts as a partially open valve and has a low discharge coef-

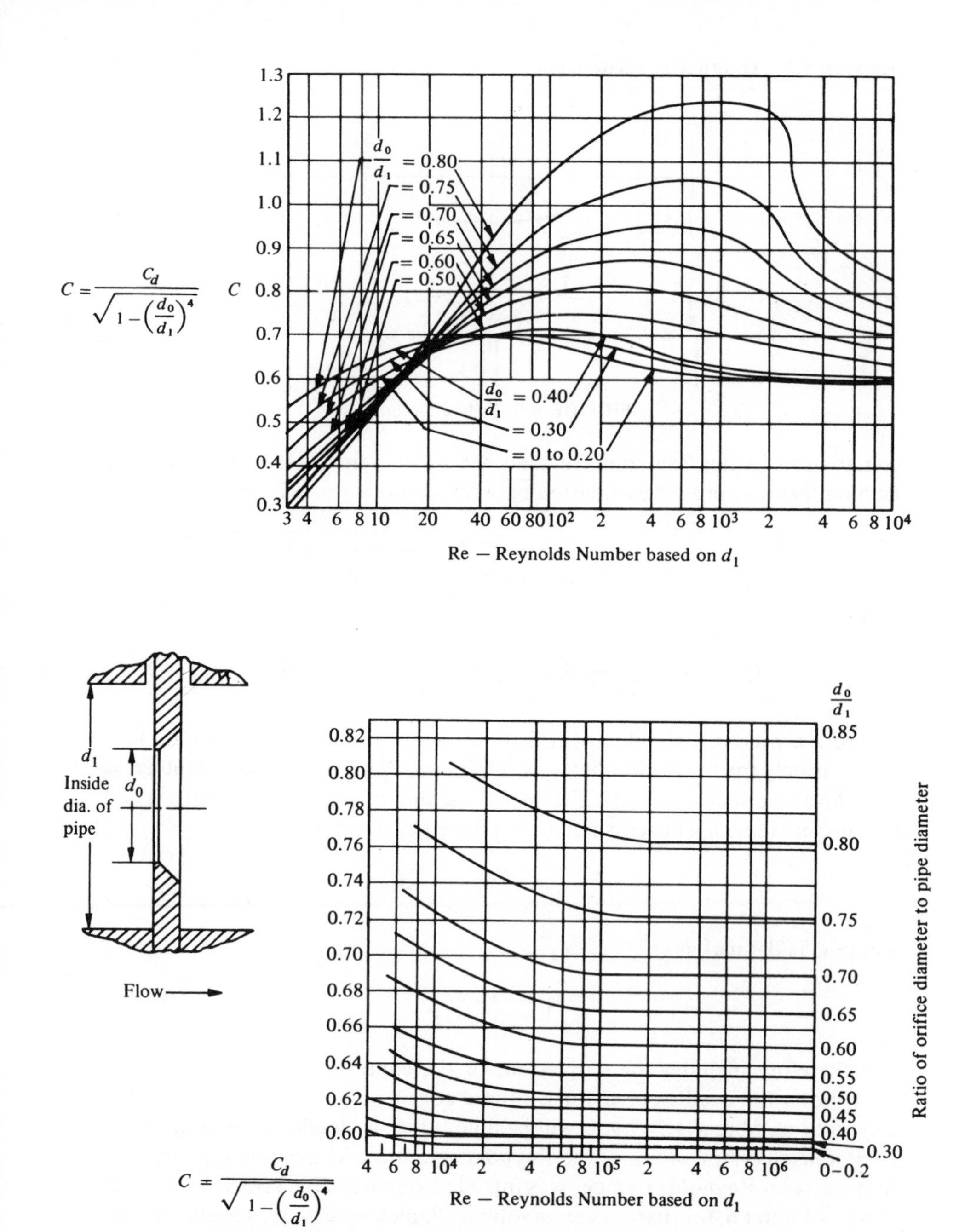

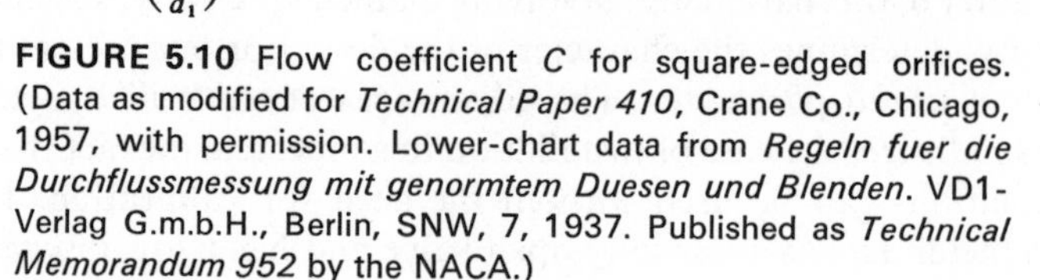

FIGURE 5.10 Flow coefficient C for square-edged orifices. (Data as modified for *Technical Paper 410*, Crane Co., Chicago, 1957, with permission. Lower-chart data from *Regeln fuer die Durchflussmessung mit genormtem Duesen und Blenden*. VD1-Verlag G.m.b.H., Berlin, SNW, 7, 1937. Published as *Technical Memorandum 952* by the NACA.)

ficient and consequently a relatively high head loss. In addition, errors can be caused due to poor placement leading to the orifice being eccentric to the inside diameter of the pipe, inaccurate location of the pressure taps, and a rough edge on the orifice opening.

Some of the problems associated with the orifice meter can be alleviated by using a nozzle, as shown in Figure 5.11. The coefficient C is defined as for the orifice meter, and it will be seen that the high value of this coefficient indicates relatively low losses in the nozzle. The manner in which C is defined as well as the location of the pressure taps for this flow nozzle leads to values of C that can exceed unity.

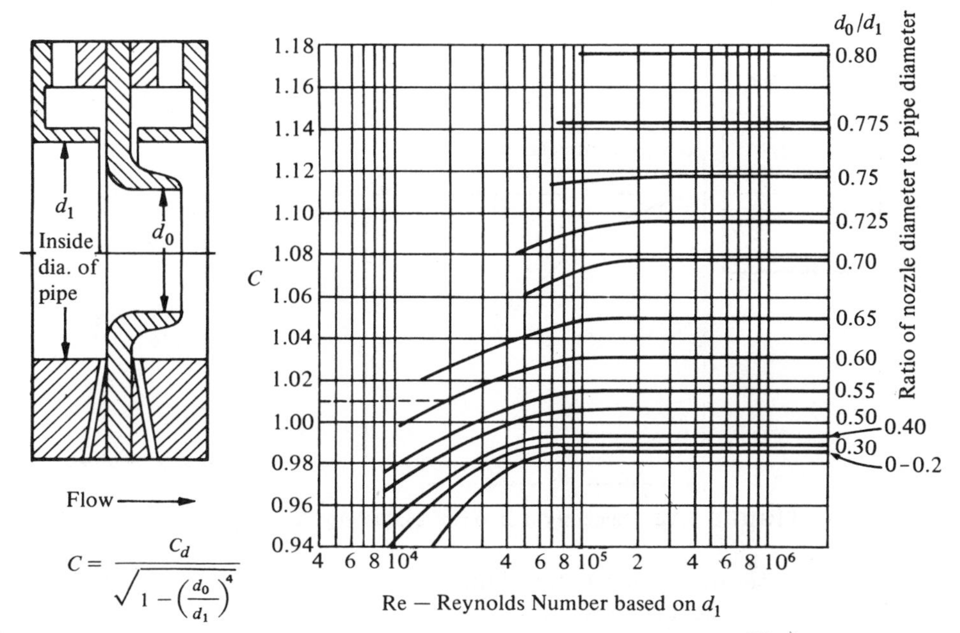

FIGURE 5.11 Flow coefficient C for nozzles. (Data as modified for *Technical Paper 410*, Crane Co., Chicago, 1957, with permission. Data from *Regeln fuer die Durchflussmessung mit genormtem Duesen und Blenden,* VD1-Verlag G.m.b.H., Berlin, SNW, 7, 1937. Published as *Technical Memorandum 952* by the NACA.)

The measurement of the three orifice coefficients (either singly or deriving the third from two of the others) presents an interesting problem in experimental fluid mechanics. We can readily visualize weighing the efflux for a given period of time and thus obtaining the actual rate of discharge. Knowing the area of the opening and the height of fluid above the opening we can directly compute C_D from its definition. We can (in principle) also measure the area of the vena contracta by use of a caliper gage if proper care is taken.

This measurement yields C_C, and C_V is calculated from the values of C_D and C_C. However, this method of measuring C_C is not very accurate, and rather than using it, let us attempt to measure C_V and to compute C_C. It is possible to measure C_V in the following manner. Let the jet flow and let us measure the position of a point along the trajectory downstream of the vena contracta. We can accomplish this by using a beam and sharp-edged pointers or else by simply noting the point at which the jet strikes a plate placed at some convenient position, as indicated in Figure 5.12. The time for a particle to fall the distance y (neglecting air friction) is given as

$$t = \sqrt{\frac{2y}{g}}$$

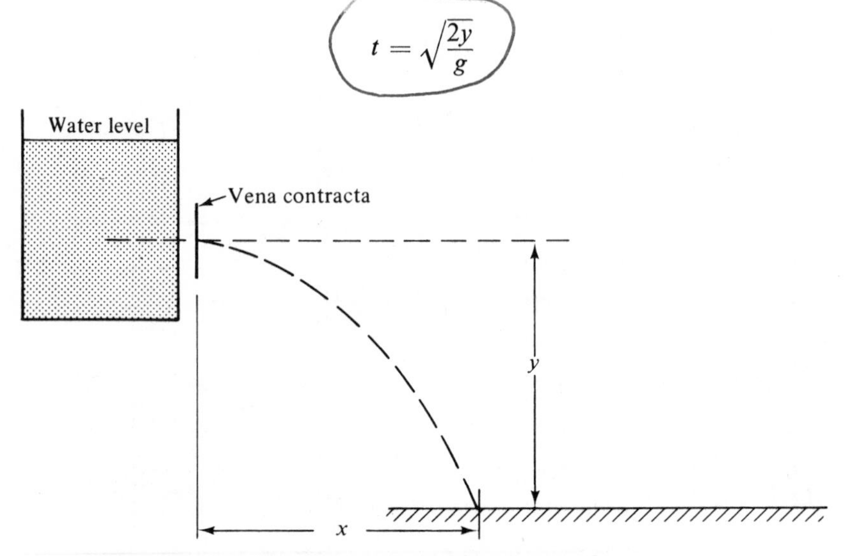

FIGURE 5.12 Trajectory method of evaluating C_v.

and the time for the particle to travel the distance x with the constant velocity of the vena contracta (V_{vc}) is

$$t = \frac{x}{V_{vc}}$$

Therefore,

$$V_{vc} = \frac{x}{\sqrt{2y/g}}$$

and

$$C_V = \frac{V_{vc}}{V_2} \tag{5.14}$$

by definition. The velocity V_2 is given by equation (5.7).

ILLUSTRATIVE PROBLEM 5.5

In a test it is found that the orifice described in Illustrative Problem 5.3 discharges 65 gal/min when subjected to a head of 16 ft of water. The values of x and y were masured as shown in Figure 5.12 and were found to be 17.25 and 5.0 ft, respectively. Determine C_C, C_V, and C_D.

Solution

For 16 ft,

$$V_2 = \sqrt{2(g)(16)} = 32.1 \text{ ft/sec}$$

From Illustrative Problem 5.3, $A_0 = 0.00546 \text{ ft}^2$. Therefore, the ideal flow is

$$(32.1)(60)(0.00546) \times \frac{1728}{231} = 78.7 \text{ gal/min}$$

From its definition,

$$C_D = \frac{65}{78.7} = 0.826$$

Using the trajectory data, we obtain

$$V_{vc} = \frac{x}{\sqrt{2y/g}} = \frac{17.25}{\sqrt{(2 \times 5)/g}} = 30.9 \text{ ft/sec}$$

and

$$C_V = \frac{30.9}{32} = 0.963$$

Having C_D and C_V,

$$C_C = \frac{C_D}{C_V} = \frac{0.826}{0.963} = 0.858$$

5.4 SIPHON

In a siphon, fluid is made to flow above the level of a source by the pressure force exerted by the atmosphere. The maximum height to which it can go (in the absence of flow losses) can be directly evaluated by considering the problem to be that of a static column of fluid where pressure due to the weight of the column plus the vapor pressure of the fluid must be balanced by the pressure of the atmosphere. For water at 60°F the maximum height to which water will flow of its own accord above a source is approximately 33.4 ft or 10.2 m at normal atmospheric pressure.

Figure 5.13 shows two flow situations. In Figure 5.13b the pipe is at a higher level than the level of the source, while in Figure 5.13a the level of fluid in the pipe is always below the level of the source. If a Bernoulli equation is written between points 1 and 2 for both of these cases, we obtain two identical equations in the absence of flow losses, yet it is known that if x is made high enough (in the limiting ideal case approximately 33.4 ft or 10.2 m for 60°F water), there will be no flow in the arrangement shown in Figure 5.13b. Thus, the height x will determine the flow that will exist in the siphon

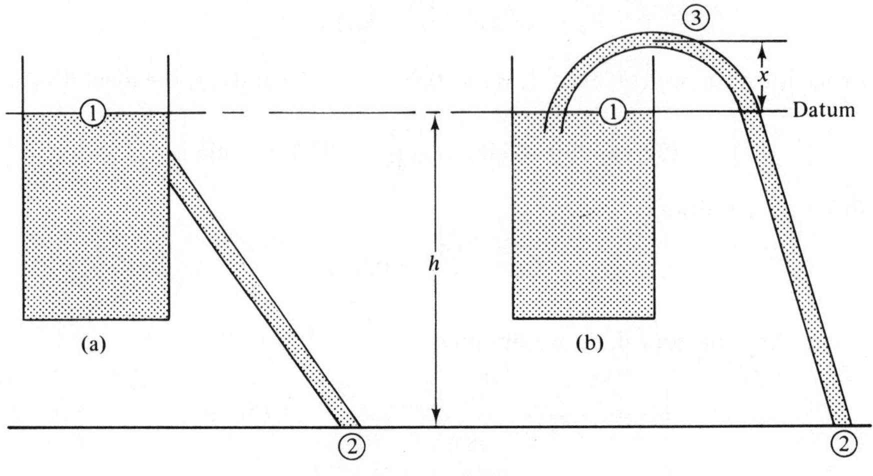

FIGURE 5.13 Siphon.

and must be investigated. Let us assume that there are flow losses in the siphon and write the Bernoulli equation between sections 1 ⟶ 3 and 3 ⟶ 2. Proceeding, we have

$$\frac{V_1^2}{2g} + Z_1 + \frac{p_1}{\gamma} = \frac{V_3^2}{2g} + Z_3 + \frac{p_3}{\gamma} + \text{losses}_{1\to 3} \tag{5.15}$$

and

$$\frac{V_3^2}{2g} + Z_3 + \frac{p_3}{\gamma} = \frac{V_2^2}{2g} + Z_2 + \frac{p_2}{\gamma} + \text{losses}_{3\to 2} \tag{5.16}$$

Simplifying these expressions and noting that the pressure at 1 and 2 is atmospheric, p_a, we obtain

$$0 + 0 + \frac{p_a}{\gamma} = \frac{V_3^2}{2g} + x + \frac{p_3}{\gamma} + \text{losses}_{1\to 3} \tag{5.17}$$

and

$$\frac{V_3^2}{2g} + x + \frac{p_3}{\gamma} = \frac{V_2^2}{2g} - h + \frac{p_a}{\gamma} + \text{losses}_{3\to2} \tag{5.18}$$

Substituting from (5.18) into (5.17), we obtain

$$\frac{p_a}{\gamma} = \frac{V_2^2}{2g} - h + \frac{p_a}{\gamma} + \text{losses}_{1\to3} + \text{losses}_{3\to2} \tag{5.19}$$

which yields

$$\frac{V_2^2}{2g} = h - \text{losses}_{1\to2} \tag{5.20}$$

which we obtain by applying the Bernoulli equation directly between sections 1 and 2. Returning to equations (5.17) and (5.18), let us obtain a solution for x, the height of the siphon:

$$x = \frac{p_a}{\gamma} - \frac{V_3^2}{2g} - \frac{p_3}{\gamma} - \text{losses}_{1\to3} \tag{5.21}$$

or

$$x = \frac{V_2^2}{2g} - \frac{V_3^2}{2g} - h + \frac{p_a}{\gamma} - \frac{p_3}{\gamma} + \text{losses}_{3\to2} \tag{5.22a}$$

If the siphon is flowing full and is of constant cross section, $V_2 = V_3$ and equation (5.22a) becomes

$$x = \frac{p_a}{\gamma} - \frac{p_3}{\gamma} - h + \text{losses}_{3\to2} \tag{5.22b}$$

From equations (5.20) through (5.22b) the siphon can be analyzed. This is best shown in the following illustrative problem.

ILLUSTRATIVE PROBLEM 5.6

A siphon is operated to discharge water at the rate of 1 ft^3/sec from a reservoir through a pipe whose inside diameter is 4 in. If the height of the siphon is 10 ft above the level in the reservoir and the discharge of the siphon is 2.02 ft below the reservoir level, determine the velocity in the pipe and the pressure at the highest point in the pipe if (a) flow losses are negligible; (b) if the losses from 1 $\longrightarrow$ 3 are equivalent to a head of 0.1 ft of water and represent one-third of the total losses in the pipe, determine the quantity of fluid flowing, the velocity in the pipe, and the pressure at the highest point in the pipe.

Solution

a. Refer to Figure 5.13b. The quantity of fluid flowing is given as 1 ft³/sec. From continuity,

$$V = \frac{Q}{A} = \frac{1}{(\pi/4)[(4)^2/144]} = 11.46 \text{ ft/sec}$$

and using equation (5.21) with no flow losses,

$$\frac{p_3}{\gamma} = \frac{p_a}{\gamma} - \frac{V_3^2}{2g} - x = \frac{14.7 \times 144}{62.4} - \frac{(11.46)^2}{2 \times 32.17} - 10$$
$$= 33.92 - 2.04 - 10 = 21.88 \text{ ft}$$

or 33.92 — 21.88 = 12.04 ft of vacuum at the summit. For the case of no flow losses, we note that this is equal to the head difference between the summit of the siphon and its exit; that is, 10 + 2.02 = 12.02 ft (closely).

b. In this part of the problem the quantity of fluid flowing is not known. We therefore apply equation (5.20) and from the statement of the problem note that losses 1 ⟶ 2 = 0.3 ft. Thus,

$$\frac{V_2^2}{2g} = h - 0.3 = 2.02 - 0.3$$

and

$$V_2 = 10.5 \text{ ft/sec}$$

The quantity flowing is therefore

$$\frac{\pi}{4} \times \frac{(4)^2}{144} \times 10.5 = 0.916 \text{ ft}^3\text{/sec}$$

Once again applying equation (5.21),

$$\frac{p_3}{\gamma} = \frac{p_a}{\gamma} - \frac{V_3^2}{2g} - x - \text{losses}_{1\to 3}$$
$$= \frac{14.7 \times 144}{62.4} - \frac{(10.5)^2}{2g} - 10 - 0.1$$
$$= 22.11$$

and this is 33.92 — 22.11 ft, or 11.81 ft, vacuum at the summit.

When analyzing siphons it is important to verify that the summit pressure p_3 is not impossible and also to check that the pipe is flowing full by independently checking the summit velocity against the discharge that would be obtained if the siphon is operated at its maximum height of atmospheric pressure minus the vapor pressure of the fluid (the pressure exerted by vapor in equilibrium with its fluid at the temperature of the fluid).

5.5 PRESSURE AND VELOCITY MEASUREMENTS[1]

The measurement of any property of a system requires that the method used should not change the magnitude of the property being measured, thereby yielding an erroneous result. There are many methods available to determine the pressure and velocity at a given location in a flow system, and the student is referred to the technical literature for detailed discussions on this subject. In particular, the latest publication of the American Society of Mechanical Engineers, *Flow Meters: Their Theory and Application*, and the *Transactions of the ASME* are recommended for this purpose. At present we shall be concerned solely with those devices to which we can apply the Bernoulli equation to evaluate the desired results.

5.5a Piezometer

The measurement of the pressure of a fluid in motion is particularly difficult since the presence of the measuring device will almost invariably alter the flow or change the magnitude of the pressure. An opening in the wall of a pipe is quite often used to measure the pressure of a flowing fluid. This static pressure is presumed to be the pressure of the undisturbed fluid, and the opening is known as a *piezometer opening*. The reading obtained with a single opening of this type will not be correct if the opening is not normal to the pipe surface or if the edges of the opening have burrs, and this method of pressure measurement is subject to large errors due to the presence of small perturbations in the installation of the piezometer opening (Figure 5.14). To obtain a more correct reading several holes are often placed around the periphery of a pipe in a given plane and connected together. This device is known as a *piezometer ring* and the pressure is read directly on a manometer attached to the common pressure connection.

When a body is submerged in a fluid such as in the case of an airfoil, it is often desirable to obtain the static pressure distribution along its boundaries. This can be determined by making piezometer openings along the surface and taking the corresponding static pressure readings by attaching tubes to these openings. In the case of an airfoil we can use these pressure readings to obtain the velocity distribution by applying the Bernoulli equation.

[1]Appendix 2-1 should be consulted in conjunction with this section.

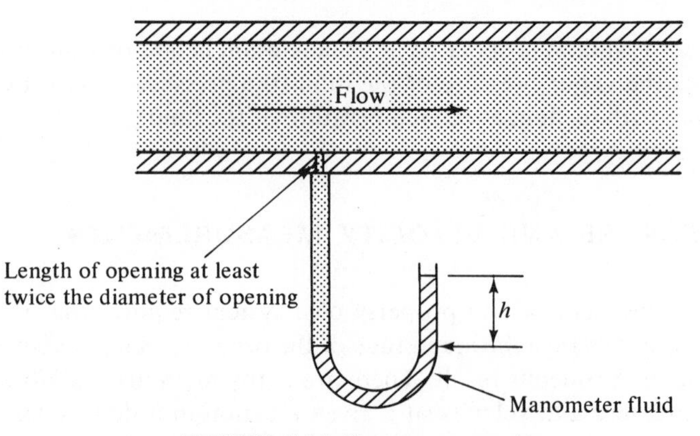

FIGURE 5.14 Piezometer opening.

5.5b The Pitot Tube and Pitot-Static Tube

Henri Pitot, a Frenchman, inserted an open-ended tube into the Seine River near Paris in the early 1700s and found that the rise of water in the vertical leg of the tube was proportional to the square of the velocity of the stream at the location at which he made his measurements. Tubes of this type are used to measure the local velocity of a flowing fluid and are called Pitot tubes in honor of Henri Pitot. Figure 5.15 shows a simple Pitot tube inserted into a stream with its open end facing into the stream flow. Both the tube and its opening are made as small as is practical to keep flow disturbances to a minimum due to the presence of the tube. The assumption is made that the streamline at the center of the Pitot tube is not disturbed by the tube. At the inlet to the tube the stream velocity is reduced to zero and the level of liquid

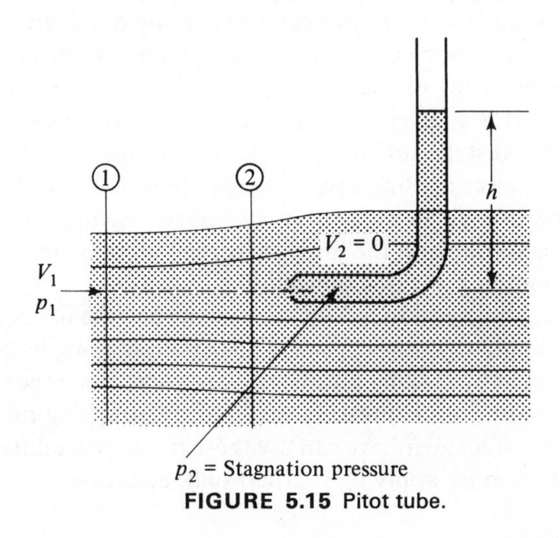

FIGURE 5.15 Pitot tube.

rises in the tube until it reaches the level h above the center line of the tube. It is assumed that the pressure at section 2 is the stagnation pressure of the streamline in question. By applying the Bernoulli equation to the streamline at the center of the tube successively at sections 1 and 2 in the flow, we obtain

$$\frac{V_1^2}{2g} + \frac{p_1}{\gamma} = \frac{p_2}{\gamma}$$

and

$$V = \sqrt{\frac{2g(p_2 - p_1)}{\gamma}} \tag{5.23}$$

The term $p_2 - p_1$ is sometimes known as the impact, dynamic, or velocity pressure of a stream and it is equal to $\frac{1}{2}\rho V^2$. In equation (5.23) the specific weight, γ, is that of the flowing stream and not that which exists in the manometer leg. If the Pitot tube is installed in an open channel, then $(p_2 - p_1)/\gamma$ is simply h, as shown in Figure 5.16. For a closed conduit it is necessary to

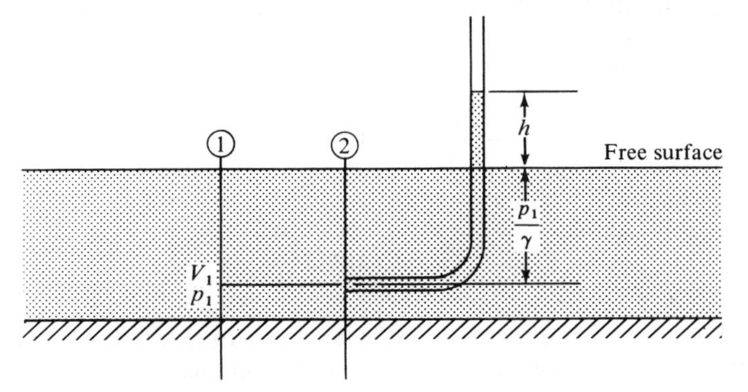

FIGURE 5.16 Pitot tube in an open channel.

evaluate $(p_2 - p_1)/\gamma$ by making a second measurement. This can be done by combining a piezometer tube with the Pitot tube or using a piezometer opening in the pipe wall. Let us assume that a piezometer opening is used in conjunction with a Pitot tube, as shown in Figure 5.17 to measure the velocity of a fluid in a pipe. Further assume that a suitable manometer fluid whose specific weight is γ_M will be used to evaluate the pressure differential between the Pitot tube and the piezometer (as shown in Figure 5.17) and let us write the Bernoulli equation between points 1 and 2:

$$\frac{V_1^2}{2g} + \frac{p_1}{\gamma} = \frac{p_2}{\gamma} \tag{5.24}$$

By evaluation of the manometer, we have

$$p_1 + \gamma h_1 + \gamma_M h_2 - \gamma(h_1 + h_2) = p_2$$

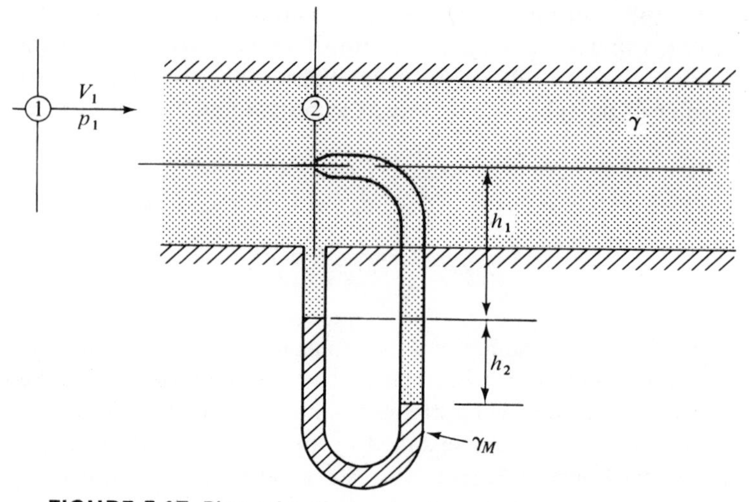

FIGURE 5.17 Pitot tube-piezometer for velocity measurement.

Simplifying yields

$$\frac{p_2 - p_1}{\gamma} = h_2\left(\frac{\gamma_M}{\gamma} - 1\right) \tag{5.25a}$$

and

$$V_1 = \sqrt{2gh_2\left(\frac{\gamma_M}{\gamma} - 1\right)} \tag{5.25b}$$

In the system shown in Figure 5.17 it is necessary to exercise care in installing the piezometer tap to avoid errors in the measurement that have been previously noted.

ILLUSTRATIVE PROBLEM 5.7

Water flows in a pipe. A Pitot-static tube uses mercury as the metering fluid and a deflection of 50 mm is noted on the manometer. Calculate the flow velocity if $\gamma_w = 9810\ \text{N/m}^3$, and the specific gravity of mercury is 13.6.

Solution

Referrning to Figure 5.16 and applying equation (5.25a), we have

$$\frac{p_2 - p_1}{\gamma} = h_2\left(\frac{\gamma_M}{\gamma} - 1\right) = 0.050\left(\frac{13.6\gamma_W}{\gamma_W} - 1\right) = 0.63\ \text{m}$$

But equation (5.24) gives us

$$\frac{V_1^2}{2g} = \frac{p_2 - p_1}{\gamma} = 0.63 \text{ m}$$

Therefore, $V_1 = 3.52$ m/s.

The arrangement shown in Figure 5.17 has the disadvantage of requiring two openings in the pipe wall, requires care in making the piezometer opening, and is usually rigid so that only a single measurement point can be obtained. To overcome these objections, the Pitot tube is combined with a static tube in a single unit called the Pitot-static tube. It usually consists of two tubes installed one inside of the other and connected to a manometer, as shown in Figure 5.18. The inner tube is open at the end and measures the total pressure of the stream, while the outer tube has holes drilled in it normal to the direction of the flow and measures the static pressure of the stream. With the manometer connections as shown in Figure 5.18, the manometer reads the dynamic pressure. Using an analysis similar to that used for the system shown in Figure 5.17, we obtain equation (5.26) for the system shown in Figure 5.18:

$$V_1 = \sqrt{2gh_2\left(\frac{\gamma_M}{\gamma} - 1\right)} \tag{5.26}$$

The accuracy of the Pitot-static tube is dependent on its construction and proper installation in the pipe. In addition, it is quite sensitive to misalign-

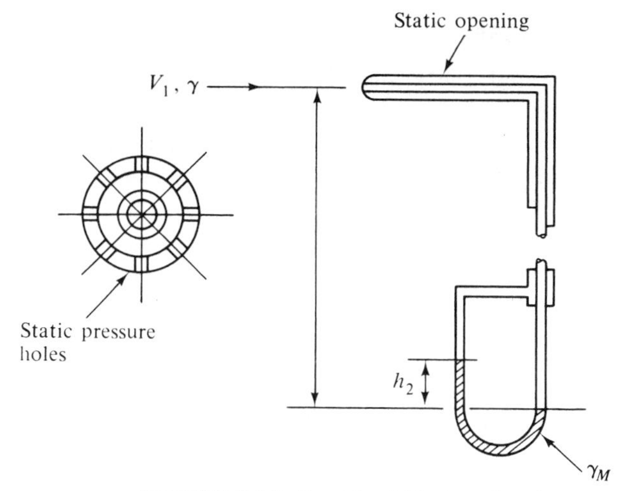

FIGURE 5.18 Combined Pitot-static tube.

ment with the flow. Although there are many Pitot-static tubes of standard size, shape, and construction that are intended to minimize these problems, it is always advisable to calibrate a particular installation. When this is done, equation (5.26) is modified to account for the nonideal action of the tube by introducing an empirical coefficient C:

$$V_1 = C\sqrt{2gh_2\left(\frac{\gamma_M}{\gamma} - 1\right)} \tag{5.27}$$

ILLUSTRATIVE PROBLEM 5.8

A stream of water flows at 3 m/s and the manometer (Figure 5.18) indicates a differential of 50 mm Hg. Determine C for this Pitot-static tube if the specific gravity of mercury is 13.6.

Solution

Using equation (5.27), we obtain

$$C^2 = \frac{V_1^2}{2gh_2[(\gamma_M/\gamma) - 1]} = \frac{3^2}{2 \times 9.81 \times 0.050(13.6 - 1)}$$
$$= 0.728$$

and $C = 0.85$.

5.5c Venturi Meter

As previously noted, the Pitot tube is used to measure the velocity at a point in a flow stream. For a permanent installation in a pipe that can be used to continuously monitor the rate of fluid with a minimum of flow losses, a venturi meter is frequently utilized. The venturi meter consists of a converging–diverging conical section of pipe arranged to give an increase in velocity and kinetic energy of the stream, thereby causing a measurable drop in pressure in the converging section. The diverging section is used to reconvert the increased kinetic energy of the stream back to pressure energy at the outlet of the meter with a minimum of turbulence and friction loss. The cone angle at the entrance to the throat is usually 20 to 30°, while the cone of the exit section is usually from 5 to 14°.

In a simplified installation, as shown in Figure 5.19, a manometer is connected to the inlet and throat of the venturi meter. Let us assume that there are no flow losses in the meter and write the Bernoulli equation for sections 1

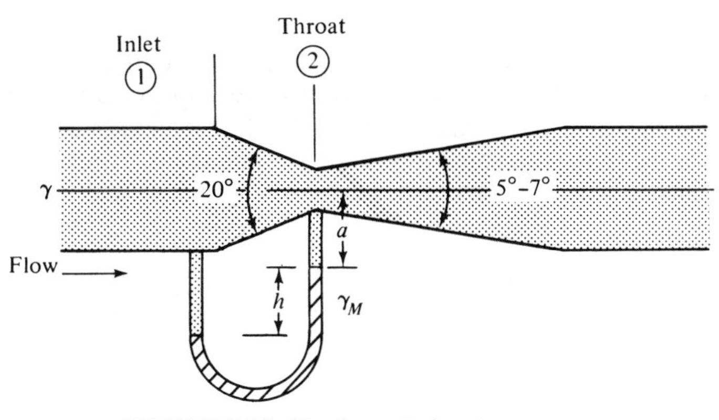

FIGURE 5.19 Simple venturi meter.

and 2; we obtain

$$\frac{p_1}{\gamma} + \frac{V_1^2}{2g} = \frac{p_2}{\gamma} + \frac{V_2^2}{2g}$$

and

$$\frac{V_2^2}{2g} - \frac{V_1^2}{2g} = \frac{p_1}{\gamma} - \frac{p_2}{\gamma} \tag{5.28}$$

The continuity equation requires that for the steady flow of an incompressible fluid $A_1V_1 = A_2V_2$. Therefore, equation (5.28) becomes

$$\frac{p_1}{\gamma} - \frac{p_2}{\gamma} = \frac{V_2^2}{2g}\left[1 - \left(\frac{A_2}{A_1}\right)^2\right] \tag{5.29}$$

and the volume rate of fluid flow (AV) is

$$AV = A_2V_2 = \left[\frac{A_2}{\sqrt{1 - (A_2/A_1)^2}}\right]\sqrt{\frac{2g(p_1 - p_2)}{\gamma}} \tag{5.30}$$

where the specific weight γ is that of the flowing fluid and not that of the manometer fluid shown in Figure 5.19. Using the manometer and a recording instrument, it is possible to obtain a continuous record of the flow in a pipe. For the case of a real fluid with friction, we introduce a coefficient C_V into equation (5.30). Therefore,

$$(AV)_{\text{actual}} = \frac{C_V A_2}{\sqrt{1 - (A_2/A_1)^2}}\sqrt{\frac{2g(p_1 - p_2)}{\gamma}} \tag{5.31}$$

In terms of the manometer shown in Figure 5.18,

$$(AV)_{\text{actual}} = \frac{C_v A_2}{\sqrt{1 - (A_2/A_1)^2}}\sqrt{2g\left(\frac{\gamma_M}{\gamma} - 1\right)h} \tag{5.32}$$

The velocity coefficient of a venturi meter is not constant and is found to be a function of the Reynolds number, the area ratio A_2/A_1, and the shape of the venturi. For greatest accuracy an in-place calibration is most desirable. Figure 5.20 shows C_V for venturi meters of different sizes but having a 2:1 diameter ratio. In most instances of practical interest C_V can be taken to be 0.9 or greater.

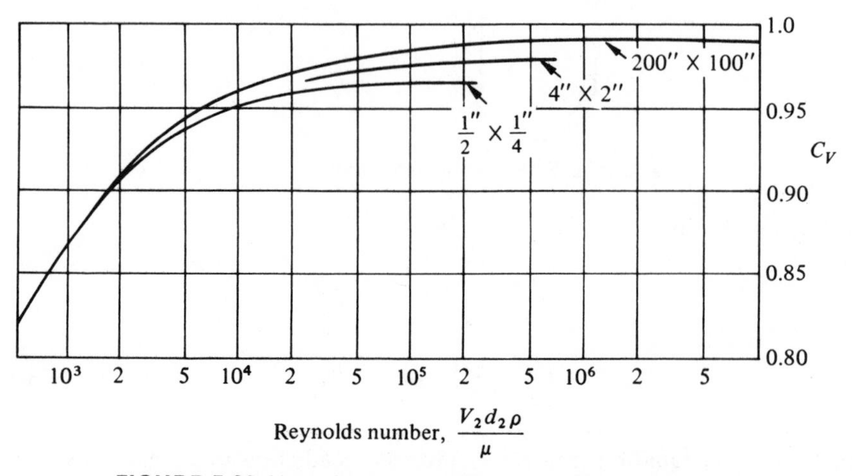

FIGURE 5.20 Venturi meter coefficients. (Reproduced with permission from *Elementary Fluid Mechanics* by J. K. Vennard, 4th ed., John Wiley & Sons, Inc., New York, 1961, p. 418.)

The venturi meter is a low-head-loss device, since its diverging downstream section past the throat of the venturi acts to recover kinetic energy. The head loss from inlet to throat can be derived from equation (5.3), yielding

$$\text{losses}_{1\to 2} = \left\{\left(\frac{1}{C_V^2} - 1\right)\left[1 - \left(\frac{A_2}{A_1}\right)^2\right]\right\}\frac{V_2^2}{2g} \tag{5.33}$$

For most installations the term in braces in equation (5.33) will be from 0.1 to 0.2.

ILLUSTRATIVE PROBLEM 5.9

A venturi meter consists of a 125-mm-diameter inlet and a 50-mm-diameter throat. If h in Figure 5.19 is 250 mm Hg (specific gravity = 13.6), determine the ideal quantity of water flowing.

Solution

From the manometer, we have

$$p_1 + \gamma_W(h + a) - \gamma_M h - \gamma_M a = p_2$$

or

$$\frac{p_1 - p_2}{\gamma_W} = h\left(\frac{\gamma_M}{\gamma_W} - 1\right)$$

where the subscripts M and W refer to mercury and water, respectively. Thus,

$$\frac{p_1 - p_2}{\gamma_W} = 0.25(13.6 - 1) = 3.15 \text{ m of water}$$

Using equation (5.30), we obtain

$$AV = \left[\frac{A_2}{\sqrt{1 - (A_2/A_1)^2}}\right]\sqrt{2g\left(\frac{p_1 - p_2}{\gamma}\right)}$$

We first evaluate

$$A_1 = (\pi/4)(0.125)^2 = 1.227 \times 10^{-2} \text{ m}^2$$

and

$$A_2 = (\pi/4)(0.050)^2 = 1.963 \times 10^{-3} \text{ m}^2.$$

The ratio

$$\frac{A_2}{A_1} = \frac{1.963 \times 10^{-3}}{1.227 \times 10^{-2}} = 0.16$$

Therefore,

$$Q = AV = \left[\frac{1.963 \times 10^{-3}}{\sqrt{1 - (0.16)^2}}\right]\sqrt{2 \times 9.81 \times 3.15}$$

$$= 1.56 \times 10^{-2} \text{ m}^3/\text{s}$$

5.6 CLOSURE

The equations developed in Chapter 5 are the necessary relations needed to solve problems involving the steady flow of an incompressible fluid. They are widely used for this purpose and will probably be retained by the student longer than most concepts regarding this subject. This chapter has been devoted to the solution of many problems that can be treated analytically. Because of nonideal flow situations, empirical coefficients have been introduced to yield a better correlation between measured and computed quantities. In each application studied in this chapter the actual physical problem was reduced to an idealized one-dimensional model in which it was unnecessary to know the details of the flow in order to obtain the desired information. It will be noted that the judicious selection of the control volume makes this approach feasible and considerably reduces the required computational effort.

PROBLEMS

Use $\gamma = 62.4$ lb/ft³ (9810 N/m³) for water unless indicated otherwise.

5.1 A jet of water issues vertically from a nozzle with a velocity of 25 ft/sec. What is the maximum height that it will reach if air resistance is neglected?

5.2 A jet of water issues vertically from a nozzle with a velocity of 10 m/s. What is the maximum height that it will reach if air resistance is neglected?

5.3 A 4-in.-i.d. pipe carries water at the rate of 300 gal/min. If the pressure in the pipe is 25 psig, determine the velocity in the pipe, the velocity head, the pressure head, and the total head referenced to a datum 10 ft below the center of the pipe.

5.4 A 75-mm-i.d. pipe carries water at the rate of 1400 litres/min. The pressure in the pipe is 210 kPa above atmospheric pressure. Determine the velocity, velocity head, pressure head, and total head if a datum is selected 3 m below the center of the pipe. $P_{atmos} = 100$ kPa.

5.5 Oil (specific weight, 50 lb/ft³) flows in a 4-in.-i.d. pipe. If the pressure in the pipe is 30 psia and the total head with respect to a reference plane 5 ft below the center of the pipe is 200 ft of oil, determine the velocity in the pipe and the volume rate of flow in gallons per minute.

5.6 A stream flows over a waterfall 150 ft high. What is the velocity with which it strikes the base of the fall if air resistance is negligible?

5.7 In Problem 5.6, assume the flow rate to be 500 ft³/sec. Determine the horsepower of the fall.

5.8 Determine the pressure at section 2 of the pipe shown in Figure P5.8 if water flows at the rate of 300 gal/min past section 1.

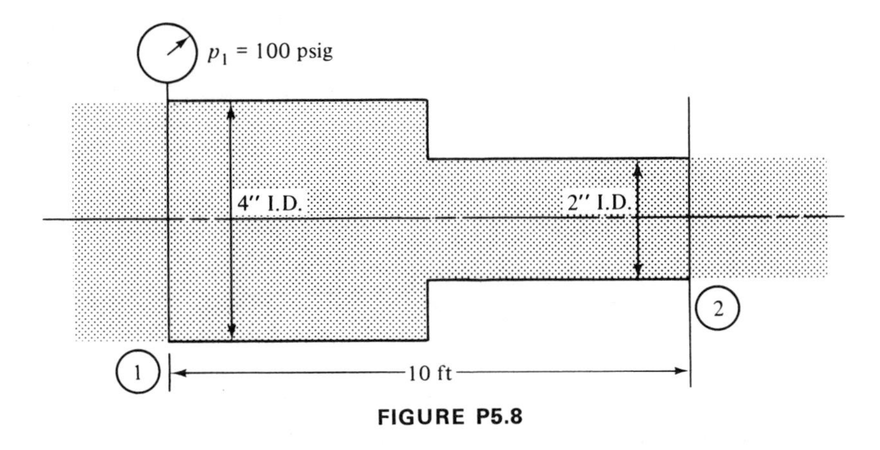

FIGURE P5.8

5.9 If the pipe in Problem 5.8 is turned vertically with the large end (section 1) at the bottom, what is the pressure at section 2?

5.10 A tank has a water level of 20 ft above a datum plane. Five feet above the datum plane there exists a 6-in. opening in the tank. If there are no flow losses, determine the velocity of the water leaving the opening and the quantity of water flowing at this instant in gallons per minute.

5.11 Assuming no flow losses, determine the velocity of the water leaving the system shown in Figure P5.11.

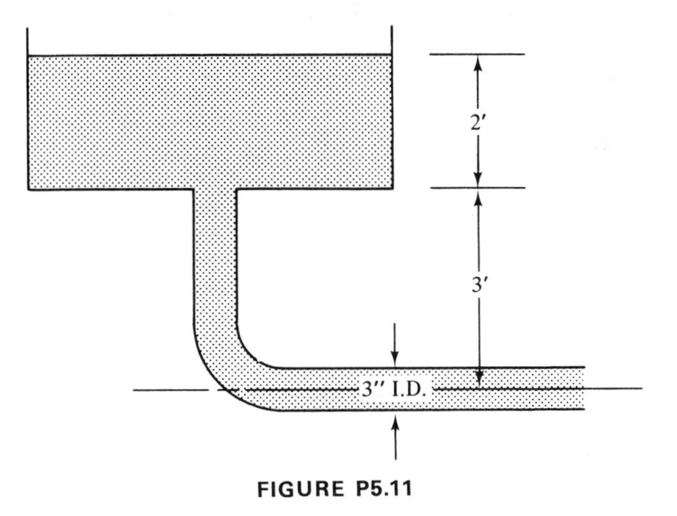

FIGURE P5.11

5.12 Calculate the quantity of water flowing in Problem 5.11 in gallons per minute.

5.13 Assume that the nozzle shown in Figure P5.13 has no losses. Determine the velocity of the jet.

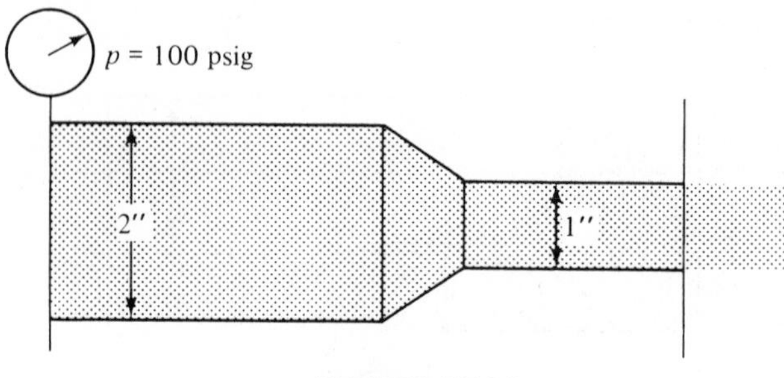

FIGURE P5.13

5.14 Solve Problem 5.10 if $C_c = 0.61$ and $C_v = 0.95$.

5.15 Solve Problem 5.13 if $C_v = 0.96$.

5.16 A large tank has a 50-mm opening located 4 m below the surface of oil having a specific gravity of 0.85. If $C_c = 0.85$ and $C_v = 0.95$, determine the volume rate of flow through the opening.

5.17 Derive an equation for the trajectory of a jet issuing from an orifice with a head of h ft and a velocity coefficient of unity. Assume resistance to be negligible.

5.18 A jet issues from an orifice and strikes horizontally 10 ft and vertically 2 ft from the vena contracta (Figure P5.18). Determine the head h on the jet.

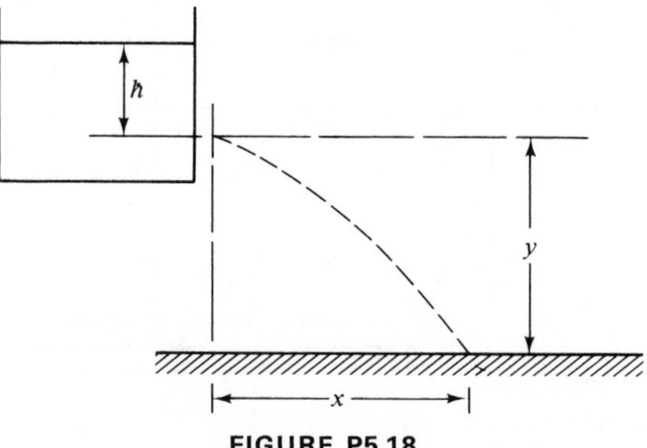

FIGURE P5.18

5.19 A Pitot tube is used to determine the velocity of a stream of water. If the specific weight of the water is 60 lb/ft³ and the water rise in a vertical tube attached to the Pitot tube is 5 in., determine the velocity of the stream at this point.

5.20 A body is immersed in a stream of water in such a manner that the water is caused to stagnate against the front face of the body. If the body is 10 ft below the surface of the stream and a Pitot tube in the face of the body has a liquid rise above the stream surface of 1 ft, calculate the stream velocity at this point.

5.21 A Pitot-static tube is directed into a stream of water flowing with a velocity of 10 ft/sec. If the manometer fluid is mercury having a specific weight of 850 lb/ft^3, determine its coefficient when the manometer differential is 1.8 in.

5.22 A siphon is used to discharge water from one tank to another. If the level in the second tank is 10 ft below the level in the first tank, what is the maximum flow rate that can be obtained through a 3-in.-i.d. tube?

5.23 Determine the volume flow through the siphon shown in Figure P5.23. Assume no flow losses.

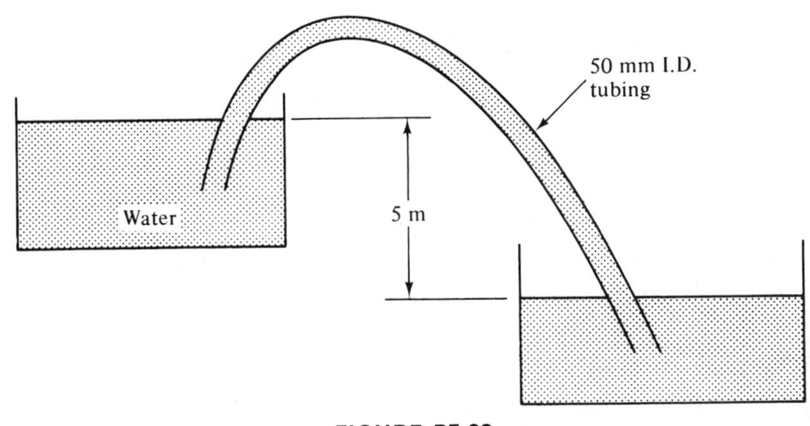

FIGURE P5.23

5.24 If the total pressure on a submerged research vessel is 220 psig at a level of 500 ft, determine its velocity. Assume the specific weight of water to be 62.4 lb/ft^3 and that it remains constant.

5.25 An airplane is designed to have a cruising speed of 500 mi/hr. If the specific weight of air is 0.075 lb/ft^3, determine the pressure between the openings of a Pitot-static tube.

5.26 A Pitot-static tube is used to measure the flow of air in a duct. If the density of the air is 0.075 lb/ft^3 and a differential manometer reads 1 in. of water ($\gamma = 62.4$), determine the air velocity.

5.27 An airplane travels at 400 mi/hr in air whose density is 0.07 lb/ft³. A Pitot-static tube is installed with the static connection on a wing and the Pitot connection facing forward into the air stream. If the air velocity relative to the wing is 480 mi/hr, determine the reading of a differential pressure gage to which this instrument is connected.

5.28 A venturi meter consists of a 4-in.-diameter inlet and a 2-in.-diameter throat (Figure P5.28). If 1 ft³/sec of water is flowing in the meter, determine h if the specific weight of mercury is 850 lb/ft³ and losses can be neglected.

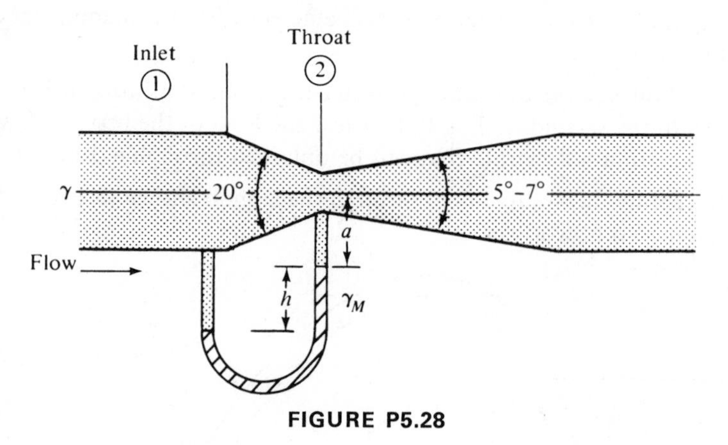

FIGURE P5.28

5.29 Solve Problem 5.28 if C_v for this meter is 0.9.

5.30 A venturi consists of a 100-mm-diameter inlet and a 40-mm-diameter throat. If h is 200 mm of mercury (sg = 13.6), what is the ideal quantity of water flowing?

5.31 Solve Problem 5.30 if C_v is 0.91.

5.32 A venturi meter is installed in a pipe so that its axis is vertical. An identical venturi meter is installed in a pipe the same size with its axis horizontal. If each meter gives the same reading on a differential manometer, is the flow the same in both pipes, assuming the density of the fluids to be the same for each installation?

5.33 Show that independent of the angle of the axis of a venturi meter, equation (5.30) or (5.31) holds.

5.34 Can the discharge coefficient of a venturi meter exceed unity? Assume that there is friction present in the meter.

REFERENCES

1. *Basic Fluid Mechanics* by J. L. ROBINSON, McGraw-Hill Book Company, New York, 1963.
2. *Engineering Applications of Fluid Mechanics* by J. C. HUNSAKER and B. G. RIGHTMIRE, McGraw-Hill Book Company, New York, 1947.
3. *Elementary Theoretical Fluid Mechanics* by K. BRENKERT, JR., John Wiley & Sons, Inc., New York, 1960.
4. *Fluid Mechanics* by V. L. STREETER, 3rd ed., McGraw-Hill Book Company, New York, 1962; 6th ed., 1979.
5. *Elementary Fluid Mechanics* by J. K. VENNARD, 3rd ed., John Wiley & Sons, Inc., New York, 1954.
6. *Mechanics of Fluids* by G. MURPHY, 2nd ed., International Textbook Co., Scranton, Pa., 1952.
7. *Fluid Mechanics for Engineers* by P. S. BARNA, Butterworth and Co. (Publishers) Ltd., London, 1957.
8. *Fluid Meters: Their Theory and Application*, 5th ed., American Society of Mechanical Engineers, New York, 1959.
9. "Flow of Fluids through Valves, Fittings, and Pipe," *Technical Paper 410*, Crane Co., Chicago, 1957.
10. *Fundamentals of Fluid Mechanics* by J. A. SULLIVAN, Reston Publishing Co., Reston, Va., 1978.
11. *Applied Fluid Mechanics* by R. L. MOTT, 2nd ed., Charles E. Merrill Publishing Co., Columbus, Ohio, 1979.

Steady Flow of Incompressible Fluids in Pipes

6.1 INTRODUCTION

Fluids are most commonly transported from one location to another by forcing them through pipes and tubes. The Bernoulli equation that was developed in Chapter 4 and extensively applied in Chapter 5 is applicable to the flow of a frictionless fluid and with modification can be applied (judiciously) to the flow of real fluids by inserting a term in the equation to account for these real effects (losses). In this chapter we shall consider the problem of the steady flow of incompressible fluids in pipes, define the flow regimes encountered and the general form of the friction equation, and give expressions for entrance, exit, enlargement, contraction, bend, and valve losses. The literature on this subject is quite extensive, and the data relating to these effects have been presented in many different (and many times inconsistent) forms. One of the most comprehensive, consistent correlations of available data has been carried out by the Hydraulic Institute and has been published as the Pipe Friction section of the *Standards of the Hydraulic Institute*. Several of the tables and charts in this chapter are taken from this standard with permission of the Hydraulic Institute, and for more complete and extensive literature references and details, the student is specifically referred to this publication in addition to the other general references given at the end of the chapter.

6.2 CHARACTER OF FLOW IN PIPES—LAMINAR AND TURBULENT

To visualize the character of a fluid flowing in a pipe, Osborne Reynolds devised a simple experiment, shown schematically in Figure 6.1. Basically, the dye is introduced into the glass tube by injecting it from a fine piece of tubing into the well-rounded entrance of the glass tube. The velocity of the test fluid is controlled by varying the height of the fluid in the glass tank and by changing the setting of the valve in the downstream portion of the glass tube.

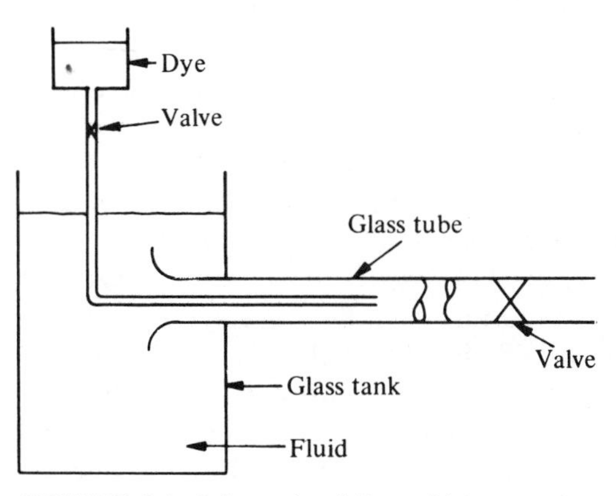

FIGURE 6.1 Schematic of Reynolds' apparatus.

At small average velocities it is found that the dye filament appears as a straight unbroken line parallel to the axis of the tube. This type of flow is known as laminar, viscous, or streamline flow and is composed of concentric cylindrical layers flowing past each other in a manner determined by the viscosity of the fluid. The fluid particles are retained in layers, and their motion is along parallel paths. As the flow rate is increased by changing the valve setting, it is found that the dye line will remain straight until a velocity is reached that causes it to waver and break into diffused patterns. The velocity at which the dye streak wavers and breaks is known as the *critical velocity*. At velocities greater than the critical velocity the colored dye filament becomes completely diffused in the main body of the fluid a short distance downstream from its point of injection. At velocities greater than the critical velocity the flow is termed *turbulent*, and the particles have a random motion that is transverse to the main flow direction and that causes the particles to intermingle in a random manner. In laminar flow the velocity of the fluid is maximum at the pipe axis and decreases to zero at the wall of the pipe, while in turbulent flow the velocity distribution is more uniform across the pipe diameter, as shown in Figure 6.2. By taking the average velocity as the char-

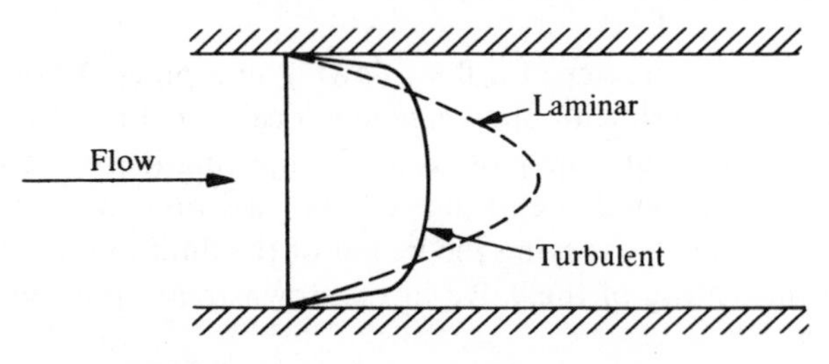

FIGURE 6.2 Velocity profiles in a pipe.

acteristic velocity and the pipe diameter as the characteristic length, Reynolds showed that the character of the flow of a fluid in a pipe was dependent on the pipe diameter, the velocity of flow, the density of the fluid, and its viscosity. The combination of these four variables yields a dimensionless parameter known as the Reynolds number, $DV\rho/\mu$.

During the course of his experiments, Reynolds was able to obtain the change from laminar to turbulent flow at Reynolds numbers as low as 1200 and higher than 40,000. However, the conditions under which these high Reynolds numbers were obtained are not ordinarily found in commercial installations. These numbers at which the transition in character from laminar to turbulent flow occur are known as the Reynolds upper critical numbers and are not normally significant in the analysis of normal pipe flow. However, if the flow is initially turbulent and the velocity of the fluid is decreased, it will be found that all initial disturbances will disappear and that the flow will become laminar. The value of this Reynolds number, known as the Reynolds lower critical number, is generally agreed to be approximately 2000. The usual piping installation will be found to undergo a transition from laminar to turbulent flow at Reynolds numbers from 2000 to 4000, with the flow always laminar below a Reynolds number of 2000 and always turbulent above a Reynolds number of 4000. Between these two values there is a region known as the *transition region* in which the flow may be either laminar or turbulent. In the transition region a disturbance will cause the character of the flow to change from laminar to turbulent.

6.3 EVALUATION OF REYNOLDS NUMBER

We have noted earlier that the Reynolds number is a dimensionless parameter involving diameter, velocity, density, and viscosity. Mathematically, the relationship is expressed as

$$\mathrm{Re} = \frac{DV\rho}{\mu} \tag{6.1}$$

Since the kinematic viscosity was defined in Chapter 2 to be the ratio of the absolute viscosity to the density, we can also write equation (6.1) as

$$\mathrm{Re} = \frac{DV}{\nu} \qquad \text{since} \qquad \nu = \frac{\mu}{\rho} \tag{6.2}$$

The evaluation of the Reynolds number from either equation (6.1) or (6.2) requires that care be exercised since the technical literature abounds with units of μ and ν that are not necessarily consistent. The following illustrative problems will help to clarify this point.

ILLUSTRATIVE PROBLEM 6.1

Water flows in a 50-mm-i.d. pipe with a velocity of 3 m/s. If the water is at 40°C, what is its Reynolds number? Use the data of Table 2.3.

Solution

From Table 2.4 at 40°C, $\rho = 992.2\ \mathrm{kg/m^3}$ and $\mu = 0.656 \times 10^{-3}\ \mathrm{N \cdot s/m^2}$. Using equation (6.1),

$$\mathrm{Re} = \frac{DV\rho}{\mu}$$

$$= \frac{0.050\ \mathrm{m} \times 3\ \mathrm{m/s} \times 992.2\ \mathrm{kg/m^3}}{0.656 \times 10^{-3}\ \mathrm{N \cdot s/m^2}} = 226\,900\ \frac{\mathrm{kg \cdot m}}{\mathrm{N \cdot s^2}}$$

Since a newton has the units of km/s^2,

$$\mathrm{Re} = 226\,900\ \frac{\mathrm{kg \cdot m}}{\mathrm{N \cdot s^2}} = 226\,900\ \frac{\mathrm{kg \cdot m}}{\mathrm{s^2}} \times \frac{1}{\mathrm{kg \cdot m/s^2}} = 226\,900$$

Note that the Reynolds number is dimensionless, *as it must be.*

ILLUSTRATIVE PROBLEM 6.2

Water flows in a 2-in.-i.d. pipe at a velocity of 30 ft/sec. If the water is at 100°F, find its Reynolds number. Use the data of Table 2.2.

Solution

From Table 2.2 at 100°F, $\rho = 1.927$ slugs/ft³, $\mu = 1.424 \times 10^{-5}$ lb sec/ft². Using equation (6.1) with all dimensions in feet yields

$$\text{Re} = \frac{DV\rho}{\mu} = \frac{2\text{ in.}/(12\text{ in./ft}) \times 30\text{ ft/sec} \times 1.927\text{ slugs/ft}^3}{1.424 \times 10^{-5}\text{ lb sec/ft}^2}$$

$$= 676{,}615 \frac{\text{slug ft}^2/\text{sec ft}^3}{\text{lb sec/ft}^2}$$

At first glance our answer would seem to be far from dimensionless. However, if we note that the slug has the units of lb sec²/ft, we obtain

$$\text{Re} = 676{,}615 \frac{\text{lb sec}^2/\text{ft} \times \text{ft}^2/\text{sec ft}^3}{\text{lb sec/ft}^2} = 676{,}615$$

which is indeed dimensionless.

ILLUSTRATIVE PROBLEM 6.3

An oil has a specific weight of 50 lb/ft³. If its viscosity is 5 cP, determine its Reynolds number if it is flowing in a 1-in.-i.d. pipe with an average velocity of 1 ft/sec.

Solution

Using the conversion from Table 2.4, 1 cP = 2.09×10^{-5} lb sec/ft², we have a viscosity of $5 \times 2.09 \times 10^{-5}$ lb sec/ft². To convert the specific weight to density requires that we divide the specific weight by g. Therefore,

$$\rho = \frac{50}{32.17}$$

$$= 1.55\text{ slugs/ft}^3$$

Using the data given, we obtain

$$\mathrm{Re} = \frac{DV\rho}{\mu} = \frac{1\ \text{in.}/(12\ \text{in./ft}) \times 1\ \text{ft/sec} \times 1.55\ \text{slugs/ft}^3}{1.045 \times 10^{-4}\ \text{lb sec/ft}^2}$$

$$= 1236$$

Using slugs = lb sec²/ft, we find Re to be dimensionless.

ILLUSTRATIVE PROBLEM 6.4

Discuss the character of the flow in Illustrative Problems 6.1, 6.2, and 6.3.

Solution

(a) Illustrative Problem 6.1; Re = 226 900. Since this is greater than 2000, the flow is turbulent.
(b) Illustrative Problem 6.2; Re = 676,615. Since this is greater than 2000, the flow is turbulent.
(c) Illustrative Problem 6.3; Re = 1236. Since this is less than 2000, the flow is laminar.

6.4 LAMINAR FLOW IN PIPES

In circular pipes where the Reynolds number is less than 2000, the flow is said to be laminar and we can characterize the flow pattern as consisting of a series of thin shells that are sliding over one another. At the center the velocity of the fluid is the greatest and at the wall the velocity is zero. This type of flow is depicted in Figure 6.3 and it can readily be analyzed mathematically. When this is done, we obtain an equation relating the pressure drop (head

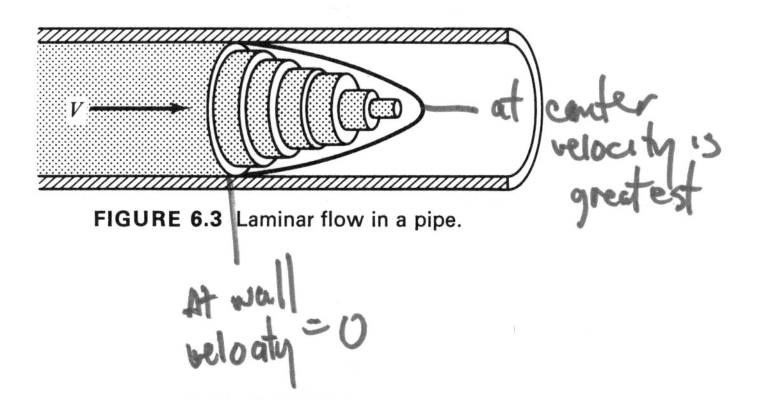

FIGURE 6.3 Laminar flow in a pipe.

loss) to the other variables involved. Equation (6.3) gives us this expression, which is known as the *Hagen–Poiseuille equation*:

$$\Delta p = \frac{128\mu LQ}{\pi D^4} \tag{6.3}$$

where Δp = pressure difference (pressure drop)
μ = viscosity
D = inside diameter of the pipe
L = length of the pipe
Q = volume rate of flow (since there is a wide variation in velocity, Q is taken to be equal to $A\bar{V}$, where $\bar{V}$ is the average velocity; for laminar flow the velocity varies parabolically from zero at the wall to a maximum value at the center of the pipe; the average velocity for this case equals one-half of the maximum velocity)

Equation (6.3) has been verified against experiments and has been found to give excellent agreement with the experimental results. It is important to note that the pressure drop in laminar flow is independent of the character (roughness) of the pipe wall. Also, because of the excellent agreement between measured and calculated results in laminar flow, the Hagen–Poiseuille relation has been used as the basis for measuring the viscosity of fluids in commercial viscosimeters.

It is sometimes more convenient to express equation (6.3) in terms of the average velocity, $\bar{V}$. Since $Q = A\bar{V} = (\pi D^2/4)\bar{V}$ we have

$$\Delta p = \frac{128\mu LA\bar{V}}{\pi D^4} = \frac{128\mu L(\pi D^2/4)\bar{V}}{\pi D^4} = \frac{32\mu L\bar{V}}{D^2} \tag{6.4}$$

In terms of a "head loss" and using $p_1 - p_2 = \Delta p$, we obtain

$$\frac{p_1 - p_2}{\gamma} = h_f = \frac{32\mu L\bar{V}}{\gamma D^2} \tag{6.5}$$

where h_f in the friction "head loss."

ILLUSTRATIVE PROBLEM 6.5

What is the pressure drop in a straight, horizontal run of 20 ft of pipe for the flow of Illustrative Problem 6.3?

Solution

Since Re is less than 2000, the flow is laminar and we can use equation (6.5). The use of equation (6.5) requires care that all units be consistent since the equation has dimensions. Proceeding with the data given,

$$h_f = \frac{32\mu L\bar{V}}{\gamma D^2} = \frac{32 \times 1.045 \times 10^{-4}\ \text{lb sec/ft}^2 \times 20\ \text{ft} \times 1\ \text{ft/sec}}{50\ \text{lb/ft}^3 \times [1\ \text{in.}/(12\ \text{in./ft})]^2}$$

and

$$h_f = 0.1926 \qquad \text{say } 0.193\ \text{ft}$$

or

$$h_f = 0.193\ \text{ft lb/lb}$$

It will be found later that the general form of equation most used to calculate the pressure loss in a pipe is known as the *Darcy–Weisbach equation*, given by

$$\frac{p_1 - p_2}{\gamma} = h_f = f\frac{L}{D}\frac{V^2}{2g} \tag{6.6}$$

where h_f is the head loss due to friction in feet or metres of fluid flowing and f is known as the *friction factor*. If we equate equations (6.5) and (6.6), we obtain

$$f\frac{L}{D}\frac{V^2}{2g} = \frac{32\mu LV}{\gamma D^2} \tag{6.7}$$

Solving for f yields

$$f = \frac{64}{DV\gamma/\mu g} = \frac{64}{DV\rho/\mu} = \frac{64}{\text{Re}} \tag{6.8}$$

where Re is based on the *average* velocity. We can therefore conclude that the friction factor in laminar flow is simply 64 divided by the Reynolds number and is independent of the pipe roughness.

ILLUSTRATIVE PROBLEM 6.6

Solve Illustrative Problem 6.5 using equation (6.8).

Solution

Earlier we had to find the Reynolds number to determine that the flow is laminar. From Illustrative Problem 6.3, Re = 1236. Therefore,

$$f = \frac{64}{\text{Re}} = \frac{64}{1236} = 0.052$$

and

$$h_f = f\,\frac{L}{D}\,\frac{V^2}{2g} = 0.052 \times \frac{20}{1/12} \times \frac{(1)^2}{2 \times 32.17} = 0.194\ \text{ft}$$

or

$$h_f = 0.194\ \text{ft lb/lb}$$

which agrees with the result of Illustrative Problem 6.5.

6.5 BOUNDARY LAYER[1]

The parabolic velocity profile of fluid flowing in laminar flow inside a pipe is due partly to the viscosity of the fluid and partly to the adhesive force between the liquid and the pipe wall. As the flow remains laminar, the velocity profile remains parabolic. When the flow becomes turbulent, the velocity decreases only very slightly from the axis of the pipe to the vicinity of the pipe wall. However, there is a steep velocity gradient in the thin layer of liquid next to the stationary boundary. This is called the *boundary layer*. Although the main body of fluid is turbulent, there still exists a thin laminar layer immediately next to the pipe wall. Some typical velocity distributions are shown in Figure 6.4.

Let us consider the flow of fluid over a flat plate held parallel to the direction of flow (see Figure 6.5). The vertical scale is purposely enlarged to show the detail of the flow pattern. When the fluid passes the leading edge of the plate, the velocity gradient and the viscous boundary shear are high. The fluid is moving in the laminar state, and the boundary layer is thin. This is called the *laminar boundary layer*. As the fluid travels farther down the stream along the plate, the retardation of fluid flow increases due to shearing force, and the boundary layer also grows in thickness. As a result, the velocity gradient gradually decreases and concurrently the boundary shear is reduced

[1]The material in this section is developed with permission of the Van Nostrand Reinhold Company from *Engineering Heat Transfer* by S. T. Hsu, Litton Educational Publishing, Inc., New York, 1963, pp. 209–211.

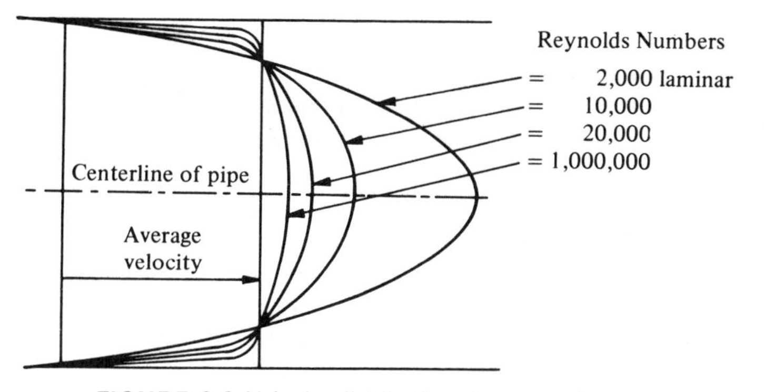

FIGURE 6.4 Velocity distributions in a smooth pipe.

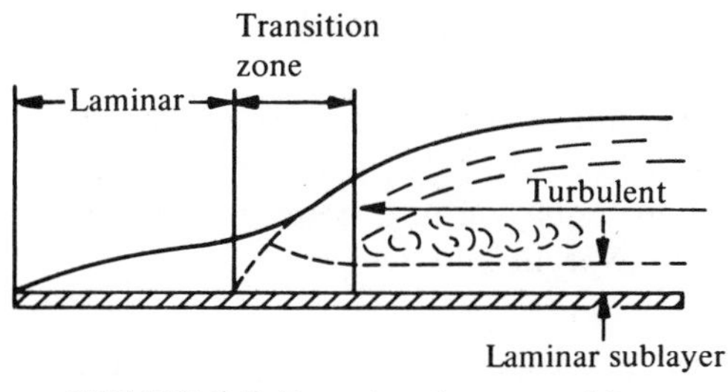

FIGURE 6.5 Boundary-layer transition.

as the thickness increases. When the boundary layer becomes thick enough, the particles begin to move out of the smooth layers or laminae, the laminar motion becomes unstable, and finally the flow becomes turbulent. However, under the turbulent boundary layer there is still a thin layer of fluid immediately next to the solid boundary, and this thin layer is flowing in the laminar regime. This layer is called the *laminar sublayer* of the turbulent boundary layer. The layer of transition from the laminar sublayer to the turbulent layer is called the *buffer layer*. Once the boundary layer is in the turbulent state, velocity distribution becomes more nearly constant in a lateral direction. A laminar boundary layer cannot change suddenly into a turbulent one, and a transition zone is found to exist between the turbulent and laminar zones. The turbulent motion is accompanied by an increase of the boundary shear and by an expansion of the thickness of the boundary layers. For a submerged plate, the transition takes place in the range of Reynolds numbers from 500,000 to 1 million, where the characteristic dimension is the distance from the leading edge of the plate. The number depends on the initial state of flow and also on the shape of the front edge and the roughness of the plate.

The thickness of the boundary layer is defined as the distance from the boundary to the point where the velocity reaches the value of the main stream velocity. It is extremely difficult actually to measure the thickness of the boundary layer, especially in turbulent flow. But the thickness of the boundary layer can be predicted by analytical methods or from the experimental results of velocity and temperature measurements. In laminar boundary layers the velocity profiles join the outside velocity curve asymptotically.

The velocity profile inside the boundary layer may be approximated by an analytical method based upon the theory of velocity distribution in pipes, and the drag force due to the boundary layer may also be derived by an analytical method. A complete discussion of these methods, however, is too advanced to fall within the scope of this book, and the student is referred to the references at the end of the chapter for further information on this subject.

6.6 FRICTION PRESSURE LOSSES IN PIPE FLOW

In Chapter 5 it was indicated that the Bernoulli equation can and has been used when a real fluid flows in a pipe. It will be recalled that a term was added to the right side of the equation to account for "head losses" that occur in actual flow situations. Specifically, we wrote

$$\frac{p_1}{\gamma_1} + \frac{V_1^2}{2g} + Z_1 = \frac{p_2}{\gamma_2} + \frac{V_2^2}{2g} + Z_2 + \text{losses}_{1\to 2} \tag{6.9}$$

For a horizontal pipe of constant diameter,

$$\text{losses}_{1\to 2} = \frac{p_1 - p_2}{\gamma} \tag{6.10}$$

That is, the loss in head between two sections of a horizontal pipe in which an incompressible fluid is flowing is simply the difference in the static pressures that exist at each of the sections.

When an incompressible fluid flows in a pipe and the flow is turbulent, it has been found experimentally that the head loss is a function of the length of the pipe, the diameter of the pipe, the surface roughness of the pipe wall, the velocity of the fluid, the density of the fluid, and the viscosity of the fluid. As noted earlier, the Darcy–Weisbach equation, equation (6.6), is most generally used to calculate the frictional losses in pipes. It is repeated here for convenience:

$$h_f = f\frac{L}{D}\frac{V^2}{2g} \tag{6.6}$$

Unfortunately, in the turbulent flow regime it is not possible to obtain an analytical solution for the friction factor as we just did for the case of laminar flow. Most of the usable data available for evaluating the friction factor in turbulent flow have been derived from experiments. Prior to 1933, most of these results were scattered throughout the technical literature. In 1933, J. Nikuradse published his work on pipes whose walls had been artificially roughened using glued sand grains of different diameters. He termed the diameter of the sand grains (e) the absolute roughness and the ratio of the sand grain diameter to the inside pipe diameter (e/D) the relative roughness. The results of Nikuradse's work is shown graphically in Figure 6.6. In the laminar zone (Re $\leqslant \sim 2100$) the friction factor is independent of the relative or absolute roughness. In the turbulent zone the friction factor is seen to be a function of the relative roughness and the Reynolds number, and it will be seen that as the Reynolds number increases, the friction factor becomes constant for each given value of the relative roughness and consequently independent of the Reynolds number.

In the same year, 1933, R. J. S. Pigott and E. Kemler published papers in which much of the existing data on friction in pipes were correlated and plotted in a diagram similar to Figure 6.6. This type of diagram, in which the logarithm of the friction factor is plotted as a function of the logarithm of the Reynolds number, is known as a Stanton diagram and is a very convenient portrayal of the data. In 1944, L. F. Moody published his paper on

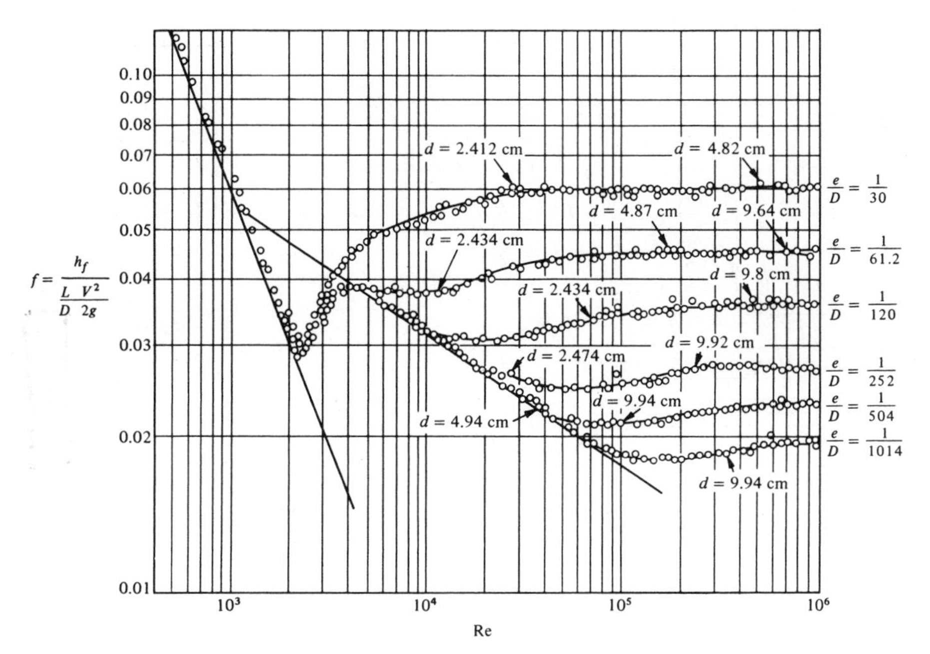

FIGURE 6.6 Nikuradse's sand-roughened pipe tests. (Reproduced by permission of the Van Nostrand Reinhold Company from *Engineering Heat Transfer* by S. T. Hsu, Litton Educational Publishing, Inc., 1963, p. 213.)

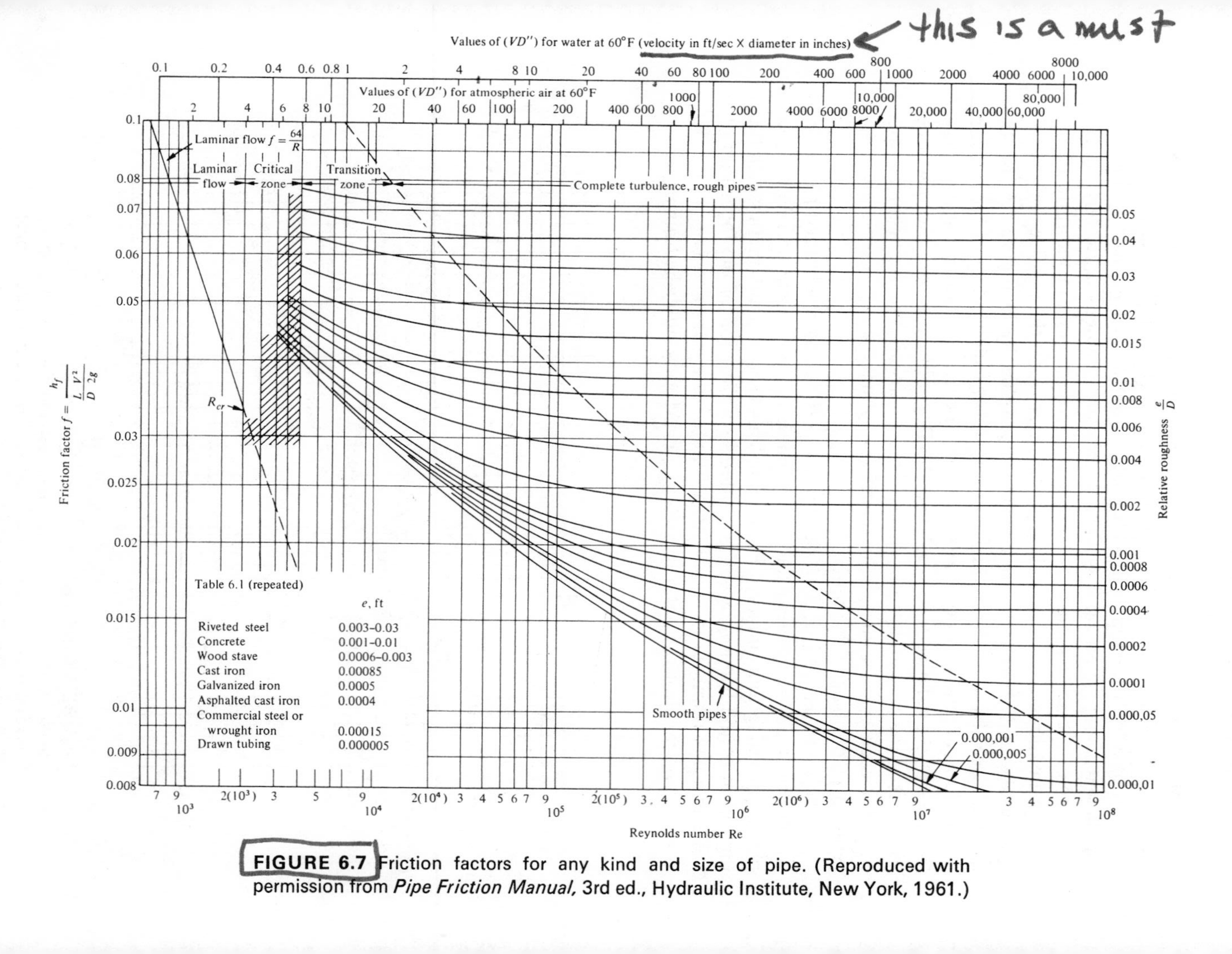

FIGURE 6.7 Friction factors for any kind and size of pipe. (Reproduced with permission from *Pipe Friction Manual,* 3rd ed., Hydraulic Institute, New York, 1961.)

friction factors in pipe flow and in this paper gave his results in the form of a Stanton diagram, which has proved to be the most widely used source of friction factors for new or clean commercial pipes. This figure is reproduced as Figure 6.7. Qualitatively, the trends in Nikuradse's data shown in Figure 6.6 and the Moody diagram shown in Figure 6.7 are in good agreement. Detailed differences exist in the critical and transition zones, as might be expected when comparing artificially roughened pipes with commercial pipes. Also, the values of the absolute roughness of pipe used by Moody and given in Table 6.1 are arbitrary and not comparable to the sand grain diameters

TABLE 6.1 Absolute Roughness of Pipes

Material	Absolute Roughness, e (ft)	Absolute Roughness, e (m)
Glass, new commercial pipe surfaces, drawn tubing (brass, copper, lead)	0.000005	1.52×10^{-6}
Commercial steel or wrought iron	0.00015	4.57×10^{-5}
Asphalted cast iron	0.0004	1.22×10^{-4}
Galvanized iron	0.0005	1.52×10^{-4}
Cast iron	0.00085	2.59×10^{-4}
Wood stave	0.0006–0.0003	1.83×10^{-4}–9.14×10^{-5}
Concrete	0.001–0.01	3.05×10^{-4}–3.05×10^{-3}
Riveted steel	0.003–0.03	9.14×10^{-4}–9.14×10^{-3}

used by Nikuradse. The solid line on the Moody diagram indicates the value of the Reynolds number at which the flow becomes independent of Reynolds number and becomes a function of e/D only. This is noted to be the zone of complete turbulence, rough pipes. Figure 6.8 gives the values of e/D for various types of pipe and also the friction factor for complete turbulence, rough pipes. It will be noted that the friction factor in the turbulent zone decreases with increasing Reynolds number until the limiting value of complete turbulence, rough pipes, is reached. The values of e/D used on the Moody diagram apply to commercial, clean pipes and are not comparable to Nikuradse's artificially roughened values. For convenience, Figure 6.9 is included, since it is a convenient presentation of both the kinematic viscosity and Reynolds number for many liquids and gases.

ILLUSTRATIVE PROBLEM 6.7

Water at 50°F is supplied from a reservoir to a 1000-ft-long, horizontal, round, concrete pipe 36 in. in diameter. Determine the frictional pressure drop in the pipe if the flow is 20,000 gal/min. Neglect minor loses.

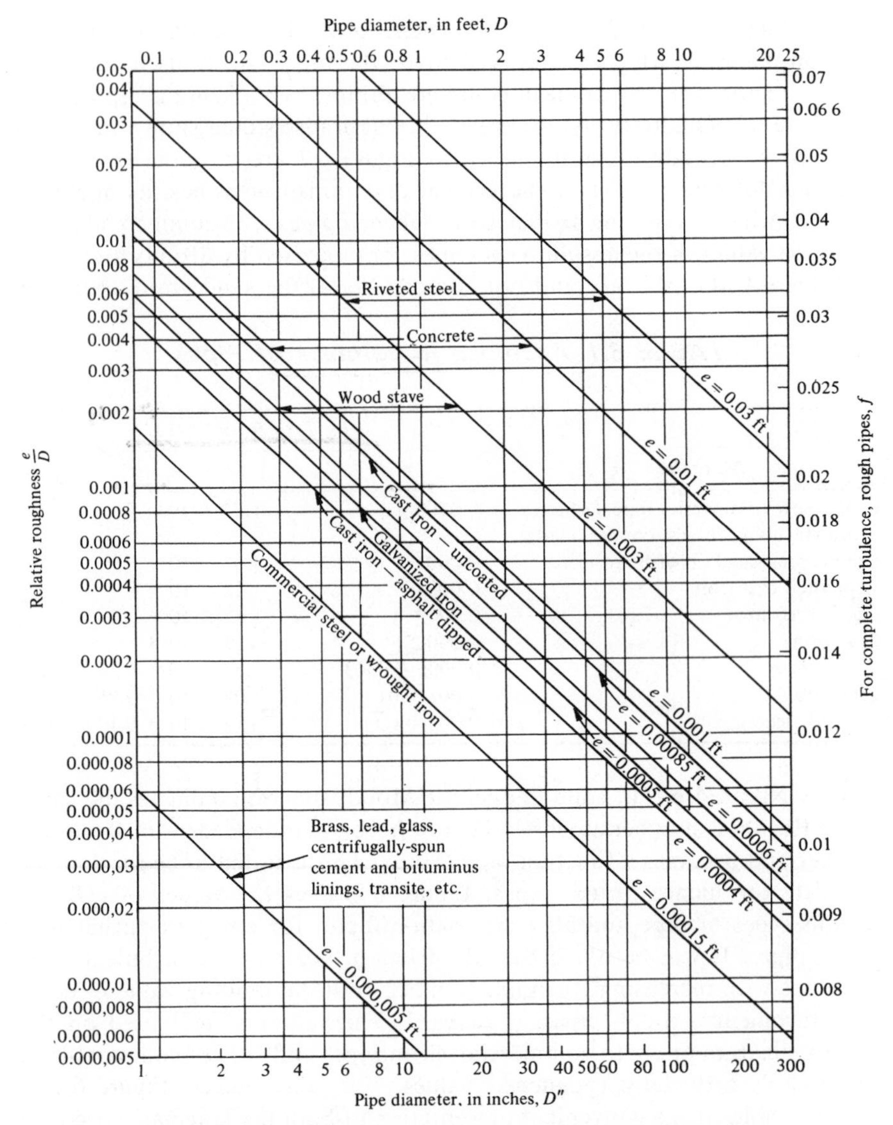

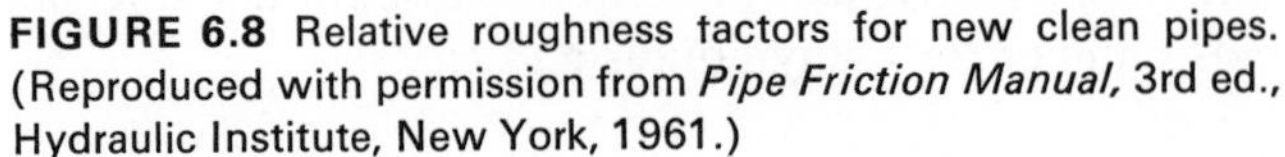

FIGURE 6.8 Relative roughness tactors for new clean pipes. (Reproduced with permission from *Pipe Friction Manual,* 3rd ed., Hydraulic Institute, New York, 1961.)

Solution

Using 231 in.^{3} in 1 gal, we obtain

$$Q = \left(\frac{20{,}000}{60}\right)\left(\frac{231}{1728}\right) = 44.6 \text{ ft}^2/\text{sec}$$

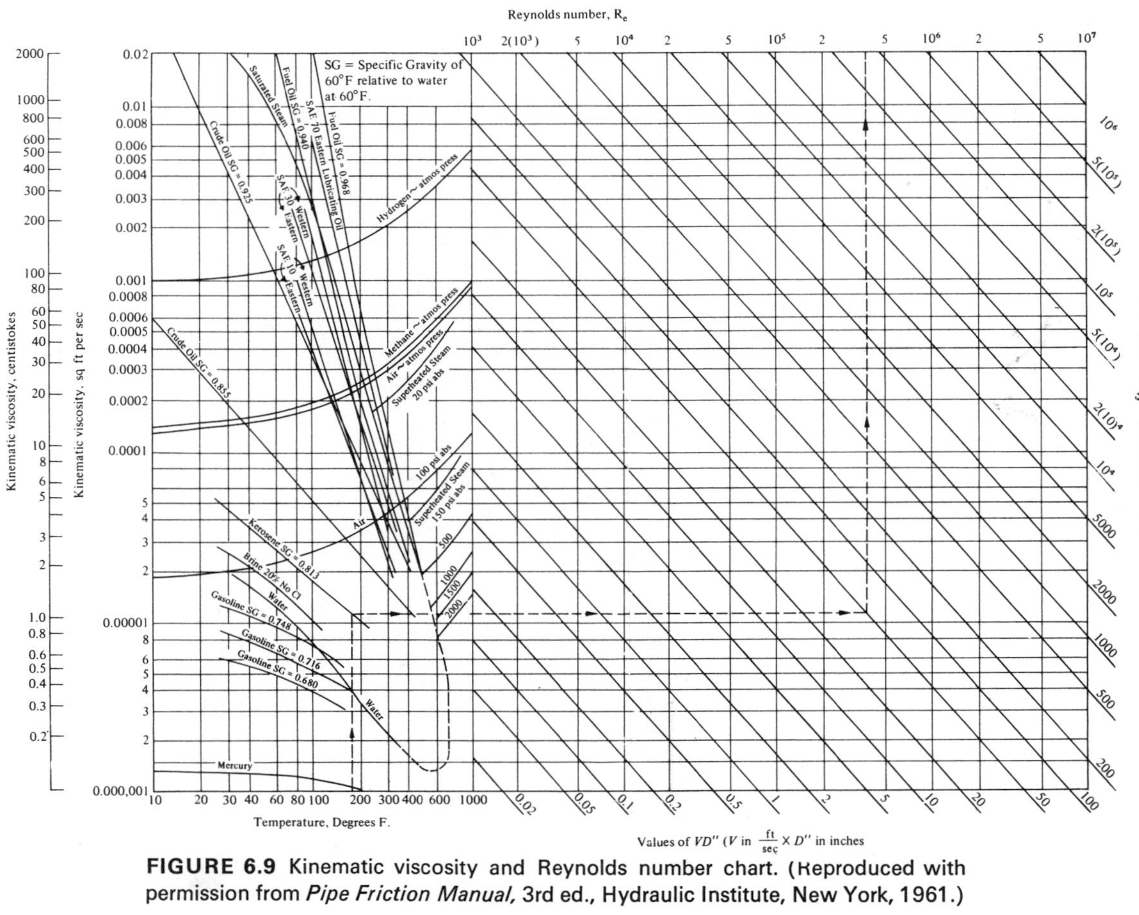

FIGURE 6.9 Kinematic viscosity and Reynolds number chart. (Reproduced with permission from *Pipe Friction Manual,* 3rd ed., Hydraulic Institute, New York, 1961.)

Since $Q = AV$,

$$V = \frac{Q}{A} = \frac{44.6}{(\pi/4)[(36^2)/144]} = 6.3 \text{ ft/sec}$$

The product of (VD) is $6.3 \times 36 = 227$. Using Figures 6.7 or 6.9, we can evaluate the Reynolds number directly:

$$\text{Re} \simeq 1.6 \times 10^6$$

The flow is therefore turbulent. The student should check the value of Re by direct calculation of $DV\rho/\mu$ or $DV\gamma/\mu g$.

From Table 6.1, e for concrete varies from 0.001 to 0.01 ft. Using the arithmetic average of 0.005,

$$\frac{e}{D} = \frac{0.005 \times 12}{36} = 0.0017$$

From Figure 6.7, $f = 0.023$.

It will be noted that at the top of the friction factor chart (Figure 6.7), there is an auxiliary scale giving values of (VD) for water and air at 60°F, where V is the velocity in ft/sec and D is the diameter in inches. If we enter this chart at (VD) of 227, we can go directly to e/D of 0.0017 and read f of 0.023.

Continuing the problem, the pressue drop due to friction is

$$h_f = f\frac{L}{D}\frac{V^2}{2g} = 0.023\left(\frac{1000}{36/12}\right)\frac{(6.3)^2}{2g} = 4.73 \text{ ft of water}$$

or

$$h_f = 4.73 \times \frac{62.4}{144} = 2.05 \text{ psi}$$

ILLUSTRATIVE PROBLEM 6.8

If the pipe of Illustrative Problem 6.1 is steel and 50 m long, what is the pressure drop?

Solution

In Illustrative Problem 6.1 we found that Re = 226 900. For commercial steel pipe, $e = 4.57 \times 10^{-5}$ m. Therefore,

$$\frac{e}{D} = \frac{4.57 \times 10^{-5}}{0.050} = 0.0009$$

From Figure 6.7, $f = 0.02$ and

$$h_f = f\frac{L}{D}\frac{V^2}{2g} = 0.02\left(\frac{50}{0.05}\right)\frac{(3)^2}{2 \times 9.81} = 9.17 \text{ m} = 9.17 \text{ N}\cdot\text{m/N}$$

or

$$h_f = 9.17 \text{ m} \times 9810 \frac{\text{N}}{\text{m}^3} = 90 \text{ kPa}$$

ILLUSTRATIVE PROBLEM 6.9

Water flows in a steel pipe whose inside diameter is 2.067 in. If the pipe is 200 ft long and a pump can furnish a head of 50 ft to overcome friction losses, determine the flow in the pipe.

Solution

In this problem it is not possible to obtain a direct solution, and it will be necessary to use successive iterations (trial and error). The procedure is to assume a value of friction factor and to solve the problem based upon this assumption. The friction factor is then evaluated using the solution based upon the initial assumption. If the calculated and assumed values agree, this is the solution. If they do not agree, a new assumption is made and the procedure is repeated.

Assume that $f = 0.02$. Therefore,

$$50 = 0.02\left(\frac{200}{2.067/12}\right)\frac{V^2}{2g}$$

and

$$V = 11.8 \text{ ft/sec}$$

$$(VD) = 11.8 \times 2.067 = 24.4$$

From Figure 6.7, Re $\simeq 2.0 \times 10^5$. From Table 6.1, $e = 0.00015$ ft. Therefore,

$$\frac{e}{D} = \frac{0.00015}{2.067/12} = 0.00087$$

and from Figure 6.7, $f = 0.021$.

The calculated value of f based upon an assumed value of f of 0.020 is

0.021. This is satisfactory and the value of 11.8 ft/sec will be used. Therefore,

$$Q = AV = 11.8 \times \frac{\pi}{4}\frac{(2.067)^2}{144} = 0.275\ \text{ft}^3/\text{sec}$$

$$\frac{0.275 \times 60 \times 1728}{231} = 123\ \text{gal/min}$$

While Figure 6.7 is convenient when solving problems in pipe friction, it is often desirable to have an explicit formula expressing friction factor as a function of the relative roughness and the Reynolds number. In 1947, Moody published such a formulation, which is sufficiently accurate for most engineering purposes, namely,

$$f \simeq 0.0055\left[1 + \left(20{,}000\frac{e}{D} + \frac{10^6}{\text{Re}}\right)^{1/3}\right] \tag{6.11}$$

Equation (6.11) is useful when problems involving pipe friction are programmed on a digital computer since an explicit solution can be obtained. It is sometimes more convenient to write it in the following form:

$$f \cong 0.0055 + 0.0055\left(20{,}000\frac{e}{D} + \frac{10^6}{\text{Re}}\right)^{1/3} \tag{6.12}$$

ILLUSTRATIVE PROBLEM 6.10

Water flows in a commercial steel pipe whose inside diameter is 75 mm. If the pipe is 100 m long and a pump can furnish a head of 20 m to overcome friction losses, determine the flow in the pipe.

Solution

As in Illustrative Problem 6.9, we cannot get an explicit solution to this problem. We will assume a friction factor of 0.02. Therefore,

$$20 = 0.02 \times \frac{100}{0.075}\frac{V^2}{2 \times 9.81}$$

and

$$V = 3.84\ \text{m/s}$$

While we could fundamentally obtain Re, it is simpler to first obtain (VD).

Therefore,

$$(VD) = (3.84 \times 3.281)\left(\frac{75}{25.4}\right) = 37.2$$

$$\frac{e}{D} = \frac{4.57 \times 10^{-5}}{0.075} = 0.000\,61$$

Entering Figure 6.7, we read $f = 0.019$. If we now use $f = 0.019$ as our second trial, $V = 3.94$ m/s and $(VD) = 38.1$. Entering Figure 6.7, we read $f = 0.019$. Using this value, $V = 3.94$ m/s and

$$Q = AV = \frac{\pi}{4}(0.075)^2 \times (3.94) = 1.74 \times 10^{-2}\ \text{m}^3/\text{s}$$

The problem of evaluating friction factors for old pipes and allowing for the deterioration of new pipes is particularly difficult since pipe deterioration is a function of the fluid and pipe chemical properties. Thus it is recommended that the Moody curve and the data presented in this chapter should not be used when estimating the pressure losses in old pipes or to allow for the aging of new pipes. If at all possible, experience with an existing installation will yield the best data for estimating the pressure drop in old or aging pipes. Data in the engineering literature are not satisfactory and may yield results that are in large error when applied to a specific situation.

6.7 OTHER LOSSES

In addition to the pressure drop incurred due to friction as a fluid flows in a pipe, other losses in pressure occur. These other losses occur at the entrance to a pipe, at abrupt changes in section in a pipe, at valves, at fittings, at bends, and at the pipe exit; they are often called minor losses. The data available in the literature on the losses in valves and fittings have been presented in two different forms. The first form expresses the fact that these losses can be expressed in terms of velocity heads of fluid flowing; that is,

$$h = K\frac{V^2}{2g} \tag{6.13}$$

The value of K increases with increasing roughness and decreases with increasing Reynolds numbers but primarily depends on the geometric shape of the valve or fitting. Figure 6.10 gives the resistance coefficients for pipe fittings and Figure 6.11 gives the resistance coefficients for valves, couplings, and unions. To appreciate that these values are approximate and to appreciate

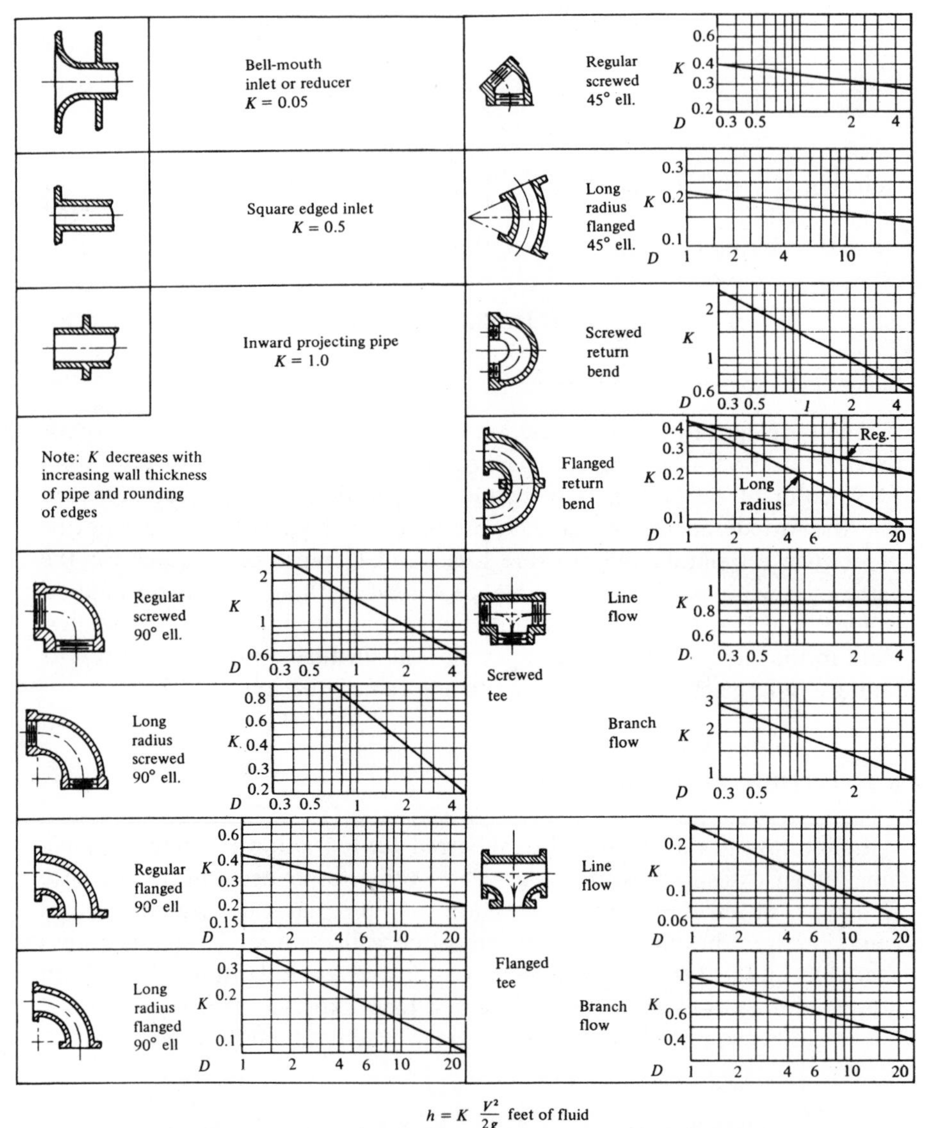

$$h = K \frac{V^2}{2g} \text{ feet of fluid}$$

FIGURE 6.10 Resistance coefficients for valves and fittings. The valve of D is nominal iron pipe size. For velocities below 15 ft/sec, check valves and foot valves will be only partially open and will exhibit higher values of K than that shown. (Reproduced with permission from *Pipe Friction Manual,* 3rd ed., Hydraulic Institute, New York, 1961.)

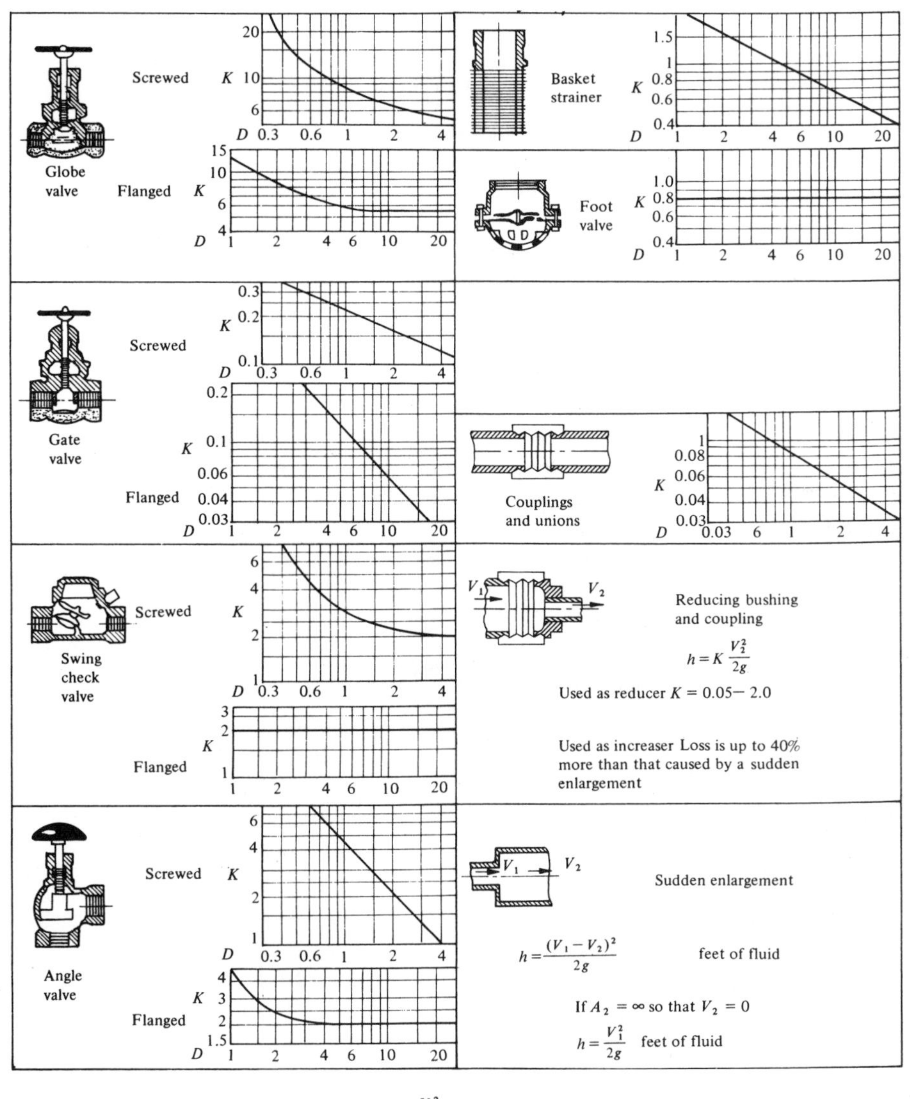

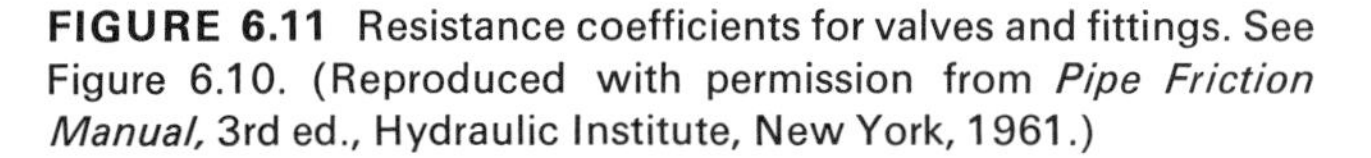

$$h = K\frac{V^2}{2g} \text{ Feet of fluid}$$

FIGURE 6.11 Resistance coefficients for valves and fittings. See Figure 6.10. (Reproduced with permission from *Pipe Friction Manual,* 3rd ed., Hydraulic Institute, New York, 1961.)

the wide variations that are found in specific fittings, Table 6.2 has been included. It is apparent from this table that, at best, the calculation of these losses is approximate and that the values given in Figures 6.10 and 6.11 for the resistance coefficients are just good approximations.

TABLE 6.2 Resistance Coefficients for Valves and Fittings†

Approximate Range of Variation for K

Fitting		**Range of Variation**
90-deg elbow	Regular screwed	±20% above 2-in. size
	Regular screwed	±40% below 2-in. size
	Long radius, screwed	±25%
	Regular flanged	±35%
	Long radius, flanged	±30%
45-deg elbow	Regular screwed	±10%
	Long radius, flange	±10%
180-deg bend	Regular screwed	±25%
	Regular flanged	±35%
	Long radius, flanged	±30%
Tee	Screwed, line, or branch flow	±25%
	Flanged, line, or branch flow	±35%
Globe valve	Screwed	±25%
	Flanged	±25%
Gate valve	Screwed	±25%
	Flanged	±50%
Check valve	Screwed	±30%
	Flanged	+200% −80%
Sleeve check valve		Multiply flanged values by 0.2–0.5
Tilting check valve		Multiply flanged values by 0.13–0.19
Drainage gate check		Multiply flanged values by 0.03–0.07
Angle valve	Screwed	±20%
	Flanged	±50%
Basket strainer		±50%
Foot valve		±50%
Couplings		±50%
Unions		±50%
Reducers		±50%

†Reproduced with permission from *Pipe Friction Manual*, 3rd ed., Hydraulic Institute, New York, 1961.

The second method of presenting these losses is to express them as equivalent lengths of pipe L_{eq} that have the same head loss for the same discharge. Therefore,

$$f\frac{L_{eq}}{D}\frac{V^2}{2g} = K\frac{V^2}{2g} \tag{6.14}$$

and

$$L_{eq} = \frac{KD}{f} \tag{6.15}$$

These minor losses are in excess of the friction loss produced in a straight pipe whose length is equal to the length of the axis of the pipe (see Table 6.3).

ILLUSTRATIVE PROBLEM 6.11

A commercial steel pipeline has an inside diameter of 6 in. and the flow is completely turbulent. If the sum of all the loss coefficients in the pipe (K) is 12, determine the equivalent length that must be added to the actual pipe length in order to cause the same resistance to flow.

Solution

For steel pipe, $e = 0.00015$ and $e/D = 0.00015/(6/12) = 0.00030$. For completely turbulent flow, $f = 0.015$. Therefore,

$$L_{eq} = \frac{12 \times (6/12)}{0.015} = 400 \text{ ft to be added}$$

In addition to valves and fittings, most pipelines have abrupt changes at entrances, exits, reducers, increasers, diffusers, and bends. One of these losses, the case of loss at a sudden enlargement, can be calculated, with the results agreeing reasonably well with experiment. When this is done, the head loss is found to be

$$h = \frac{(V_1 - V_2)^2}{2g} \tag{6.16}$$

where V_1 is the upstream velocity and V_2 the downstream velocity. Such a condition exists when a pipe exits into a large tank or reservoir. For this case equation (6.16) gives a loss equal to one velocity head in the pipe. Alternatively, we can argue that the kinetic energy in the pipe is converted to internal

TABLE 6.3 Equivalent Length of New Straight Pipe for Valves and Fittings for Turbulent Flow Only†

Fittings			Pipe Size																				
			$\frac{1}{4}$	$\frac{3}{8}$	$\frac{1}{2}$	$\frac{3}{4}$	1	$1\frac{1}{4}$	$1\frac{1}{2}$	2	$2\frac{1}{2}$	3	4	5	6	8	10	12	14	16	18	20	21
Regular 90° ell.	Screwed	Steel	2.3	3.1	3.6	4.4	5.2	6.6	7.4	8.5	9.3	11	13										
		Castiron										9.0	11										
	Flanged	Steel			0.92	1.2	1.6	2.1	2.4	3.1	3.6	4.4	5.9	7.3	8.9	12	14	17	18	21	23	25	30
		Castiron										3.6	4.8		7.2	9.8	12	15	17	19	22	24	28
Long radius 90° ell.	Screwed	Steel	1.5	2.0	2.2	2.3	2.7	3.2	3.4	3.6	3.6	4.0	4.6										
		Castiron										3.3	3.7										
	Flanged	Steel			1.1	1.3	1.6	2.0	2.3	2.7	2.9	3.4	4.2	5.0	5.7	7.0	8.0	9.0	9.4	10	11	12	14
		Castiron										2.8	3.4		4.7	5.7	6.8	7.8	8.6	9.6	11	11	13
Regular 45° ell.	Screwed	Steel	0.34	0.52	0.71	0.92	1.3	1.7	2.1	2.7	3.2	4.0	5.5										
		Castiron										3.3	4.5										
	Flanged	Steel			0.45	0.59	0.81	1.1	1.3	1.7	2.0	2.6	3.5	4.5	5.6	7.7	9.0	11	13	15	16	18	22
		Castiron										2.1	2.9		4.5	6.3	8.1	9.7	12	13	15	17	20
Teeline flow	Screwed	Steel	0.79	1.2	1.7	2.4	3.2	4.6	5.6	7.7	9.3	12	17										
		Castiron										9.9	14										
	Flanged	Steel			0.69	0.82	1.0	1.3	1.5	1.8	1.9	2.2	2.8	3.3	3.8	4.7	5.2	6.0	6.4	7.2	7.6	8.2	9.6
		Castiron										1.9	2.2		3.1	3.9	4.6	5.2	5.9	6.5	7.2	7.7	8.8
Teebranch flow	Screwed	Steel	2.4	3.5	4.2	5.3	6.6	8.7	9.9	12	13	17	21										
		Castiron										14	17										
	Flanged	Steel			2.0	2.6	3.3	4.4	5.2	6.6	7.5	9.4	12	15	18	24	30	34	37	43	47	52	62
		Castiron										7.7	10		15	20	25	30	35	39	44	49	57
180° return bend	Screwed	Steel	2.3	3.1	3.6	4.4	5.2	6.6	7.4	8.5	9.3	11	13										
		Castiron										9.0	11										
	Regular flanged	Steel			0.92	1.2	1.6	2.1	2.4	3.1	3.6	4.4	5.9	7.3	8.9	12	14	17	18	21	23	25	30
		Castiron										3.6	4.8		7.2	9.8	12	15	17	19	22	24	28
	Long radius flanged	Steel			1.1	1.3	1.6	2.0	2.3	2.7	2.9	3.4	4.2	5.0	5.7	7.0	8.0	9.0	9.4	10	11	12	14
		Castiron										2.8	3.4		4.7	5.7	6.8	7.8	8.6	9.6	11	11	13
Globe valve	Screwed	Steel	21	22	22	24	29	37	42	54	62	79	110										
		Castiron										65	86										
	Flanged	Steel			38	40	45	54	59	70	77	94	120	150	190	260	310	390					
		Castiron										77	99		150	210	270	330					
Gate valve	Screwed	Steel	0.32	0.45	0.56	0.67	0.84	1.1	1.2	1.5	1.7	1.9	2.5										
		Castiron										1.6	2.0										
	Flanged	Steel								2.6	2.7	2.8	2.9	3.1	3.2	3.2	3.2	3.2	3.2	3.2	3.2	3.2	3.2
		Castiron										2.3	2.4		2.6	2.7	2.8	2.9	2.9	3.0	3.0	3.0	3.0
Angle valve	Screwed	Steel	12.8	15	15	15	17	18	18	18	18	18	18										
		Castiron										15	15										
	Flanged	Steel			15	15	17	18	18	21	22	28	38	50	63	90	120	140	160	190	210	240	300
		Castiron										23	31		52	74	98	120	150	170	200	230	280
Swing check valve	Screwed	Steel	7.2	7.3	8.0	8.8	11	13	15	19	22	27	38										
		Castiron										22	31										
	Flanged	Steel			3.8	5.3	7.2	10	12	17	21	27	38	50	63	90	120	140					
		Castiron										22	31		52	74	98	120					
Coupling or union	Screwed	Steel	0.14	0.18	0.21	0.24	0.29	0.36	0.39	0.45	0.47	0.53	0.65										
		Castiron										0.44	0.52										
	Bell mouth inlet	Steel	0.04	0.07	0.10	0.13	0.18	0.26	0.31	0.43	0.52	0.67	0.95	1.3	1.6	2.3	2.9	3.5	4.0	4.7	5.3	6.1	7.6
		Castiron										0.55	0.77		1.3	1.9	2.4	3.0	3.6	4.3	5.0	5.7	7.0
	Square mouth inlet	Steel	0.44	0.68	0.96	1.3	1.8	2.6	3.1	4.3	5.2	6.7	9.5	13	16	23	29	35	40	47	53	61	76
		Castiron										5.5	7.7		13	19	24	30	36	43	50	57	70
	Reentrant pipe	Steel	0.88	1.4	1.9	2.6	3.6	5.1	6.2	8.5	10	13	19	25	32	45	58	70	80	95	110	120	150
		Castiron										11	15		26	37	49	61	73	86	100	110	140
	Sudden enlargement		$h = \frac{(V_1 - V_2)^3}{2g}$ feet of liquid; if $V_2 = 0$, $h = \frac{V_1^2}{2g}$ feet of liquid																				

†Reproduced with permission from *Pipe Friction Manual*, 3rd ed., Hydraulic Institute, New York, 1961.

energy and does not appear again as a pressure head. In this sense one velocity head is "lost" at the exit of the pipe. For a sudden contraction Figure 6.12 gives the loss coefficient. It will be noted that if this curve is applied to the sharp entrance of a pipe from a large reservoir, the loss coefficient is half a velocity head. As will be also seen from Figure 6.10, this loss can be greatly reduced by rounding off the entrance to the pipe.

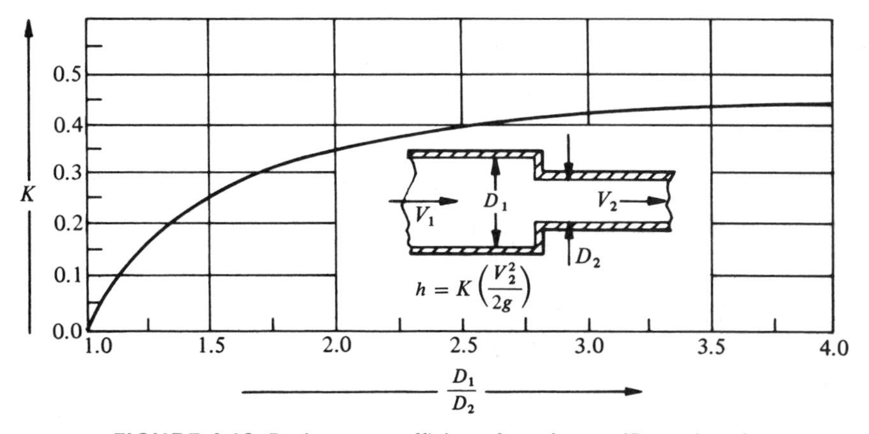

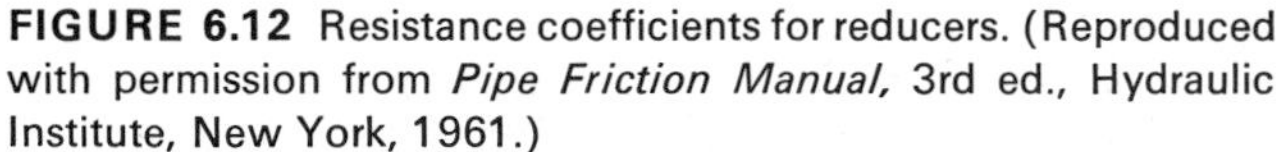

FIGURE 6.12 Resistance coefficients for reducers. (Reproduced with permission from *Pipe Friction Manual,* 3rd ed., Hydraulic Institute, New York, 1961.)

In addition to the foregoing, Figures 6.13 and 6.14 can be used to obtain the resistance coefficient, K, for increasers, diffusers, and bends. It should be noted that many of these losses can be decreased dramatically by good design techniques. In large air-handling systems, vanes are frequently used to decrease the turbulence that occurs in elbows and bends. Figure 6.15 shows the effect of using vanes in such a system. These vanes serve to decrease the turbulence and concurrently they also reduce the noise caused by the air flowing in the ducts. The same principle of good design also applies to fluid systems.

ILLUSTRATIVE PROBLEM 6.12

A large open reservoir is connected to a steel tube whose inside diameter is 2 in. A screwed gate valve is placed in the tube near its outlet, and the tube is 50 ft long. If the level of water in the tank is 50 ft above the outlet of the tube, determine the rate at which the water flows when the valve is opened. Assume that the loss coefficients for bends and elbows add up to 3.0.

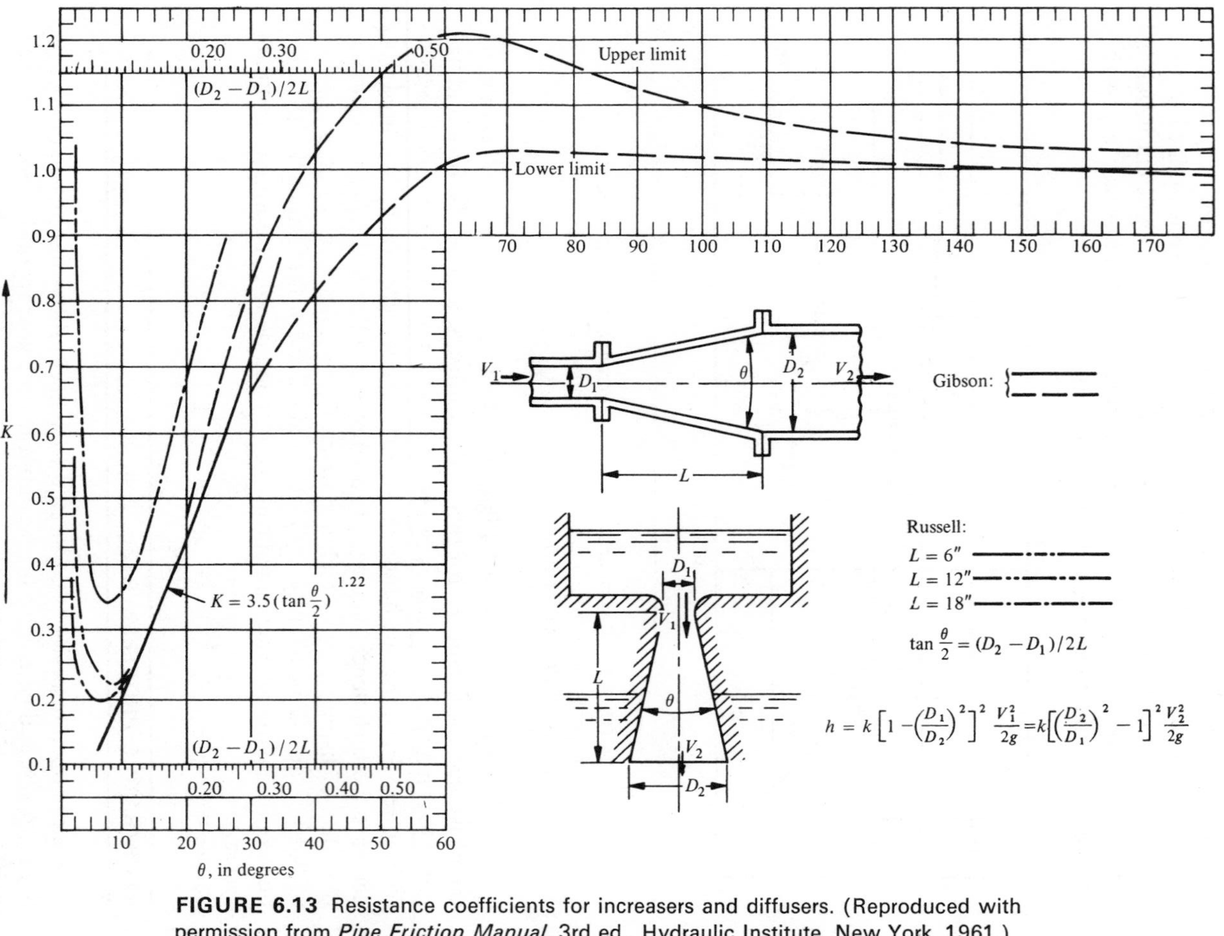

FIGURE 6.13 Resistance coefficients for increasers and diffusers. (Reproduced with permission from *Pipe Friction Manual,* 3rd ed., Hydraulic Institute, New York, 1961.)

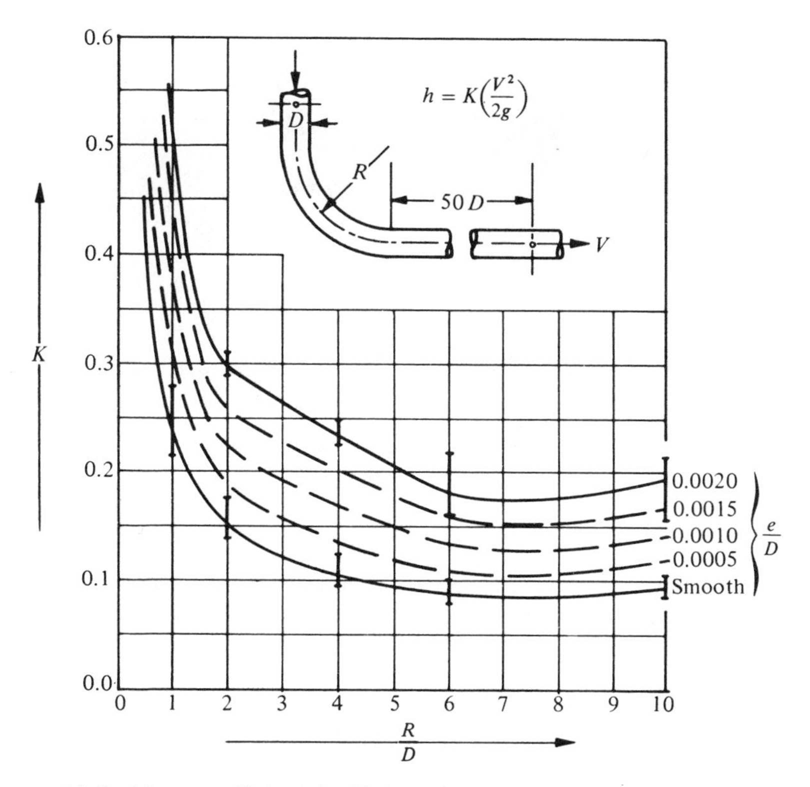

(a) Resistance coefficients for 90 degree bends of uniform diameter.

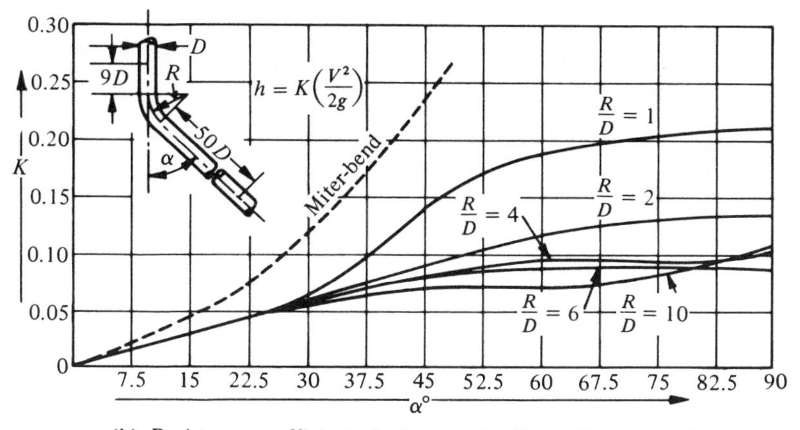

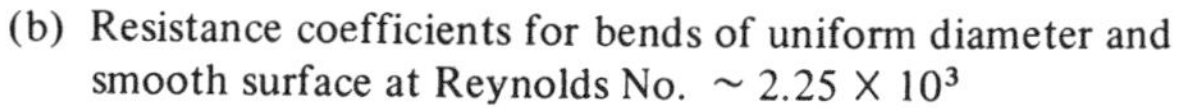
(b) Resistance coefficients for bends of uniform diameter and smooth surface at Reynolds No. ~ 2.25×10^3

FIGURE 6.14 (a) Resistance coefficients for 90° bends of uniform diameter. (b) Resistance coefficients for bends of uniform diameter and smooth surface at Reynolds number 2.25×10^3. (Reproduced with permission from *Pipe Friction Manual*, 3rd ed., Hydraulic Institute, New York, 1961.)

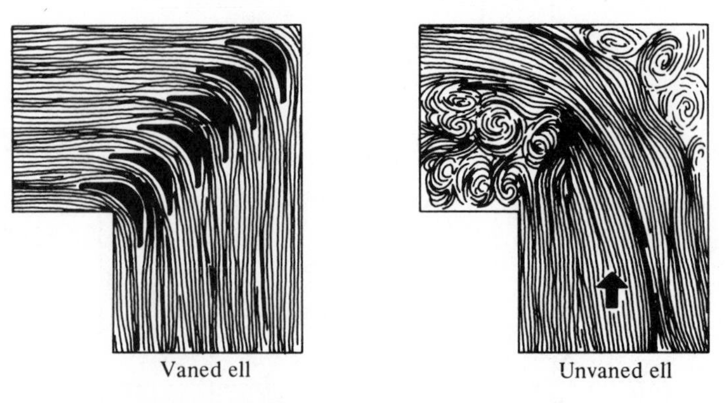

FIGURE 6.15 Effect of vanes on turbulence.

Solution

Writing a Bernoulli equation between sections 1 and 2 in Figure 6.16,

$$\frac{p_1}{\gamma} + \frac{V_1^2}{2g} + Z_1 = \frac{p_2}{\gamma} + \frac{V_2^2}{2g} + Z_2 + \text{losses}_{1\to2}$$

Since we have atmospheric pressure at sections 1 and 2 and V_1 can be taken to be zero,

$$\frac{V_2^2}{2g} + \text{losses}_{1\to2} = Z_1 - Z_2$$

where we have already termed $V_2^2/2g$ to be the exit loss. Inserting an entrance

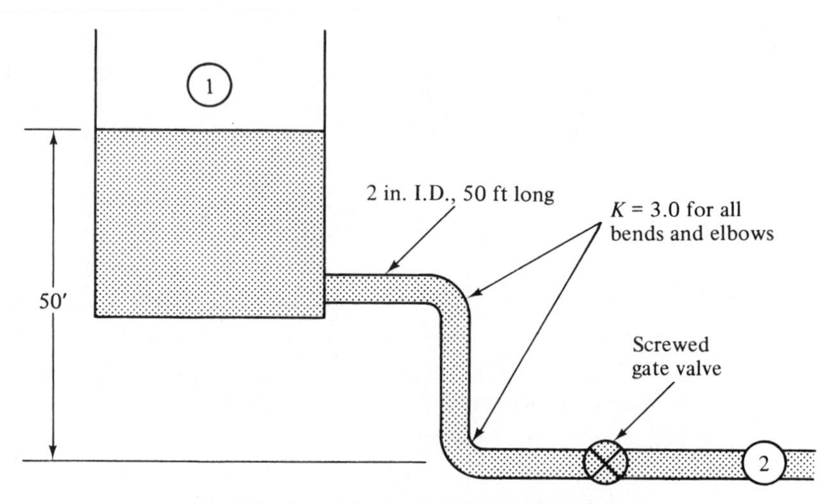

FIGURE 6.16 Illustrative Problem 6.12.

loss, a loss term for the valve, and a friction loss term as well as the other losses, we have

$$\frac{V_2^2}{2g}\left(1 + f\frac{L}{D} + 3.0 + 0.5 + 0.17\right) = Z_1 - Z_2 = 50$$

In general, f is a function of the velocity V, and an explicit solution cannot be obtained for this problem. However, let us assume that the flow is fully turbulent:

$$\frac{e}{D} = \frac{0.00015}{2/12} = 0.0009$$

and f for fully developed turbulent flow is 0.019. Therefore,

$$\frac{V_2^2}{2g}\left(1 + \frac{0.019 \times 50}{2/12} + 3.0 + 0.5 + 0.17\right) = 50$$

and

$$V_2 = 17.7 \text{ ft/sec}$$

At this point it is necessary to check the Reynolds number to verify the assumption of fully turbulent flow. Since $(VD) = 17.7 \times 2 = 35.4$, $\text{Re} \simeq 2.0 \times 10^5$ from Figure 6.13. From Figure 6.7 we have $f = 0.021$. As a second trial assume that $f = 0.021$, and solving we have $V_2 = 17.1$ ft/sec. For this velocity $\text{Re} \simeq (17.1/17.7) \times 2 \times 10^5 = 1.9 \times 10^5$ and $f = 0.021$. Therefore, the solution is that water flows at the rate of 17.1 ft/sec in the pipe.

ILLUSTRATIVE PROBLEM 6.13

Air flows in a $\frac{1}{2}$-in. schedule 80 steel pipe (i.d. = 0.546 in.) from a large storage tank. The piping system will contain twelve 90-deg regular screwed elbows, two screwed gate valves, one screwed globe valve, three screwed tees with line flow, and 200 ft of straight pipe. If 11.3 lb/min flows and the air has a specific weight of 5.36 lb/ft³, determine the pressure drop in the system. Assume that the tank pressure is 1000 psig, that air temperature is 1000°F, and that the pressure drop in the pipe is less than 10% of the tank pressure. If the pressure drop exceeds 10% of the tank pressure, it is incorrect to apply the equations for incompressible flow to this flow situation.

Solution

The velocity in the pipe is

$$\dot{W} = \gamma A V; \qquad \frac{11.3}{60} = 5.36 \times \frac{\pi}{4}\frac{(0.546)^2}{144} \times V$$

and

$$V = \frac{11.3}{60}\left(\frac{4}{\pi}\right)\frac{144}{(0.546)^2 5.36} = 21.5 \text{ ft/sec}$$

$$(VD) = 21.5 \times 0.546 = 11.8$$

From Figure 6.9, Re $\simeq 8 \times 10^3$ for air at 100 psia and 1000°F. Since the viscosity of a gas is primarily dependent on its temperature, the viscosity at 1014.7 psia and 1000°F will be the same as at 100 psia and 1000°F. The only term in the Reynolds number that changes due to pressure is the specific weight. Therefore, the specific weight (and consequently the Reynolds number) at 1014.7 psia will exceed the value at 100 psia by the ratio of 1014.7 : 100. For the conditions of this problem,

$$\text{Re} \simeq 8 \times 10^3 \times \frac{1014.7}{100} \simeq 8 \times 10^4$$

For commercial steel pipe,

$$e = 0.00015 \text{ ft}$$

Therefore,

$$\frac{e}{D} = \frac{0.00015}{0.546/12} = 0.0033$$

From Figure 6.7, $f = 0.028$.

Expressing the friction loss in terms of velocity heads yields

$$K_f = f\frac{L}{D}$$

or

$$K_f = 0.028\frac{200}{0.546/12} = 123.0$$

For twelve 90° regular screwed elbows in a $\frac{1}{2}$-in. pipe size, $K = 2$ for each elbow.

For two screwed gate valves, $K = 0.32$ per valve.

For one screwed globe valve, $K = 14$.

For three screwed tees with line flow, $K = 0.9$/tee.

For entrance, $K = 0.5$.

For exit, $K = 1$.

Adding all of these terms, we have

$$K = 123.0 + 12(2) + 2(0.32) + 14 + 3(0.9) + 1 + 0.5 = 165.8$$

Therefore,

$$h_f = 165.8\left[\frac{(21.5)^2}{2g}\right]\frac{5.36}{144} = 44.2 \text{ psi}$$

Since this figure is within the 10% limit the assumption of incompressible flow is valid for this problem.

Before continuing, two cautions must be noted at this time. The tables and charts in this chapter for the calculation of friction and minor losses are based upon nominal IPS (iron pipe size). In all of the illustrative problems, we have used the actual inside diameter, thereby introducing a small error. In most instances the error introduced is well within the limits of accuracy of the data. For certain cases (particularly small pipe sizes) this may not be true, and the data given in Appendix B for pipe sizes should be used.

The second caution concerns the use of the Bernoulli equation. Because of the nonuniform velocity distribution across any section of a pipe, a correction should be made in the velocity head terms. Again, in most cases of concern to engineers this correction can be neglected. However, in laminar flow the correction factor is 2. For turblent flow it varies from 1.01 to about 1.10 and is usually neglected, except for very precise work.

6.8 NONCIRCULAR PIPE SECTIONS

The problem of determining the pressure drop in pipes having noncircular cross sections would at first appear to be quite difficult. However, if the shear stress at the wall of the noncircular pipe is equated to the shear stress at the wall of an equivalent circular pipe, we can obtain an "equivalent" diameter for the noncircular pipe. The equivalent diameter D_{eq} is given by equation (6.17),

$$D_{eq} = \frac{4A}{P} \tag{6.17}$$

where A is the cross-sectional flow area and P is the wetted perimeter (i.e., the perimeter in contact with the fluid). The calculation of pressure drop in a noncircular pipe involves the calculation of the equivalent diameter and using it to obtain the friction factor. Care should be exercised when using equation (6.17) for shapes that are widely different from being nearly circular. Also, equation (6.17) gives better results when applied to turbulent flow situations, but if used for laminar flow situations, large errors can be introduced.

ILLUSTRATIVE PROBLEM 6.14

Three pipes, one in the shape of a square of side L, another an equilateral triangle of side L, and the third a rectangle whose width is half of its depth, are to be used in a flow system. Determine the equivalent diameter of each of these pipes.

Solution

1. *Square:*

$$A = L^2 \qquad P = 4L$$

$$D_{eq} = \frac{4A}{P} = \frac{4(L^2)}{4L} = L$$

The equivalent diameter of a square is simply the dimension of any side.

2. *Equilateral triangle:*

$$A = \frac{\sqrt{3}}{4}L^2 \qquad P = 3L$$

$$D_{eq} = \frac{4A}{P} = \frac{4\sqrt{3}\,L^2}{4(3L)} = \frac{L}{\sqrt{3}}$$

3. *Rectangle:*

$$A = \frac{L^2}{2} \qquad P = \frac{L}{2} + \frac{L}{2} + L + L = 3L$$

$$D_{eq} = \frac{4A}{P} = \frac{4(L^2/2)}{3L} = \frac{2}{3}L$$

6.9 INTERSECTING PIPES—PARALLEL FLOW

When pipes are connected in parallel, as shown in Figure 6.17, we can state that the sum of the flows in the three branches must equal the total flow in the main, and loss in pressure between sections 1 and 2 must be the same regardless of the path taken. It is interesting to note the analogy that this problem has with a direct-current network in which the circuit elements are resistors connected in parallel. We can compare the current in the electrical circuit with the quantity of fluid flowing in the pipes; the resistance (in ohms) of the electrical circuits has its counterpart in the flow resistance in the pipes, and

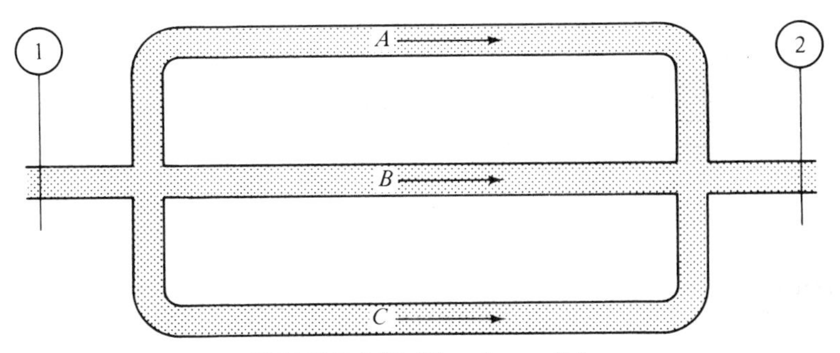

FIGURE 6.17 Pipes in parallel.

the potential drop (voltage) in the electrical circuit is analogous to the pressure drop in the pipe. This type of electrical analogy has provided the basis for the solution of many fluid flow problems on analog computers.

If we consider any of the branches, such as branch A between points 1 and 2 in Figure 6.17, we can write the following:

$$Q_A = A_A V_A \tag{6.18}$$

and

$$V = K'_A \sqrt{h} \tag{6.19}$$

where h is the loss of head between sections 1 and 2, K' is $\sqrt{2g/K_{eq}}$, and K_{eq} is the sum of all the loss coefficients in branch A. For each of the other branches, we can write similar relations. Thus we can write for the flow in the main, Q,

$$Q = Q_A + Q_B + Q_C$$

or

$$Q = (A_A K'_A + A_B K'_B + A_C K'_C)\sqrt{h} \tag{6.20}$$

Equation (6.20) can be used to obtain either the flow in the main or the flow in any of the branches. The following example will illustrate this application.

ILLUSTRATIVE PROBLEM 6.15

A steel horizontal branch piping system is as shown in Figure 6.18. Neglect minor losses and determine the pressure drop from 1 to 2 if water at 68°F is flowing into the system at section 1 at the rate of 750 gal/min.

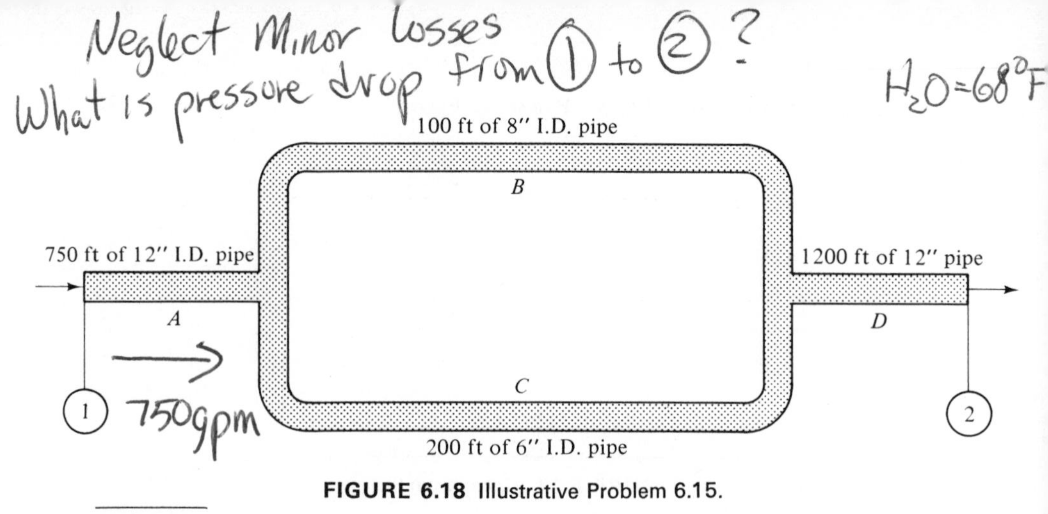

FIGURE 6.18 Illustrative Problem 6.15.

Solution

The pressure drop in branches A and D can be determined explicitly as follows: For branch A,

$$750 \text{ gal/min} \times \frac{231}{1728} = 100.5 \text{ ft}^3/\text{min}$$

$$\text{flow area} = \frac{\pi}{4}(1)^2 = 0.7854 \text{ ft}^2$$

$$\text{velocity} = \frac{Q}{A} = \frac{100.5}{60 \times 0.7854} = 2.13 \text{ ft/sec}$$

$$(VD) = 2.13 \times 12 = 25.5$$

$$e = 0.00015 \qquad \frac{e}{D} = 0.00015$$

From Figure 6.7,

$$\text{Re} = 1.7 \times 10^5 \qquad f = 0.017$$

$$\Delta p_f = f\frac{L}{D}\frac{V^2}{2g}\frac{\gamma}{144} = 0.017 \times \frac{750}{1} \times \frac{2.13^2}{2g} \times \frac{62.4}{144} = 0.39 \text{ psi}$$

For branch D, all calculations are the same as for branch A with only the length changed. Therefore,

$$\Delta p_f = 0.39 \times \frac{1200}{750} = 0.62 \text{ psi}$$

Branches B and C cannot be solved explicitly since the quantity flowing in each branch is unknown. However,

$$Q_B + Q_C = 750 \text{ gal/min}$$

Thus,

$$Q_B = 750 - Q_C \text{ gal/min}$$

$$Q_B = (750 - Q_C)\frac{231}{1728}\left(\frac{1}{60}\right) \text{ ft}^3/\text{sec}$$

and

$$Q_B = 1.67 - 0.00223 Q_C \text{ ft}^3/\text{sec}$$

However,

$$V_B = \frac{Q_B}{A_B} = \frac{Q_B}{(\pi/4)[(8/12)^2]} = 2.87 Q_B$$

Then

$$V_B = 4.79 - 0.0064 Q_C$$

We also have the condition that

$$(\Delta p_f)_B = (\Delta p_f)_C$$

As a first approximation, let $f_B = f_C$. Then

$$\left(\frac{L}{D}\right)_B \frac{V_B^2}{2g} = \left(\frac{L}{D}\right)_C \frac{V_C^2}{2g}$$

Substituting for V_B and rearranging, we have the following quadratic equation for Q_C:

$$Q_C^2 + 204 Q_C - 76{,}000 = 0$$

and

$$Q_C \simeq 190 \text{ gal/min}$$

Therefore,

$$Q_B \simeq 560 \text{ gal/min}$$

Now check branch C:

$$V = \frac{190 \times 231}{1728 \times 60}\frac{1}{0.196} = 2.16 \text{ ft/sec}$$

$$(VD) = 6 \times 2.16 = 13$$

$$\frac{e}{D} = \frac{0.00015}{12} = 0.0003$$

$$\text{Re} = 9 \times 10^4$$

$$f = 0.02$$

$$(\Delta p_f)_C = 0.02 \times \frac{200}{1/2} \times \frac{(2.16)^2}{2g} = 0.58 \text{ ft}$$

Now check branch B:

$$V_B = \frac{560}{60} \times \frac{231}{1728} \frac{1}{(\pi/4)[(8/12)^2]} = 3.59 \text{ ft/sec}$$

$$(VD) = 3.59 \times 8 = 28.7$$

$$\frac{e}{D} = \frac{0.00015}{8/12} = 0.000225$$

$$\text{Re} = 2 \times 10^5$$

$$f = 0.0178$$

$$(\Delta p_f)_B = 0.0178 \times \frac{(3.59)^2}{2g} \times \frac{100}{8/12} = 0.53 \text{ ft}$$

At this point we note that the two pressure drops are not equal (as they must be), and if the desired accuracy warrants the effort, another iteration can be made using the results of the foregoing solution. The quantity of water flowing in branch C will be less than 190 gal/min, and in branch B it will be greater than 560 gal/min. The total pressure drop between 1 and 2 will be, closely,

$$0.39 + 0.62 + \frac{0.55(62.4)}{144} = 1.25 \text{ psi}$$

The preceding problem is a relatively simple problem, and it can be seen that a considerable expenditure of time and effort is required before a solution can be obtained. For more complex problems the time required to obtain a solution in the manner indicated is prohibitive. Numerical procedures have been devised to decrease the effort in problems of this type, but these procedures can become quite lengthy, too. Fortunately, these problems can be programmed relatively easily for solution on both analog and digital computers, and at present almost all complex problems of this type are solved using computers. It is important that the student understand the basis of this class of problem so that the necessary inputs for the computerized solution can be obtained.

6.10 CLOSURE

In this chapter we have placed the emphasis on the practical aspects of calculating the pressure drop in pipes and ducts when the fluid flowing is incompressible. While the approach to this problem has been on a rational basis, it has been necessary to invoke empirical correlations to permit the engineer to make quantitative evaluations in commercial piping systems.

Even with the best available data it is possible to find deviations in excess of 30% between calculated and measured results for new or clean commercial pipes. For old or severely corroded pipes it is almost impossible to obtain a usable engineering correlation for pressure drop.

Pressure changes occur at changes in section, valves, bends, branches, fittings, inlets, and outlets and must be included with the friction pressure drops in any calculation of pump size or head requirements. For these losses (termed other or minor losses) the available correlations leave much to be desired, and in the worst instance a variation of +80% −200% can be found for a flanged check valve. It is evident that the designer of a piping system must exercise a great deal of judgment in evaluating the pressure drop and must allow for this lack of accurate data in sizing pumps or establishing the head on a given system. The novice or student should be aware of the inadequacies of our knowledge in this branch of fluid mechanics and should not depend blindly on a calculated flow rate or pressure drop.

PROBLEMS

6.1 Water flows in a 2-in.-i.d. pipe at the rate of 10 ft/sec. Determine the Reynolds number if the temperature of the water is 100°F. Use Table 2.3.

6.2 Water at 20°C flows in a pipe whose diameter is 50 mm. If the velocity is 5 m/s, what is the Reynolds number? Use Table 2.4.

6.3 Oil flows in a pipe that has an inside diameter of 1.5 in. If the oil has a specific weight of 45 lb/ft^3 and a viscosity of 3 cP, determine its Reynolds number if the velocity is 14 ft/sec.

6.4 Show that for a given volumetric flow that the pressure drop in a pipe is inversely proportional to the fourth power of the diameter in laminar flow and inversely proportional to the fifth power of the diameter for turbulent flow.

6.5 What is the least value of friction factor that you would expect in laminar flow?

6.6 Oil flows in a square pipe that is 100 mm on a side. If the Reynolds number is 1000 and the velocity of the oil is 0.5 m/s, determine the pressure drop per metre of pipe.

6.7 Water flows in a 5-in.-diameter pipe in laminar flow. If the Reynolds number is 1200 and the velocity is 0.1 ft/sec, what is the pressure drop per 100 ft of pipe?

6.8 Oil flows through a small size tube having a $\frac{1}{4}$-in. i.d. It is desired that the flow shall always be laminar. If the viscosity of the oil is 1 P and its weight is 52 lb/ft^3, determine the maximum velocity in the tube. Also determine the average velocity.

6.9 Assuming that the concept of an equivalent diameter can be applied to an annulus between two concentric tubes having diameters D_2 and D_1, determine the equivalent diameter of such an annulus in terms of D_2 and D_1, assuming that the fluid wets both surfaces.

6.10 Based upon the results of Problem 6.9 ($D_{eq} = D_2 - D_1$), calculate the pressure drop per foot of an annulus having D_2 equal to 1 in. and D_1 equal to $\frac{3}{4}$ in. if the fluid flowing is an oil with a viscosity of 90 cP and a specific weight of 50 lb/ft^3. The average flow velocity is 10 ft/sec.

6.11 A fluid flows in a horizontal pipe at the rate of $\frac{1}{4}$ ft^3/sec, and the friction head loss is measured to be 25 ft. If the pipe is 1000 ft long and has an inside diameter of 2 in., determine whether the flow is laminar or turbulent.

6.12 Oil is being pumped from a truck to a tank 10 ft higher than the truck through a 2-in.-i.d. pipe 100 ft long (Figure P6.12). Assume that the pump outlet pressure is 15 psig and determine the flow rate in gallons per minute. The oil has an absolute viscosity of 105 cP and a specific weight of 56 lb/ft^3. (*Hint*: Assume the flow to be laminar and check after solving the problem.)

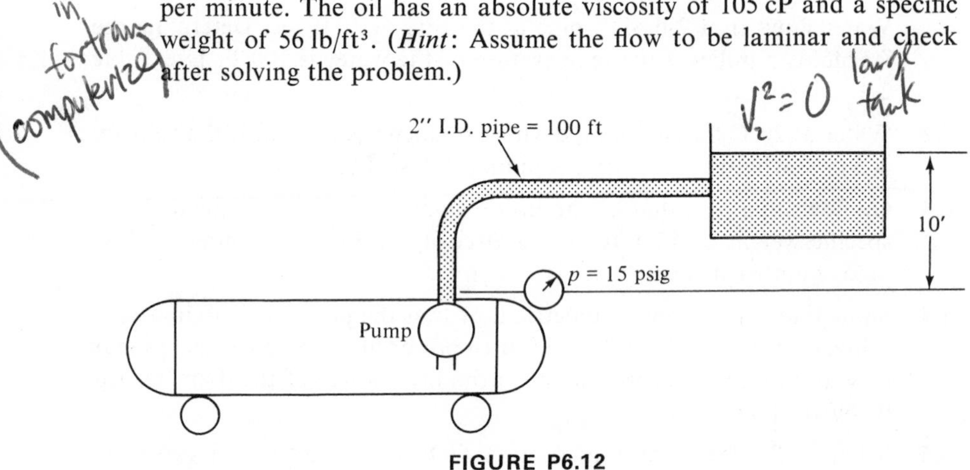

FIGURE P6.12

6.13 A company wishes to pump crude oil from a storage tank to its plant 35 mi away (Figure P6.13). Assume that an 18-in.-i.d. pipeline is to be used and that the plant is at an elevation of 50 ft above the tank. The oil has a viscosity of 150 cP and a specific weight of 52 lb/ft^3. Determine the required pressure at the pump if the average velocity in the pipe is to be limited to 1.5 ft/sec.

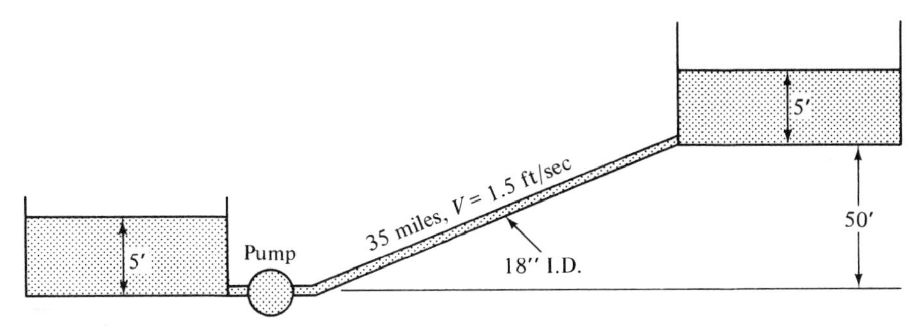

FIGURE P6.13

6.14 Determine the pressure drop in a 2-in.-i.d. pipe that is 1000 ft long in which oil flows at an average velocity of 3 ft/sec if the specific weight of the oil is 51 lb/ft^3 and its viscosity is 56 cP.

6.15 Oil flows in a 3-in.-i.d. pipe. If the absolute viscosity of the oil is 92 cP and its specific weight is 50 lb/ft^3, determine the flow rate, the Reynolds number, and the pressure drop per foot length if the maximum velocity is 4 ft/sec.

6.16 A commercial steel pipe whose inside diameter is 6 in. is used to deliver water at the rate of 700 gal/min to a chemical plant. Determine the friction loss per mile of pipe. Assume water at 68°F.

6.17 In Problem 6.16, determine the friction loss per mile if there are two regular flanged elbows, three wide-open flanged gate valves, a flanged check valve, and four 90° flanged long-radius elbows.

6.18 An 8-in.-i.d. pipe is 2500 ft long and carries water from a pump to a reservoir whose water surface is 200 ft above the pump. If 4 ft^3/sec is being pumped, determine the gage pressure at the discharge of the pump. Assume that the pipe is rough cast iron and neglect minor losses.

6.19 Water at 68°F flows in a horizontal steel pipe at the rate of 15 gal/min. If the pipe has an inside diameter of 3 in., is 500 ft long, has three 90° regular elbows, four fully open gate valves, three regular 180° radius bends, and a check valve, determine the pressure drop in the line. All fittings are flanged.

6.20 Water at 68°F flows in a vertical square steel pipe whose sides are 4 in. (Figure P6.20). Determine the difference in pressure between two static pressure taps located 100 ft apart if the average water velocity is 15 ft/sec and the flow is up.

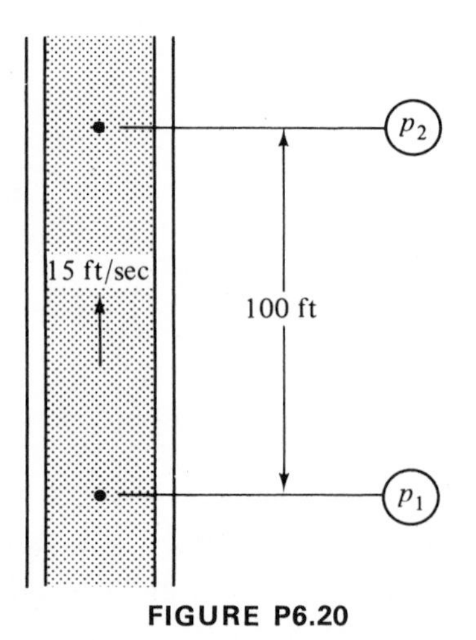

FIGURE P6.20

6.21 The flow in Problem 6.20 is reversed. Determine the pressure difference between the taps.

6.22 Oil having a viscosity of 50 cP and a specific weight of 52 lb/ft^3 flows upward in a 1-in.-i.d. pipe with a velocity of 0.1 ft/sec. Determine the difference in pressure between two static taps located 100 ft apart.

6.23 A diffuser is placed in a pipeline to connect a 6-in.-i.d. pipe to a 12-in.-i.d. pipe. If the total included angle of the diffuser is 30° and the flow is from the 6- to the 12-in. pipe at 10 ft/sec, determine the pressure drop in the diffuser.

6.24 If the diffuser included angle is 7° in Problem 6.23, determine the pressure drop in the diffuser.

6.25 Air at 50 psia and 100°F flows in a horizontal pipe that has a 6-in. i.d. It is desired to limit the pressure drop due to the pressure losses to 10% of the initial absolute air pressure. What is the maximum velocity in the pipe if minor losses are neglected and the pipe is 500 ft long? (Use $\bar{\gamma} = 0.229$ lb/ft^3.)

6.26 An abrupt contraction is placed in a horizontal pipeline that has an inside diameter of 4 in. If the contracted flow diameter is 2 in., determine the pressure loss for 68°F water flowing at 10 ft/sec in the 4-in. pipe.

6.27 In Problem 6.26 the flow is reversed. Determine the pressure drop if the velocity in the 4-in. line is kept at 10 ft/sec.

6.28 Water at 20°C flows in an abrupt contraction in a pipe whose diameter goes from 100 mm to 50 mm. If the velocity before the contraction is 3 m/s, what is the pressure loss?

6.29 If the flow is reversed in Problem 6.28, what will the pressure drop be? Both velocities are to be taken to remain the same.

6.30 Water is supplied from a reservoir to a 500-ft-long horizontal round concrete pipe 36 in. in diameter. What head is required to cause a flow of 20,000 gal/min? Neglect entrance and exit losses.

6.31 Compare the results of Problem 6.30 with the results obtained if the entrance and exit losses had not been neglected. What do you conclude from the comparison?

6.32 If water at 68°F flows in a 2-in.-i.d. pipe that is 20 ft long, determine the pressure drop if the flow is 250 gallons per minute. Neglect minor losses.

6.33 Water at 68°F flows in a 2-in.-i.d. pipe at the rate of 250 gal/min. If the pipe has a globe valve and a check valve and it delivers water to a tank that is 60 ft above the pipe inlet, determine the pressure difference between the pipe inlet and outlet. All fittings are screwed; neglect inlet and outlet losses. The total pipe length is 250 ft.

REFERENCES

1. *Pipe Friction Manual*, 3rd., Hydraulic Institute, New York, 1961.
2. "Fluid Mechanics" by A. D. KRAUS, *Electro-Technology* (New York), **69**, April 1962, p. 120.
3. "Flow of Fluids through Valves, Fittings and Pipe," *Technical Paper 410*, Crane Co., Chicago, 1957, p. 179.
4. "Flow Losses in Abrupt Enlargements and Contractions" by R. P. BENEDICT, N. A. CARLUCCI, and S. D. SWETZ, *ASME Paper 65-WA/PTC-1.*
5. *Engineering Heat Transfer* by S. T. HSU, Litton Educational Publishing, Inc., New York, 1963.
6. "Friction Factors for Pipe Flow" by L. F. MOODY, *Transactions of the ASME*, **66**, 1944, p. 671; also *Mechanical Engineering*, **69**, December 1947, p. 1005.
7. "The Flow of Fluids in Closed Conduits" by R. J. S. PIGOTT, *Mechanical Engineering*, **55**, No. 8, August 1933, p. 497.
8. "A Study of Data on the Flow of Fluids in Pipes" by E. KEMLER, *Transactions of the ASME*, **55**, 1933, HYD-55, p. 7.
9. "Laws of Flow in Rough Pipes" by J. NIKURADSE, *NACA Technical Memorandum 1292*, November 1950.

10. *Engineering Applications of Fluid Mechanics* by J. C. HUNSAKER and B. G. RIGHTMIRE, McGraw-Hill Book Company, New York, 1947.
11. *Thermodynamics of Fluid Flow* by N. A. HALL, Prentice-Hall, Inc., Englewood Cliffs, N.J., 1951.
12. *Introduction to Gas Dynamics* by R. M. ROTTY, John Wiley & Sons, Inc., New York, 1962.
13. *Fluid Mechanics* by R. C. BINDER, 4th ed., Prentice-Hall, Inc., Englewood Cliffs, N.J., 1962.
14. *Fluid Mechanics* by V. L. STREETER, 3rd ed., McGraw-Hill Book Company, New York, 1962; 6th ed., 1979.
15. *Elementary Fluid Mechanics* by J. K. VENNARD, John Wiley & Sons, Inc., New York, 1961.
16. *Applied Fluid Mechanics* by R. L. MOTT, 2nd ed., Charles E. Merrill Publishing Co., Columbus, Ohio, 1979.
17. *Fundamentals of Fluid Mechanics* by J. A. SULLIVAN, Reston Publishing Co., Reston, Va., 1978.
18. *Engineering Fluid Mechanics* by J. A. ROBERSON and C. T. CROWE, Houghton Mifflin Company, Boston, 1975.

Dynamic Forces

7.1 INTRODUCTION

Human beings have been familiar with the tremendous forces that fluid streams can exert since ancient times. The awesome power of the wind in a hurricane, the destruction caused by tidal waves, the cresting and flooding of rivers and so forth have all been described in great detail. It is also known that these forces have been utilized to humanity's benefit. Windmills, waterwheels, jets for excavation, and so on, are just some of the recorded efforts made to utilize the dynamic forces exerted by fluid streams. In recent years the application of the principles of fluid mechanics involving dynamic forces has led to the new technology known as *fluidics*. "Fluidics" can be defined to be the technology of using streams of gas or other fluids to perform such logic and control functions as amplification, sensing, logic switching computation, and control.

In this chapter our attention will be primarily directed to the computation of these forces using both the energy (Bernoulli) equation and Newton's second law in its appropriate form.

7.2 FORCE, MASS, AND ACCELERATION

Newton's second law of motion states that for a constant mass (m) the resultant force acting on a system is simply the product of its mass and its acceleration. Mathematically,

$$F = ma \tag{7.1}$$

If we now recall that acceleration (a) is the rate at which velocity changes with respect to time,

$$a = \frac{\Delta V}{\Delta t} \tag{7.2}$$

where $\Delta V = V_2 - V_1$ is the change in velocity, and Δt is the change in time during which the velocity was changing. Using equation (7.2) in conjunction with equation (7.1) gives us

$$F = \frac{m\,\Delta V}{\Delta t} \tag{7.3}$$

Equation (7.3) can be used in its present form or recast into two other useful forms. The first is

$$F\,\Delta t = m\,\Delta V = m(V_2 - V_1) \tag{7.3a}$$

The left side of equation (7.3a) is called the *impulse* imparted to the body and the right side of this equation is termed the change of *momentum* of the body.

The other form of equation (7.3) that will be useful to us is

$$F = \frac{m}{\Delta t}\Delta V = \dot{m}\,\Delta V = \dot{m}(V_2 - V_1) \tag{7.3b}$$

where we have used the symbol $\dot{m}$ for the term $m/\Delta t$, which is the time rate of flow. In SI units $\dot{m}$ is given in kg/s and in English units $\dot{m}$ is given in slugs/sec. Equation (4.1c) permits us to express $\dot{m}$ in terms of fluid properties,

$$\dot{m} = \rho A V \tag{4.1c}$$

Therefore, we can write equation (7.3b) as

$$F = \rho A V(V_2 - V_1) \tag{7.4}$$

But $AV = Q$; therefore,

$$F = \rho Q(V_2 - V_1) = \frac{\gamma}{g}Q(V_2 - V_1) \tag{7.5}$$

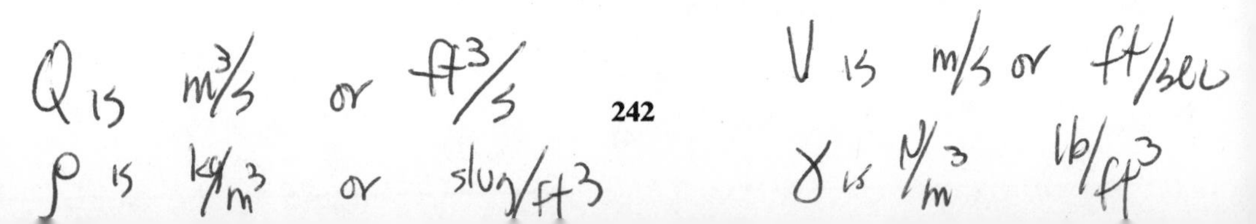

The units of Q are m^3/s or ft^3/sec, ρ is kg/m^3 or $slug/ft^3$, V is in m/s or ft/sec, and γ is in N/m^3 or lb/ft^3.

It must be noted at this time that *force*, *impulse*, and *momentum* are vectors. Therefore, it is necessary to write equations (7.3), (7.4), and (7.5) in each of the three coordinates, x, y, and z. If the force in the x direction is desired, the x velocities are used. Similarly, for the y and z directions, the appropriate velocities must be used.

ILLUSTRATIVE PROBLEM 7.1

In a turbojet engine, air enters and is compressed, fuel is added and burned, energy is removed by a turbine, and finally the combustion gases are exhausted. For such an engine let V_1 be the inlet air velocity relative to the engine, $\dot{m}_a$ the mass rate of air flow, $\dot{m}_f$ the mass rate of fuel flow, and V_2 the exhaust velocity relative to the engine. Assume that the inlet pressure, atmospheric pressure, and the exhaust pressure are all equal and that the inlet and exit areas are equal and determine the thrust of the engine. Figure 7.1 shows an aircraft turbine.

FIGURE 7.1 Aircraft gas turbine. (Courtesy of Pratt & Whitney Div. of United Aircraft Corp.)

$\dot{m}_a$ = mass rate air flow

$\dot{m}_f$ = mass rate fuel flow

Solution

Consider the control volume shown in Figure 7.2. Since V_1 and V_2 are specified relative to the engine, we may consider the engine fixed and the inlet and exhaust stream velocities as absolute. The momentum in is $\dot{m}_a(V_1)$

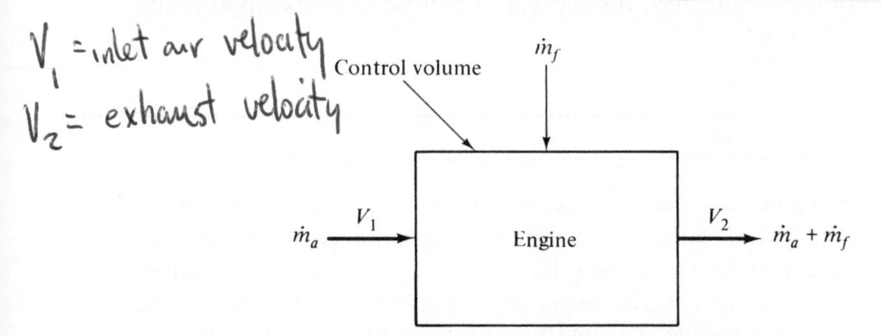

FIGURE 7.2 Illustrative Problem 7.1.

and the momentum out is $(\dot{m}_a + \dot{m}_f)V_2$. Since the pressures and areas are assumed to be equal, they will cancel out on both sides of the control volume and we do not have to consider them. Therefore,

$$F_x = (\dot{m}_a + \dot{m}_f)V_2 - \dot{m}_a(V_1)$$

In words, the thrust in the direction of the gas flow is simply the rate of change of momentum into and out of the control volume.

ILLUSTRATIVE PROBLEM 7.2

A 75-mm-i.d. pipe carries water at the rate of 0.2 m³/s. If there is a 180° bend in the horizontal plane, determine the thrust exerted by the water on the pipe. Use $\rho = 1000$ kg/m³. Neglect any pressure drop in the bend.

Solution

Figure 7.3a shows a sketch of the bend, and Figure 7.3b shows a free-body diagram of the bend. Consider x to be positive to the right. Since Q is specified as 0.2 m³/s,

$$V = \frac{Q}{A} = \frac{0.2 \text{ m}^3/\text{s}}{[\pi(0.075)^2/4] \text{ m}^2} = 45.3 \text{ m/s}$$

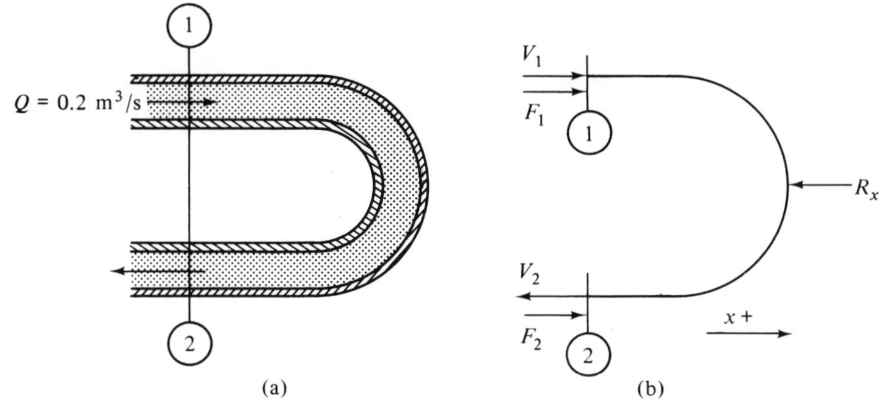

FIGURE 7.3 Illustrative Problem 7.2.

The reaction R_x due to the dynamic motion of the water is given by equation (7.5) as

$$R_x = \rho Q(V_2 - V_1)$$

Since $V_1 = 45.3$ m/s, V_2 must equal -45.3 m/s (since the stream leaves the 180° bend facing in the negative direction). It is most important to keep in mind at all times that *velocity is a vector*! Therefore,

$$R_x = 1000 \text{ kg/m}^3 \times 0.2 \text{ m}^3\text{/s} \times (-45.3 - 45.3) \text{ m/s}$$
$$R_x = -18\,120 \text{ kg}\cdot\text{m/s}^2 = -18\,120 \text{ N}$$

The negative sign indicates that the reaction force is to the left on the water. Note also that this solution does not take into account pressure differences between the inlet and the outlet of the bend. The 18 120-N force is the force required to alter the direction of flow of the water, and it is sometimes called *dynamic thrust*.

7.3 DEFLECTION OF STREAMS BY STATIONARY BODIES

As was seen in Illustrative Problem 7.2, the alteration of the direction of flow of a fluid or of its velocity means that a net force must have been exerted on the fluid stream. Concurrently, an equal and opposite force must have been exerted by the fluid stream on the body causing the velocity change. If the deflector is fixed, it is necessary for the structure to be capable of resisting the resultant force acting on it, while the same force exerted on a moving deflector (vane) is capable of doing useful work. The ability of a fluid stream to do work on a moving vane is the basis upon which the theory of turbomachinery is based.

Let us consider the situation of a fixed vane, shown in Figure 7.4, and apply the impulse–momentum relations developed earlier. The fluid stream of cross-sectional area A_0 with an initial velocity V_0 is deflected through the

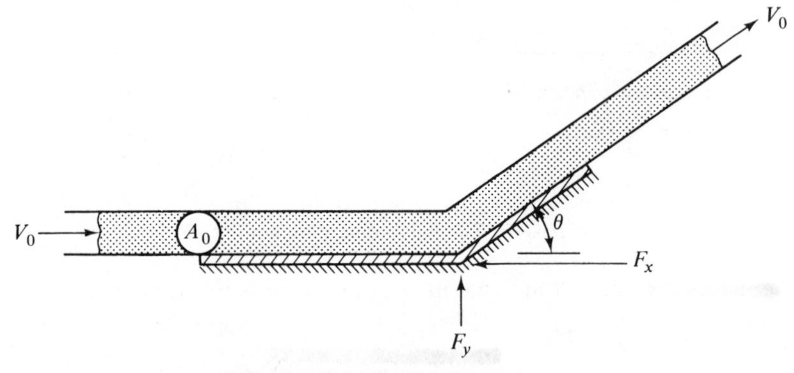

FIGURE 7.4 Fixed deflector vane.

angle θ. Assume the frictional resistance of the stream on the vane to be negligible and the velocity of the stream to be uniform throughout. For a free jet the pressure at both the inlet to the vane and the outlet of the vane must be equal since they must equal the surrounding (atmospheric) pressure. Therefore, the speed (not velocity) of the jet must be the same at inlet as at outlet; the vane only changes the direction of the stream. The force exerted by the vane on the stream in the horizontal direction is

$$F_x = \dot{m}(V_x - V_0) = \dot{m}(V_0 \cos\theta - V_0) \tag{7.6}$$

and

$$F_x = -\rho A V_0^2(1 - \cos\theta) = -\rho Q V_0(1 - \cos\theta) \tag{7.7}$$

In the vertical direction,

$$F_y = \dot{m}(V_y - 0) = \dot{m}(V_0 \sin\theta) \tag{7.8}$$

and

$$F_y = \rho A V_0^2 \sin\theta = \rho Q V_0 \sin\theta \tag{7.9}$$

The resultant force R is given by

$$R = \sqrt{F_x^2 + F_y^2} = \rho A V_0^2 \sqrt{2(1 - \cos\theta)} \tag{7.10}$$

or

$$R = \rho Q V_0 \sqrt{2(1 - \cos\theta)} \tag{7.10a}$$

The derivation of equations (7.6) through (7.10) has assumed that the difference in elevation between the ends of the vane is negligible. Note, too, that the forces given by these equations are the forces exerted on the fluid stream. The forces on the fixed vane will be equal and opposite to these forces.

ILLUSTRATIVE PROBLEM 7.3

Water is deflected by a 90° bend that is in a horizontal plane (both ends at the same elevation); see Figure 7.5. If the stream is 3 in. in diameter, γ is

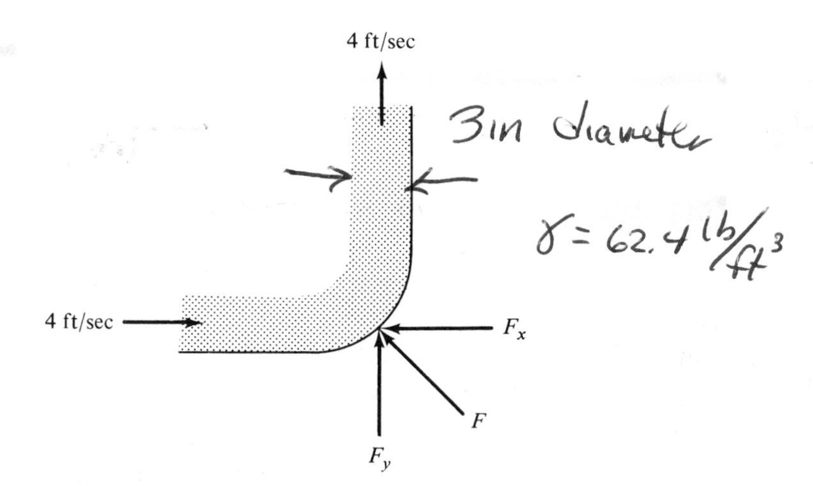

FIGURE 7.5 Illustrative Problem 7.3.

62.4 lb/ft³, and the stream velocity is 4 ft/sec, determine the forces exerted on the stream by the vane due to the dynamic action of the stream.

Solution

$$A = \frac{\pi}{4} D^2 = \frac{\pi}{4}\left(\frac{3}{12}\right)^2 = 0.0491 \text{ ft}^2$$

Using equation (7.7), we obtain

$$F_x = -\rho A V_0^2 (1 - \cos\theta) = \frac{-62.4}{32.17}(0.0491)(4)^2(1 - \cos 90) = 1.52 \text{ lb}$$

$$\left(\frac{\gamma}{g} = \rho, \cos 90 = 0\right)$$

From equation (7.9),

$$F_y = \rho A V_0^2 \sin\theta = \frac{62.4}{32.17}(0.0491)(4)^2 = 1.52 \text{ lb} \qquad (\sin 90 = 1)$$

and the resultant force is found from equation (7.10) as

$$R = \sqrt{F_x^2 + F_y^2} = \sqrt{(-1.52)^2 + (1.52)^2} = 2.15 \text{ lb}$$

ILLUSTRATIVE PROBLEM 7.4

A jet directs a stream of water against a fixed vane as indicated in Figure 7.4. If the angle θ is 45° and the jet has a velocity of 10 m/s, determine the magnitude of the forces on the vane. Also determine the magnitude and direction of the resultant of these forces. The density of water can be taken as 1000 kg/m³ and Q is 0.1 m³/s.

Solution

Applying equations (7.7) and (7.9), we have

$$F_x = -\rho Q V_0(1 - \cos\theta) = -1000 \text{ kg/m}^3 \times 0.1 \text{ m}^3\text{/s} \times 10 \text{ m/s}\,(1 - 0.7071)$$

$$F_x = -292.9 \text{ N}$$

and

$$F_y = \rho Q V_0 \sin\theta = 1000 \text{ kg/m}^3 \times 0.1 \text{ m}^3\text{/s} \times 10 \text{ m/s} \times 0.7071$$

$$F_y = 707.1 \text{ N}$$

Notice that these forces are on the fluid. For the vane,

$$F_x = 292.9 \text{ N}$$

$$F_y = -707.1 \text{ N}$$

Therefore,

$$R = \sqrt{F_x^2 + F_y^2} = \sqrt{(292.9)^2 + (-707.1)^2} = 765.4 \text{ N}$$

We can determine the line of action of this resultant from Figure 7.6. Therefore,

$$\theta = \arccos\frac{292.9}{765.4}$$

$$= 67.5°$$

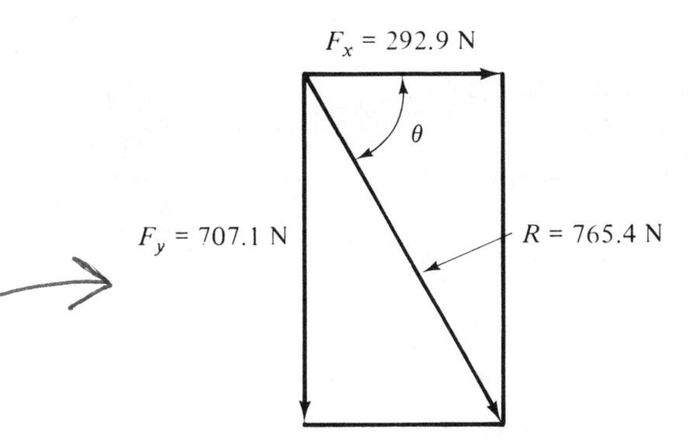

FIGURE 7.6 Illustrative Problem 7.4.

ILLUSTRATIVE PROBLEM 7.5

If the 180° bend of Illustrative Problem 7.2 is also subjected to a difference in pressure at inlet and outlet due to the pressure drop in the bend, what is the total force on the pipe? Assume that the pressure at the inlet is 250 kPa and the pressure at the outlet is 240 kPa.

Solution

The dynamic forces remain the same. In addition, there is a force at the inlet acting to the right and equal to pA. Therefore, at the inlet,

$$F_{x_1} = pA = 250\,000 \times \frac{\pi}{4}(0.075)^2 = 1104.5 \text{ N}$$

At the outlet there is also a pressure force to the right equal to pA. Therefore,

$$F_{x_2} = pA = 240\,000 \times \frac{\pi}{4}(0.075)^2 = 1060.3 \text{ N}$$

The *total* force on the pipe due to both pressure differences and change in velocity is

$$F_{x_1} + F_{x_2} - R_x = 1104.5 + 1060.3 - (-18\,120) = 20\,284.8 \text{ N}$$

where the value of R_x is obtained from Illustrative Problem 7.2.

In Chapter 5 we studied nozzles in some detail. Let us at this time look at a simple nozzle, as shown in Figure 7.7. The basic purpose of this device is to

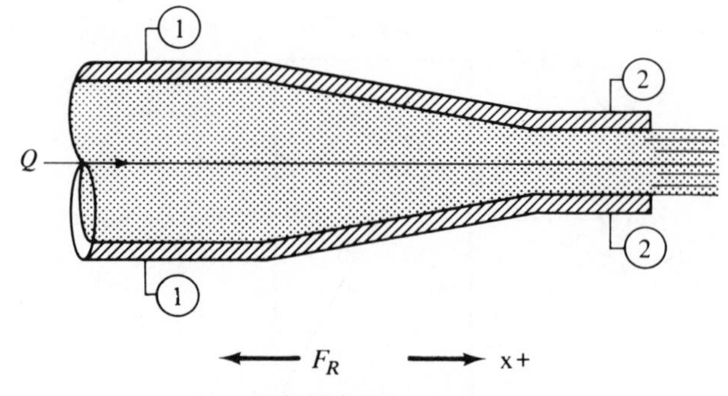

FIGURE 7.7 Nozzle.

increase the velocity at its outlet by decreasing the area. Consider the impulse–momentum relation (repeated here for convenience),

$$F_x = \rho Q(V_2 - V_1) \tag{7.5}$$

where F_x is the net force acting in the x direction. At the inlet there is a pressure force equal to p_1A_1 and at the outlet the pressure force is p_2A_2. These pressure forces act opposite to each other. If we denote the force exerted *by the nozzle* on the fluid as F_R, equation (7.5) applied to this device becomes

$$p_1A_1 - p_2A_2 - F_R = \rho Q(V_2 - V_1) \tag{7.11}$$

Solving for F_R, we obtain

$$F_R = p_1A_1 - p_2A_2 - \rho Q(V_2 - V_1) \tag{7.12}$$

ILLUSTRATIVE PROBLEM 7.6

A pump discharges 500 gpm through a 4-in.-diameter pipe. At the end of the pipe there is a nozzle with an outlet diameter of 1 in. What is the force exerted by the fluid on the hose? Take $p_1 = 100$ psig, p_2 to be atmospheric pressure, and the specific weight of water to be 62.4 lb/ft³.

Solution

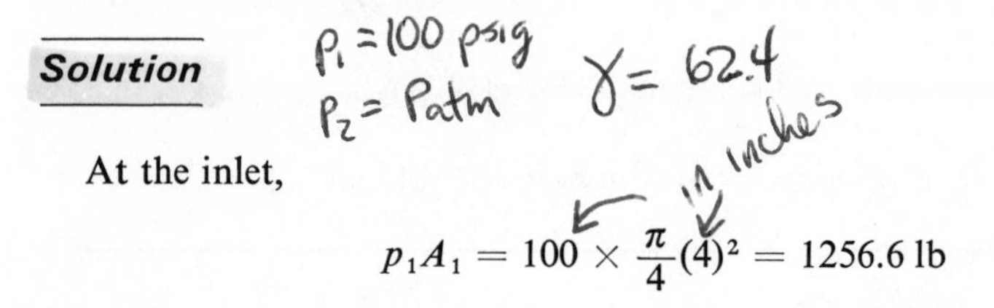

At the inlet,

$$p_1A_1 = 100 \times \frac{\pi}{4}(4)^2 = 1256.6 \text{ lb}$$

At the outlet, $p_2A_2 = 0$, since p_2 is atmospheric pressure which we take as our base for pressure differences. The velocity at each section is found from the continuity relation, $Q = AV$ or $V = Q/A$.

$$Q = 500 \text{ gpm} \times \frac{231 \text{ in.}^3/\text{gal}}{1728 \text{ in.}^3/\text{ft}^3} \times \frac{1}{60 \text{ sec/min}} = 1.11 \text{ ft}^3/\text{sec}$$

$$V_1 = \frac{1.11}{(\pi/4)[(4)^2/144]} = 12.7 \text{ ft/sec}$$

and

$$V_2 = \frac{1.11}{(\pi/4)[(1)^2/144]} = 203.5 \text{ ft/sec}$$

Applying equation (7.12), we obtain

$$F_R = p_1A_1 - p_2A_2 - \rho Q(V_2 - V_1)$$

$$= 1256.6 - 0 - \frac{62.4}{32.17} \times 1.11(203.5 - 12.7)$$

$$= 845.8 \text{ lb}$$

Since this is the force on the fluid, the force on the nozzle will be equal to the force on the fluid and directed opposite to the direction of flow.

7.4 MOVING VANES

If the vane is moving with the velocity u in the same direction as the initial velocity of the stream, we have the situation shown schematically in Figure 7.8. At the inlet the absolute velocity of the stream is V_0 and its velocity relative to the vane is $V_0 - u$. The relative velocity is turned by the vane through the angle θ without a change in its magnitude. This relative velocity

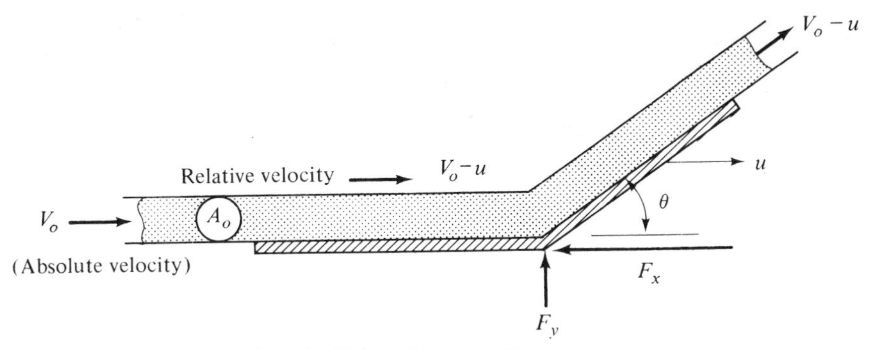

FIGURE 7.8 Moving deflector vane.

vector $V_0 - u$ is now added vectorially to u to give the final absolute velocity leaving the vane. Figure 7.9 shows the vector diagram at the outlet of the vane, where V_1 is the resultant absolute velocity at the outlet. The component of the absolute velocity in the horizontal direction at the outlet is $(V_0 - u)\cos\theta + u$ and at the inlet it is V_0. Therefore,

$$F_x = \dot{m}[(V_0 - u)\cos\theta + u - V_0] = \dot{m}(V_0 - u)(1 - \cos\theta) \qquad (7.13)$$

and

$$F_y = \dot{m}[(V_0 - u)\sin\theta] \qquad (7.14)$$

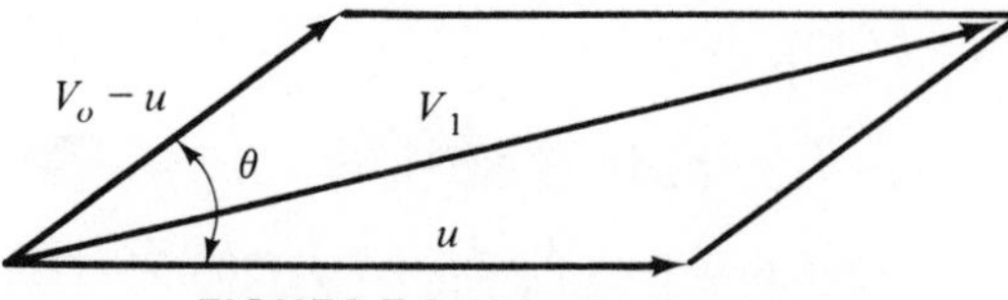

FIGURE 7.9 Velocity diagram.

Notice from equations (7.13) and (7.14) that if $V_0 - u$ is replaced by the relative velocity, the problem of the moving vane is reduced to the problem of the fixed vane with a velocity equal to the relative velocity.

ILLUSTRATIVE PROBLEM 7.7

If the vane in Illustrative Problem 7.3 has a velocity of 2 ft/sec to the right, determine the forces acting on the stream.

Solution

Refer to sketch of Illustrative Problem 7.3. Using the data of Illustrative Problem 7.3 and equations (7.13) and (7.14) gives us

$$V_0 - u = 4 - 2 = 2 \text{ ft/sec}$$

Therefore,

$$F_x = -\frac{62.4 \times 0.0491 \times 4}{32.17}(2 - 0) = -0.76 \text{ lb}$$

$$F_y = \frac{62.4 \times 0.0491 \times 4}{32.17}(2) = 0.76 \text{ lb}$$

and

$$R = \sqrt{(-0.76)^2 + (0.76)^2} = 1.07 \text{ lb}$$

ILLUSTRATIVE PROBLEM 7.8

If the stream in Illustrative Problem 7.3 is turned through an angle of 180° (a U bend), determine the forces acting on the stream.

Solution

Refer to Figure 7.10. Using the absolute velocities from the diagram we obtain

$$\begin{aligned} F_x &= \dot{m}[-(V_0 - u) + u - V_0] = \dot{m}[-2(V_0 - u)] \\ &= -\dot{m}2(V_0 - u) \\ &= -\frac{62.4 \times 0.0491 \times 4}{32.17}(2)(2) = -1.56 \text{ lb} \end{aligned}$$

and

$$F_y = 0$$

The resultant force is therefore 1.56 lb acting to the left.

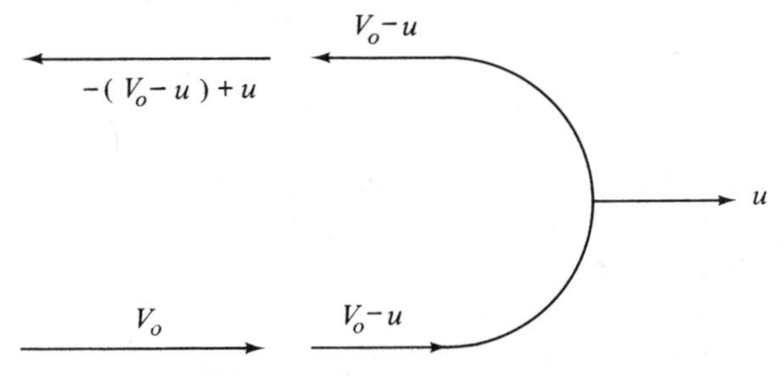

FIGURE 7.10 Illustrative Problem 7.8.

7.5 POWER

In mechanics, *work* is defined as the product of a force times the displacement (distance moved) in the direction of the force. The student will note that we have already used this definition in Chapter 4. At this point we will define the term *power* to be the rate at which work is being done. Since work is expressed as either ft lb or N·m, power is expressed as ft lb/sec or N·m/s. The term *horsepower* (hp) is also a defined term that equals 550 ft lb/sec in the English system or 746 N·m/s or 746 W in the SI system.

In order to determine the horsepower required or delivered in a given situation, it will first be necessary to define the system. As our first example, let us consider a force F acting on a solid body moving with a velocity of V ft/sec or V m/s, as shown in Figure 7.11. The rate of doing work by the applied force will be the product of the force and the displacement per unit time, which equals FV. The horsepower being supplied by the applied force is given by

$$\text{hp} = \frac{FV}{550} \quad \text{(English units)} \tag{7.15}$$

or

$$\text{hp} = \frac{FV}{746} \quad \text{(SI units)} \tag{7.15a}$$

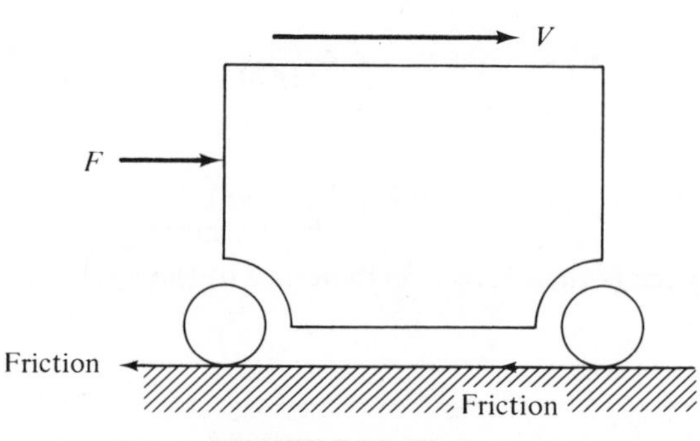

FIGURE 7.11 Work.

Equations (7.15) and (7.15a) refer to a constant force acting on a body that is moving with a constant velocity. Note that this power is needed to replenish the power lost by friction between the wheels and the ground. An interesting instance of this concept arises when we consider the power that a stream transfers to a moving blade. The power transferred from a stream to a moving vane is the product of $F_x(u)$. If the turbomachine is frictionless, its output is this product: namely,

$$p = F_x(u) = \dot{m}(V_0 - u)(1 - \cos\theta)u \tag{7.16}$$

Examination of equation (7.16) shows that no power output will be obtained if the blade is either stationary or operating at $u = V_0$. If equation (7.16) is plotted as a function of u, it is found that P is a maximum when $u = V_0/2$. If u is kept constant and P is plotted as a function of θ, P is found to be maximum when $\theta = 180°$. Using both of these conditions, we find that the maximum power transferred, P, is exactly equal to the power of the free jet. A blade operated at half the stream speed with the stream turned through 180° will have an efficiency of energy transfer theoretically equal to 100%.

Let us now consider a pump that is required to pump a quantity of fluid against a given "head." This terminology is commonly used, but it really should be worded: "How much energy must be imparted to a fluid to pump it from one location to another?" If we review Chapter 4 at this point, we will see that the weight flow rate multiplied by the total head gives us power in ft lb/sec or N·m/s. The weight flow rate is given in units of ft lb/sec or N·m/s and in either set of units equals the volume flow rate, Q, multiplied by the specific weight. Therefore, power is the product of head multiplied by the volume flow rate and the specific weight of the fluid. In terms of horsepower,

$$\text{hp} = \frac{(E_p \text{ or } E_T) \times Q \times \gamma}{550} \tag{7.17}$$

or

$$\text{hp} = \frac{(E_p \text{ or } E_T) \times Q \times \gamma}{746} \tag{7.17a}$$

where E_p and E_T are defined by equations (5.4) and (5.5) of Chapter 5. Note that if we replace the (E_p or E_T) terms by a simple term called "head," the head may represent the potential, pressure or velocity terms from the Bernoulli equation either singly or in any combination.

ILLUSTRATIVE PROBLEM 7.9

A pump is required to pump water from a large reservoir to a point located 20 m above the reservoir (Figure 7.12). If 0.05 m³/s of water having a density of 1000 kg/m³ is pumped through a 50-mm pipe, how much power is required to be delivered to the water by the pump? Neglect all flow losses in the pipe.

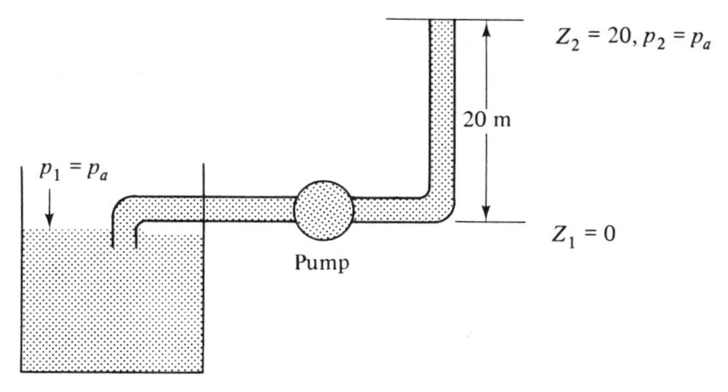

FIGURE 7.12 Illustrative Problem 7.9.

Solution

Let us first write a Bernoulli equation between sections 1 and 2,

$$\frac{p_1}{\gamma_1} + \frac{V_1^2}{2g} + Z_1 + E_p = \frac{p_1}{\gamma_2} + \frac{V_2^2}{2g} + Z_2$$

since there are no losses in the pipe. Therefore, E_p is

$$E_p = \frac{p_2 - p_1}{\gamma} + \frac{V_2^2 - V_1^2}{2g} + (Z_2 - Z_1)$$

but $p_2 = p_1 = p_a$, $V_1 = 0$, and $Z_1 = 0$. E_p is thus given by

$$E_p = \frac{V_2^2}{2g} + Z_2$$

Since $Q = AV$, we have

$$V = \frac{0.05}{\pi[(0.05)^2/4]} = 25.5 \text{ m/s}$$

Using this value of V,

$$E_p = \frac{(25.5)^2}{2 \times 9.81} + 20 = 53.1 \text{ m}$$

We can now obtain the horsepower required as follows:

$$\text{hp} = \frac{E_p \times Q \times \gamma}{746}$$

550 (for British)

$$= \frac{53.1 \times 0.05 \times 1000 \times 9.81}{746} = 34.9 \text{ hp}$$

ILLUSTRATIVE PROBLEM 7.10

A hydraulic turbine is connected as shown in Figure 7.13. How much horsepower will it develop? Use 1000 kg/m³ for the density of water. Neglect the flow losses in the system.

Solution

Again let us write a Bernoulli equation, but this time for a turbine.

$$\frac{p_1}{\gamma_1} + \frac{V_1^2}{2g} + Z_1 = \frac{p_2}{\gamma_2} + \frac{V_2^2}{2g} + Z_2 + E_T$$

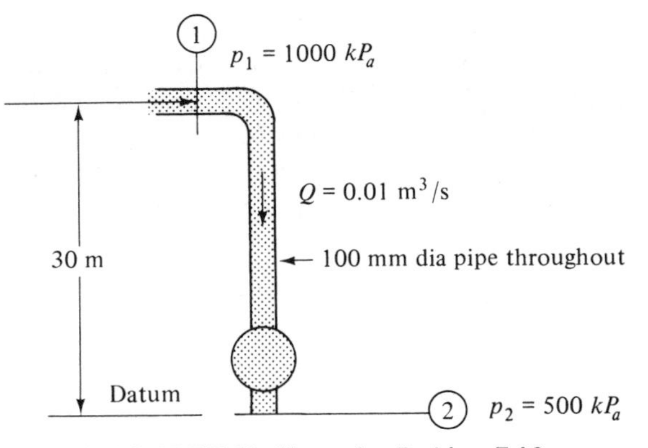

FIGURE 7.13 Illustrative Problem 7.10.

Since the pipe diameter is constant, $V_1 = V_2$, $Z_2 = 0$, and $Z_1 = 30$ m. Therefore, E_T is given as

$$E_T = \frac{p_1 - p_2}{\gamma} + (Z_1 - Z_2)$$

For the data given,

$$E_T = \frac{(1000 - 500)1000}{1000 \times 9.81} + (30 - 0) = 81 \text{ m}$$

and the horsepower is given by

$$\text{hp} = \frac{E_T Q \gamma}{746}$$

$$= \frac{81 \times 0.01 \times 1000 \times 9.81}{746} = 10.65 \text{ hp}$$

7.6 FAN OR PROPELLER

The function of a fan or propeller is to transform efficiently its rotary motion into forward linear motion of an airstream to provide an increased air velocity. The blades of such a device can be considered to be a number of airfoils of very short span set end to end that must absorb the power furnished to the propeller by its driving motor. The theory of the action of these small airfoils is known as the Drzewiecki theory or blade element theory and its development will be found in standard texts on aerodynamics or propellers. For the present we shall consider a somewhat simplified treatment in which the fan or propeller will be considered as a disk that exerts a uniform pressure or thrust on the airstream without causing any rotation of the fluid.

The only result of the thrust is to increase the momentum of the air. The additional assumption will be made that the pressure at stations far enough in front of and in back of the propeller is atmospheric and that the flow is streamline from entrance to exit of the device.

Before proceeding with the analysis, let us first recast the Bernoulli equation into a more convenient form. Repeating equation (5.2), we have

$$\frac{p_1}{\gamma} + \frac{V_1^2}{2g} + Z_1 = \text{constant} \tag{5.2}$$

Multiplying equation (5.2) by γ and noting that $\gamma = \rho g$,

$$p_1 + \frac{V_1^2 \gamma}{2g} + \gamma Z_1 = p_1 + \frac{\rho V_1^2}{2} + \gamma Z_1 = \text{constant} \tag{7.18}$$

For problems in gas flow and gas dynamics the elevation term is usually negligible when compared to the other terms, which leads to a form of the Bernoulli equation sometimes called the aerodynamic form of the Bernoulli equation: namely,

$$p + \frac{\rho V^2}{2} = \text{constant} \tag{7.19}$$

If a body is immersed in a fluid such as is shown in Figure 7.14, there will be a point A on the body where the velocity of the fluid is zero. Since the fluid is stagnant at this point, it is known as the *stagnation point* and the pressure at this point will exceed the pressure of the surrounding stream by $\rho V^2/2$ as determined by equation (7.19).

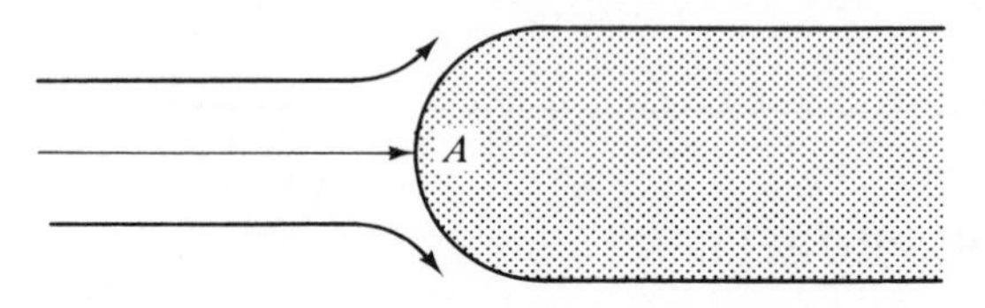

FIGURE 7.14 Streamlines around a submerged body.

We are now ready to consider an idealized fan or propeller as shown in Figure 7.15. As was stated earlier, the basic action of the fan or propeller is to impart a momentum to the fluid, thereby developing a propulsive thrust. For our analysis we will assume the density of the air to remain essentially constant and we will also assume that there are no flow losses in the device.

If we consider sections ② and ③ and write the momentum relation between these sections, we obtain

$$F = (p_3 - p_2)A = \rho Q(V_4 - V_1) \tag{7.20}$$

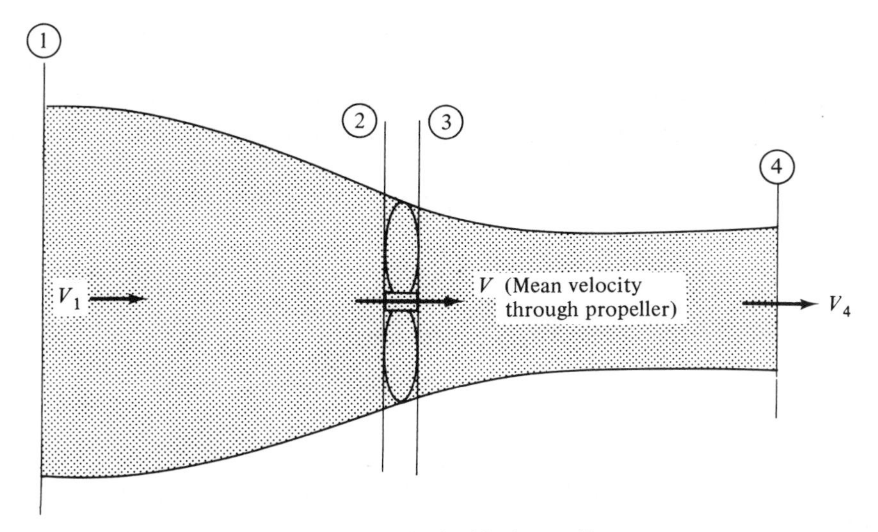

FIGURE 7.15 The ideal propeller.

where A is the plane area swept out by the propeller. Since $Q = AV$,

$$(p_3 - p_2)\not{A} = \rho \not{A} V(V_4 - V_1) \tag{7.21}$$

The assumption that there are no energy losses in this device permits us to write the energy equation in aerodynamic form between sections 1 and 2, and then between sections 3 and 4.

$$p_1 + \frac{\rho V_1^2}{2} = p_2 + \frac{\rho V_2^2}{2} \tag{7.22}$$

and

$$p_3 + \frac{\rho V_3^2}{2} = p_4 + \frac{\rho V_4^2}{2} \tag{7.23}$$

Since p_1 and p_4 are the pressures of the undisturbed stream, $p_1 = p_4$. We can therefore solve for $p_3 - p_2$ from equations (7.22) and (7.23) as

$$p_3 - p_2 = \tfrac{1}{2}\rho(V_4^2 - V_1^2) \tag{7.24}$$

But equation (7.21) also gives us $p_3 - p_2$. Therefore,

$$\tfrac{1}{2}\rho(V_4^2 - V_1^2) = \rho V(V_4 - V_1)$$

which yields V as

$$V = \frac{V_1 + V_4}{2} \tag{7.25}$$

From equation (7.25) we conclude that the velocity through the propeller area is the arithmetic mean of the initial and final velocities with half of the total change in velocity occurring *before* the air passes through the propeller and half being given to the air *after* it passes through the propeller.

The useful work done on the airstream is the product of the thrust and velocity. We can obtain the power input to the propeller using equation (7.17):

$$P = Q\gamma E_p \tag{7.26}$$

E_p is obtained from the Bernoulli equation as

$$E_p = \frac{V_4^2 - V_1^2}{2g} \tag{7.27}$$

Using equations (7.26) and (7.27) as well as equation (7.25) yields the following:

$$P = Q\gamma\left(\frac{V_4^2 - V_1^2}{2g}\right)$$

but $\gamma = \rho g$. Therefore,

$$P = Q\rho\frac{(V_4^2 - V_1^2)}{2} = Q\rho(V_4 - V_1)\frac{(V_4 + V_1)}{2} \tag{7.28}$$

and

$$P_{\text{input}} = Q\rho(V_4 - V_1)V \tag{7.29}$$

The power in the airstream is the the force at the inlet multiplied by the velocity of the stream at the inlet, giving us

$$P_{\text{airstream}} = FV_1$$

But F is given by equation (7.20) as $\rho Q(V_4 - V_1)$. Thus, the power in the airstream is

$$P_{\text{airstream}} = \rho Q(V_4 - V_1)V_1 \tag{7.30}$$

If we now define the efficiency of the propeller as the ratio of the power in the airstream to the power input by the propeller, we obtain

$$\eta = \left(\frac{V_1}{V}\right)100\% \tag{7.31}$$

At this point let us define $V_4 - V_1$ as ΔV. Using this definition and equations

(7.25) and (7.31) gives us the propeller efficiency as

$$\eta = \left(\frac{V_1}{V_1 + \Delta V/2}\right) 100\% \tag{7.32}$$

Equation (7.32) expresses the efficiency of an ideal fan or propeller in which the only energy loss considered is the loss in kinetic energy due to the change in velocity through the device. An actual propeller will have an efficiency less than that given by equation (7.32) because of the fact that the flow is not streamline and that there are friction losses in the blades, pulsations due to the finite number of blades, tip losses, and stream rotation after the stream leaves the blades. Propellers have actual efficiencies of 80 to 90%. The high efficiency of fans and propellers have made their application to wind chargers for power generation very attractive for alternate power sources.

ILLUSTRATIVE PROBLEM 7.11

What is the maximum efficiency of a propeller if the air speed of a plane is 150 mi/hr and the airstream velocity leaving the propeller is 200 mi/hr?

Solution

The increment in velocity across the entire unit is 200 − 150 = 50 mi/hr. Since half of this is added before the propeller, the velocity at the propeller is 175 mi/hr. Therefore, $\Delta V/2 = 25$ mi/hr and

$$\eta = \left(\frac{V_1}{V_1 + \Delta V/2}\right) 100 = \left(\frac{150}{150 + 25}\right) 100$$
$$= 85.7\%$$

ILLUSTRATIVE PROBLEM 7.12

An airplane is traveling at 300 mph and has a propeller that is 6 ft in diameter. If the specific weight of air is 0.0755 lb/ft², determine the thrust provided by the propeller if it is 90% efficient.

Solution

Since

$$\eta = \frac{V_1}{V_1 + \Delta V/2}$$

and V_1 is given as 300 mi/hr or 440 ft/sec,

$$0.9 = \frac{440}{440 + \Delta V/2}$$

$$0.9\left(440 + \frac{\Delta V}{2}\right) = 440$$

$$0.9\frac{(\Delta V)}{2} = 440 - 396 = 44$$

and

$$\Delta V = 97.8 \text{ ft/sec} \qquad \text{and} \qquad \frac{\Delta V}{2} = 48.9 \text{ ft/sec}$$

Therefore, the velocity at the propeller is $440 + 48.9 = 488.9$ ft/sec. At the exit, $V_4 = 488.9 + 48.9 = 537.8$ ft/sec. From equation (7.20),

$$F = \rho_1 A_1 V_1 (V_4 - V_1) = \rho_1 A_1 V_1 (\Delta V)$$

Since the propeller is 6 ft in diameter and γ is 0.0755 lb/ft^3,

$$F = \frac{0.075}{32.17} \times \frac{\pi}{4}(6)^2 \times 488.9(97.8)$$

$$= 3152 \text{ lb}$$

7.7 CLOSURE

In Chapter 5 we considered certain fluid dynamic applications that could be solved using the energy equation or Bernoulli equation. However, there are many instances where it is necessary to invoke other considerations in order to obtain solutions to problems in fluid mechanics. In this chapter we have applied Newton's second law in the form of the impulse–momentum relation to the solution of certain problems that are frequently encountered in technology. These problems have all dealt with forces exerted on or by fluids in motion. It will be noted that the basic procedure used in earlier chapters was followed in this chapter: namely, draw a sketch, label all quantities, write the applicable equation or equations, and solve the problem. It is a procedure that works and, unfortunately, many students have found that there is no shortcut to this procedure in problem solving. When one of these steps is omitted, there is usually found to be an error in the solution of a problem. Also, *always* pay attention to physical units, or else something will surely be wrong in your problem solutions.

PROBLEMS

7.1 A body of 10 kg experiences a force of 5 N for a period of 20 s. What will be its change in velocity?

7.2 Show that the thrust developed by a turbojet engine can be approximately written as

$$F = (1 + R)V_2\dot{m}_A$$

where R is the fuel–air ratio, $\dot{m}_f/\dot{m}_a$. State all your assumptions.

7.3 A turbojet engine is tested on a fixed test stand. Air having a specific weight of 0.075 lb/ft³ enters the engine at the rate of 7500 ft³/sec. Fuel is added to the engine at the rate of 1 lb of fuel for each 150 lb of air. Determine the thrust produced by the engine if the exhaust velocity is found to be 820 ft/sec. Neglect pressure differences at the inlet and outlet of the engine and take the inlet area to be 40 ft².

7.4 If the engine described in Problem 7.3 operates with 250 m³/s air at entry having a specific weight of 11.78 N/m³ and uses 1 kg of fuel for each 150 kg of air, what thrust will it produce? The exit velocity is found to be 300 m/s and the inlet and outlet pressures are to be taken to be equal. The inlet area of the engine is 4 m².

7.5 Use the results of Problem 7.2 to solve Problem 7.3. Compare your answers.

7.6 Use the results of Problem 7.2 to solve Problem 7.4. Compare your answers.

7.7 A jet of water 3 in. in diameter strikes a flat plate placed normal to the direction of flow. If the plate is stationary, calculate the force on it if the initial velocity of the jet is 40 ft/sec. Use 62.4 lb/ft³ for the specific weight of water.

7.8 A jet of water (density equal to 1000 kg/m³) issues from a nozzle having a diameter of 50 mm. What force will this water jet exert on a plate placed normal to the direction of flow if the velocity of the jet is 15 m/s?

7.9 A stream of water impinges on a flat plate. If the stream exerts a force of 1000 N on the plate, what is the velocity of the stream? Assume the fluid density to be 1000 kg/m³ and the stream to have a diameter of 75 mm.

7.10 A stream of air has a horizontal velocity of 60 ft/sec. What force is required to deflect this stream of air 180° if it has a cross-sectional area of 20 ft²? The specific weight of air = 0.0765 lb/ft³.

7.11 If air has a specific weight of 12.02 N/m^3 and an airstream has a velocity of 30 m/s, what force is required to deflect this stream 180 deg? Assume that the cross-sectional area of the stream is 3 m^2.

7.12 A 90° pipe bend in a horizontal plane carries water at a velocity of 20 ft/sec. If the pipe has an inside diameter of 5 in., what force is exerted on the bend by the dynamic action of the stream? The specific weight of water is 62.4 lb/ft^3.

7.13 A stream of water issues from a hose that has a 20-mm inside diameter. If 400 litres/min is delivered, what is the reaction force on the hose? Use $\gamma = 9810$ N/m^3 and consider that the jet is brought to rest.

7.14 What is the reaction of the water on the vane shown in Figure P7.14 if the jet has a velocity of 20 m/s? Use $\rho = 1000$ kg/m^3 and $Q = 0.2$ m^3/s.

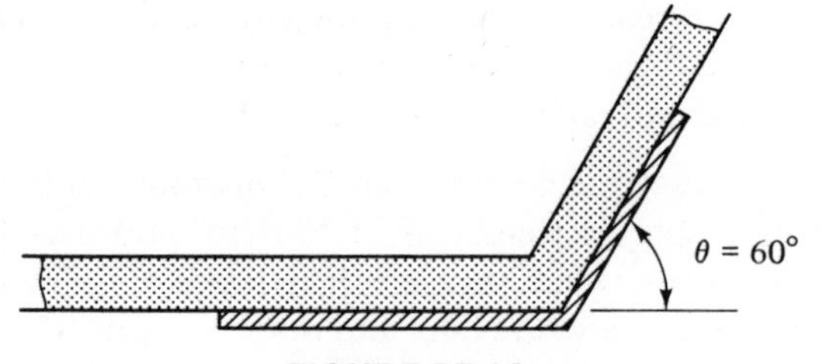

FIGURE P7.14

7.15 Solve Problem 7.14 if the angle of the vane is 120°.

7.16 Solve Problem 7.10 if friction reduces the leaving velocity to 50 ft/sec.

7.17 A stream of water strikes a plate as shown in Figure P7.17. What force does it exert on the plate? Assume that the water ($\gamma = 62.4$ lb/ft^3) issues from a 3-in.-diameter nozzle at the rate of 60 ft/sec.

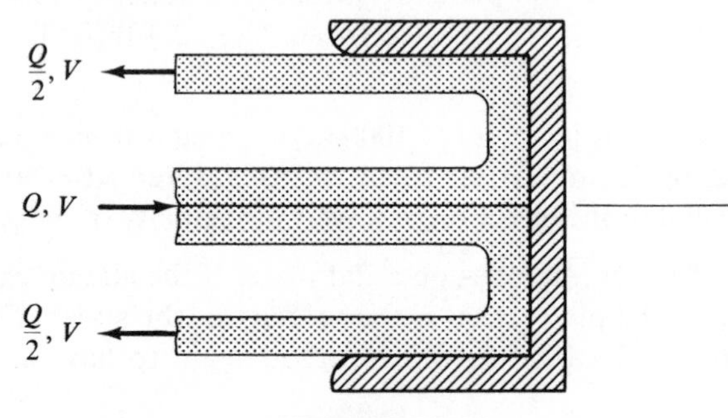

FIGURE P7.17

7.18 A pump discharges 200 gal/min through a 3-in.-diameter pipe (Figure P7.18). At the end of the pipe there is a nozzle having an outlet diameter

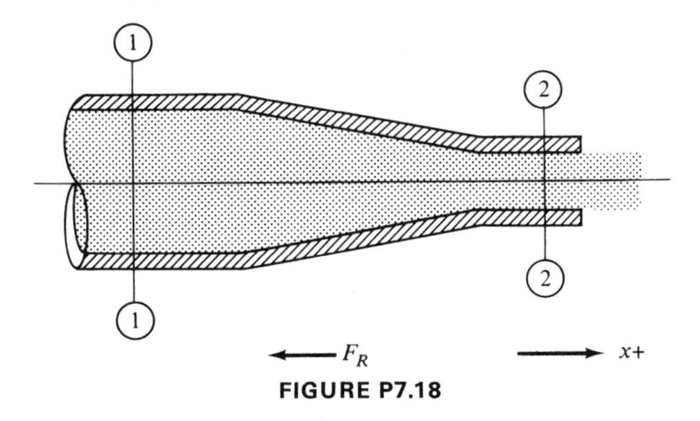

FIGURE P7.18

of 1 in. If $\gamma = 62.4$ lb/ft³, $p_1 = 50$ psig, and p_2 is atmospheric pressure, what force does the fluid exert on the hose?

7.19 A fluid stream exerts a force of 500 N on a vane that is moving in the direction of the stream with a velocity of 10 m/s. What power does the stream deliver to the vane?

7.20 A boat is propelled steadily at 20 mi/hr by a 25-hp motor. What thrust does the propeller exert on the boat?

7.21 Solve Problem 7.11 if the vane has a velocity of 5 m/s to the right.

7.22 Solve Problem 7.12 if the bend has a velocity of 5 ft/sec to the right.

7.23 Solve Problem 7.14 if the vane is moving to the right with a velocity of 5 m/s.

7.24 A pump is required to deliver 100 gal/min of water having $\gamma = 62.4$ lb/ft³ against a total head of 100 ft (ft lb/lb). What is the power required?

7.25 What horsepower must the pump shown in Figure P7.25 deliver to a fluid having $\gamma = 60$ lb/ft³? The pump delivers 40 gal/min.

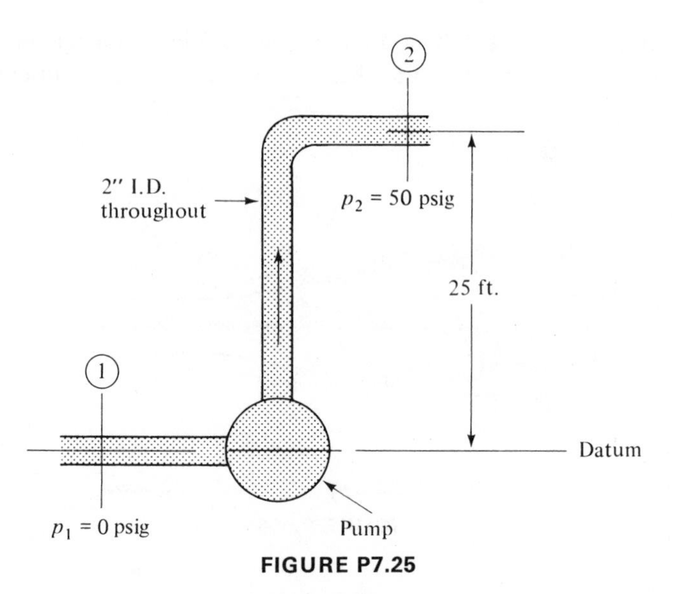

FIGURE P7.25

7.26 If there are flow losses in Problem 7.25 that are equivalent to 3 ft (ft lb/lb), what power must the pump deliver?

7.27 If the system shown in Problem 7.25 is a turbine with flow downward, what power will the fluid deliver to the turbine?

7.28 Solve Problem 7.26 if the pump is replaced by a turbine and the flow losses remain at 3 ft.

7.29 The pump shown in Figure P7.29 delivers water from the lower tank to

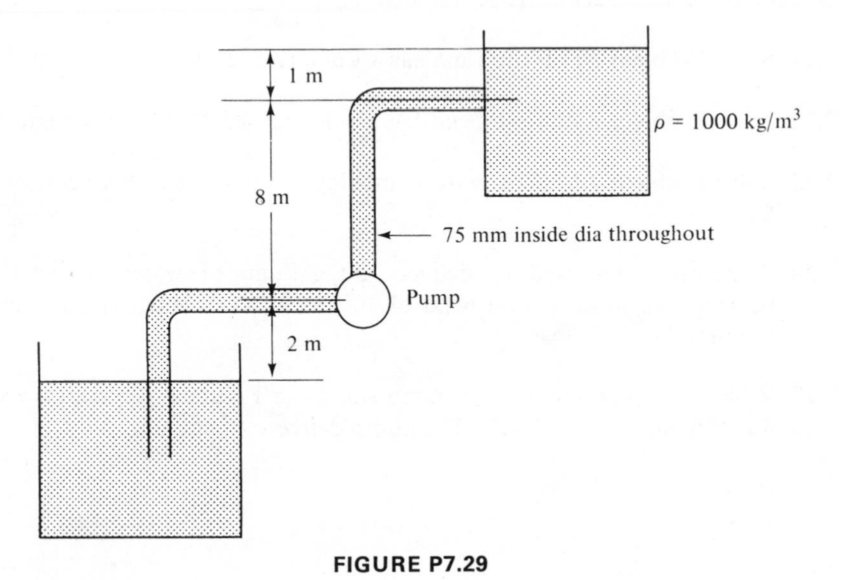

FIGURE P7.29

the upper tank. If 0.2 m^3/s is flowing, determine the power delivered to the fluid by the pump.

7.30 If there is a loss in energy between the inlet and outlet of the system shown in Problem 7.29 of 0.75 m, what power must the pump deliver?

7.31 An airplane travels at 300 mi/hr. If atmospheric pressure is 14.6 psia and the specific weight of air is 0.075 lb/ft^3, determine the stagnation pressure exerted on the airplane.

7.32 An airplane travels at 480 km/h. If atmospheric pressure is 100 kPa and the specific weight of air is 11.78 N/m^3, determine the stagnation pressure on the airplane.

7.33 If a propeller on an airplane going at 200 mi/hr causes the air to leave with an absolute velocity of 280 mi/hr, determine its efficiency.

7.34 A propeller 5 ft in diameter is mounted on a plane traveling at 350 mi/hr. If the airstream leaves the propeller with an absolute velocity of 410 mi/hr, determine the efficiency of the propeller and the thrust imparted to the aircraft. Assume that $\gamma = 0.075$ lb/ft^3.

7.35 A propeller is 2 m in diameter and is mounted on an aircraft going at 500 km/h. The airstream leaves the propeller with a velocity of 650 km/h. Determine the efficiency of the propeller and the thrust. Assume that the γ of air $= 11.8$ N/m^3.

7.36 If a propeller has an efficiency of 90%, determine its thrust if it is 6 ft in diameter and is mounted on a plane traveling at 275 mi/hr. Use $\gamma = 0.075$ lb/ft^3.

7.37 Determine the horsepower absorbed by a propeller if it is on a plane traveling at 240 mi/hr, is 7 ft in diameter, is 80% efficient, and $\gamma = 0.075$ lb/ft^3.

7.38 A propeller is 2.5 m in diameter and can exert a thrust of 7 kN on an airplane traveling at 360 km/h. What is its efficiency? $\gamma = 11.8$ N/m^3.

7.39 A propeller 7 ft in diameter is capable of exerting a 1500-lb thrust on an airplane traveling at 280 mi/hr. Determine its efficiency. Use $\gamma = 0.075$ lb/ft^3.

REFERENCES

1. *Elements of Practical Aerodynamics* by B. JONES, 4th ed., John Wiley & Sons, Inc., New York, 1950.
2. *Basic Fluid Mechanics* by J. L. ROBINSON, McGraw-Hill Book Company, New York, 1963.

3. *Engineering Applications of Fluid Mechanics* by J. C. HUNSAKER and B. G. RIGHTMIRE, McGraw-Hill Book Company, New York, 1947.
4. *Elementary Theoretical Fluid Mechanics* by K. BRENKERT, JR., John Wiley & Sons, Inc., New York, 1960.
5. *Fluid Mechanics* by V. L. STREETER, 3rd ed., McGraw-Hill Book Company, New York, 1962; 6th ed., 1979.
6. *Elementary Fluid Mechanics* by J. K. VENNARD, 3rd ed., John Wiley & Sons, Inc., New York, 1954.
7. *Mechanics of Fluids* by G. MURPHY, 2nd ed., International Textbook Company, Scranton, Pa., 1952.
8. *Fluid Mechanics for Engineers* by P. S. BARNA, Butterworth & Co. (Publishers) Ltd., London, 1957.
9. *Fluid Mechanics* by R. C. BINDER, 4th ed., Prentice-Hall, Inc., Englewood Cliffs, N.J., 1962.
10. *Applied Fluid Mechanics* by R. L. MOTT, 2nd ed., Charles E. Merrill Publishing Co., Columbus, Ohio, 1979.
11. *Engineering Fluid Mechanics* by J. A. ROBERSON and C. T. CROWE, Houghton Mifflin Company, Boston, 1975.
12. *Fundamentals of Fluid Mechanics* by J. A. SULLIVAN, Reston Publishing Co., Reston, Va., 1978.

Open Channel Flow

8.1 INTRODUCTION

The study of the fluid mechanics of rivers, culverts, canals, and conduits flowing partially full comes under the general heading of *open channel flow*. In this type of flow one surface is free and is not confined or in contact with a wall. Since the flow channel may be irregular and have a varying cross section and depth, the character of the flow differs from one location to another. The general considerations developed for the flow of fluids in pipes are applicable to open channel flow, but the presence of the free surface and the varying conditions in the channel make the analytical determination of the flow in open channels very difficult.

To approach this subject in a logical manner let us first identify and categorize the many possible flow regimes in a channel. The flow may be classified as being steady (independent of time) or unsteady (time dependent) and uniform or nonuniform; in addition, these flows can be further classified as tranquil or rapid. Steady uniform flow occurs in channels that are very long and whose depth and slope are constant along the length of the channel.

In this type of flow, the slope of the free surface is found to be parallel to the slope of the bed of the channel. The flow in this case is also said to be tranquil if the velocity is low enough for a small wave to be able to travel upstream causing upstream conditions to be governed by downstream conditions; the flow is said to be rapid (or shooting) if the stream velocity is so high that a small wave cannot travel upstream. If the flow is such that the stream velocity is just equal to the velocity of a small wave, it is said to be critical.

Steady nonuniform flow occurs where the channel or its depth (or both) change from section to section. In this case the velocity must change from section to section, and the flow can change from tranquil to rapid or from rapid to tranquil at different cross sections along the length of the channel.

8.2 UNIFORM STEADY FLOW

Where fluid is flowing uniformly and steadily in an open channel, the rate of flow of fluid past any cross section as well as the cross section will be constant at all times and locations. Thus, every section of the channel will appear to be the same as every other section. The energy that provides the driving force for the flow is the change in potential energy of the channel as it slopes down toward the discharge end. As is shown in Figure 8.1, the channel bed has a constant slope angle of α. Since the flow is uniform and the channel has a constant depth, $V_1 = V_2$ and $V_1^2/2g = V_2^2/2g$. Therefore, the water surface

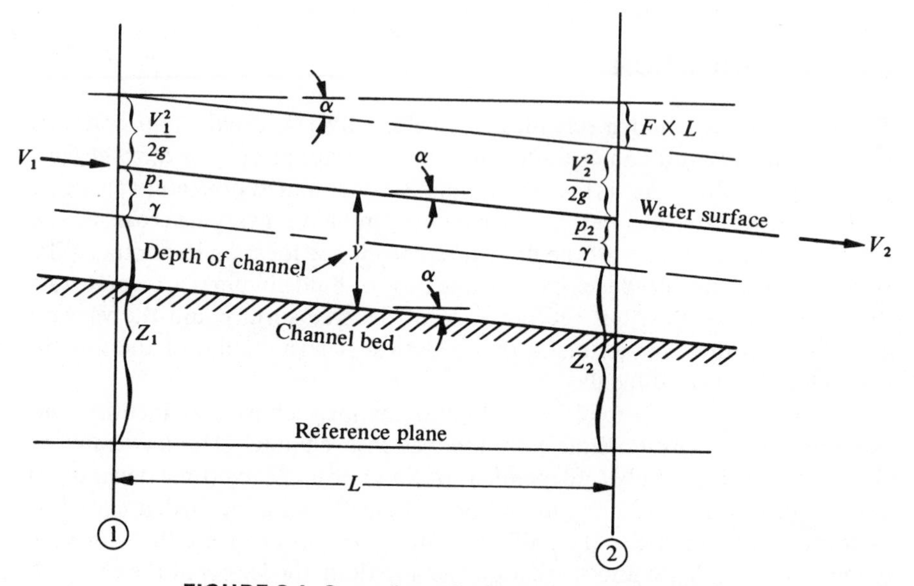

FIGURE 8.1 Open channel with uniform steady flow.

must be parallel to the bed of the channel and have the slope α. If the water has the specific energy $V_1^2/2g + p_1/\gamma + Z_1$ at the inlet, it will have as its specific energy $V_2^2/2g + p_2/\gamma + Z_2 + (FL)$ at the outlet, where F is the loss in energy due to friction per unit length of channel and L the length of the channel. The slope F of the energy line is numerically equal to the slope of the free surface, which in turn is numerically equal to the slope of the channel bed. Thus, these three lines (planes) must be parallel and are shown to be parallel in Figure 8.1.

With the foregoing in mind, let us consider the case of a rectangular channel with uniform steady flow. A frictional force will exist on the fluid due to the shear in the fluid layers adjacent to the channel wall. Denoting the shear stress by τ and the wetted perimeter by P, the shearing force equals τPL. If the flow is uniform, this shearing force must equal the component of the weight of the fluid in a direction parallel to the flow. This becomes (see Figure 8.2b) equal to $(\gamma LA) \sin \alpha$, where γLA is simply the weight of the fluid. Equating these forces yields

$$\tau PL = \gamma LA \sin \alpha \tag{8.1}$$

and for the shear stress,

$$\tau = \frac{A}{P}(\gamma \sin \alpha) \tag{8.2}$$

The ratio A/P is called the *hydraulic radius* and is denoted by the symbol R. Therefore,

$$R = \frac{\text{flow area}}{\text{wetted perimeter}} = \frac{A}{P} \tag{8.3}$$

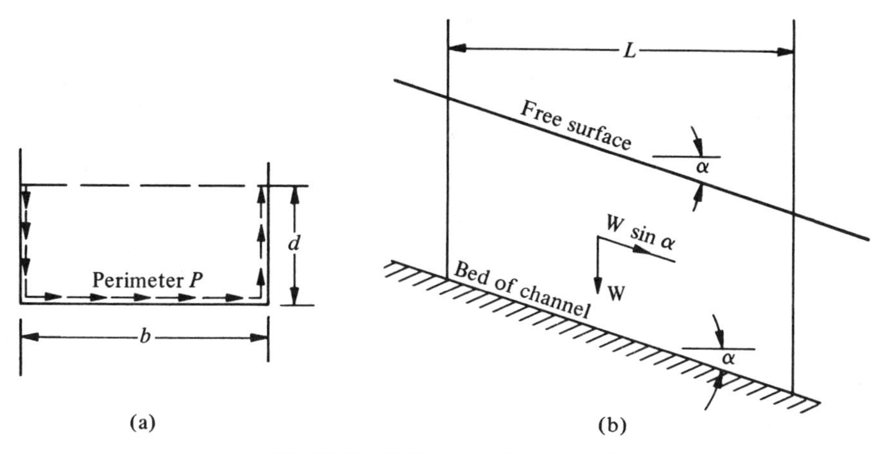

FIGURE 8.2 Rectangular channel.

From our previous study of flow in pipes, it can be stated that the shear stress will also be some function of $\rho V^2/2$. Using the mathematical symbol Ψ to denote "a function of,"

$$\tau = \Psi \frac{\rho V^2}{2} \tag{8.4}$$

If equations (8.2), (8.3), and (8.4) are combined, we will arrive at a new combined function, and using C to denote the resulting combined function,

$$V = C\sqrt{RS} \tag{8.5}$$

Equation (8.5) is also known as the *Chezy equation*, and the student will note that the coefficient C is not a constant, and we would expect it to be a function of the Reynolds number, the relative roughness of the channel, and a factor depending on the form of the channel. Manning has empirically determined C as

$$C = \frac{1.486}{n} R^{1/6} \tag{8.6}$$

where n incorporates the roughness of the channel. Combining (8.5) and (8.6), we obtain an equation known as *Manning's formula*:

$$V = \frac{1.486}{n} R^{2/3} S^{1/2} \tag{8.7}$$

In SI units equation (8.7) can be written as

$$V = \frac{1.00}{n} R^{2/3} S^{1/2} \tag{8.8}$$

Values for n for various channel surfaces are tabulated in Table 8.1. Also, it should be remembered that

$$Q = AV \tag{8.9}$$

The student should note the Manning's formula is empirical and is useful only over certain ranges of operation of channels. Other correlations have been proposed but these are no more accurate than equation (8.7) and in practice Manning's formula [equation (8.7)] is usually used. It should be further noted that n should have the units of $\text{ft}^{1/6}$ or $\text{m}^{1/6}$ for dimensional consistency. However, Manning's equation was not dimensionally consistent, and n is taken to be dimensionless in his empirical equation for conversion to SI. In the English system V will be in ft/sec and R will be in feet. In SI units,

TABLE 8.1 Flow of Water in Open Channels, Average Values of Roughness Coefficient (n), for Use in Manning's Formula

Type of Open Channel	(n)
Smooth concrete	0.014
Planed timber, asbestos pipe	0.012
Lined cast iron, wrought iron, welded steel	0.015
Vitrified sewer pipe, ordinary concrete, good brickwork, unplaned timber	0.016
Clay sewer pipe, cast iron pipe, cement lining	0.015
Riveted steel, average brickwork	0.018
Rubble masonary, smooth earth	0.025
Firm gravel, corrugated pipe	0.025
Natural earth channels (good condition)	0.025
Natural earth channels (stones and weeds)	0.032
Channels cut in rock, winding river with pools, shoals	0.040
Sluggish river (rather weedy)	0.055

V will be in m/s and R will be in metres. S, the slope, is in ft/ft or m/m and is dimensionless.

ILLUSTRATIVE PROBLEM 8.1

Determine the velocity of flow for a square, smooth concrete channel flowing full if each side is 2 m and if the slope is 1 m in 1000 m (Figure 8.3).

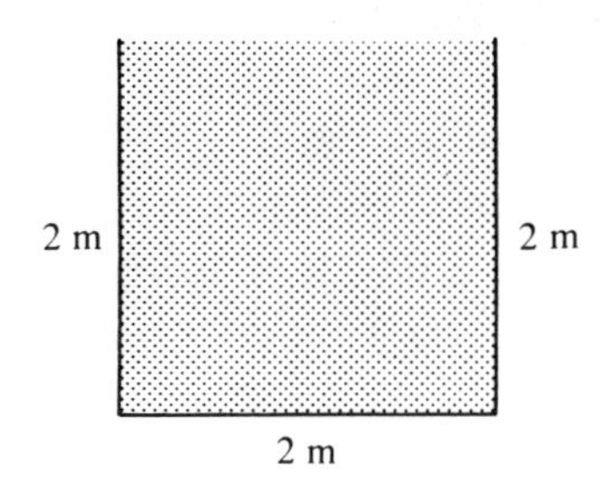

FIGURE 8.3 Illustrative Problem 8.1.

Solution

As shown in the sketch, $A = 2\text{ m} \times 2\text{ m} = 4\text{ m}^2$ and $P = 6\text{ m}$. Therefore,

$$\text{hydraulic radius} = R = \frac{A}{P} = \frac{4\text{ m}^2}{6\text{ m}} = \frac{2}{3}\text{ m}$$

From Table 8.1, $n = 0.014$. Using equation (8.8),

$$V = \frac{1.00}{n} R^{2/3} S^{1/2}$$

$$= \frac{1.00}{0.014} \left(\frac{2}{3}\right)^{2/3} (0.001)^{1/2}$$

Therefore,

$$V = 1.72 \text{ m/s}$$

If the volume flow (the discharge) is desired,

$$Q = AV = 4 \text{ m}^2 \times 1.72 \text{ m/s} = 6.88 \text{ m}^2/\text{s}$$

ILLUSTRATIVE PROBLEM 8.2

A concrete trench, square in cross section, is to be built to carry away 2250 gal/min of water. The trench will be 1000 ft long, and the change in elevation between the ends is 10 ft. What must be the dimensions if the trench is designed to operate half-full?

Solution

Denoting the width of the channel as b, the height of the liquid is $b/2$ (Figure 8.4). The area of the channel is $b(b/2) = b^2/2$. The wetted perimeter is $(b/2) + (b/2) + b$, which is $2b$. Therefore,

$$\text{hydraulic radius} = R = \frac{A}{P} = \frac{b^2/2}{2b} = \frac{b}{4}$$

The slope S is given as 10 ft in 1000 ft or $10/1000 = 0.01$. The quantity of water flowing is given as 2250 gal/min. Since 1 gal is 231 in.^{3}, 2250 gal/min

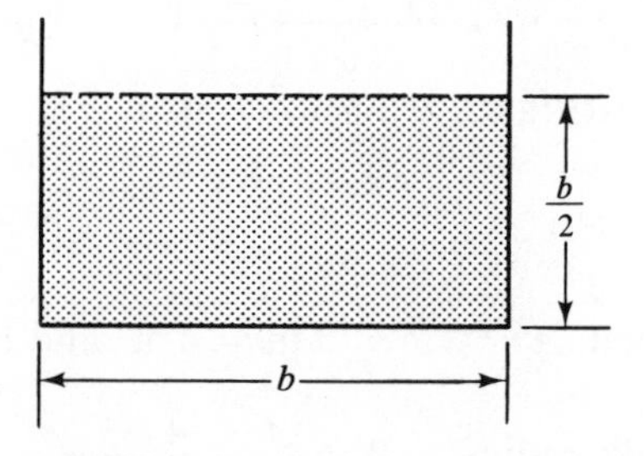

FIGURE 8.4 Illustrative Problem 8.2.

is also

$$\text{flow rate} = Q = \frac{\text{gal}}{\text{min}} \times \frac{1}{\text{sec/min}} \times \frac{\text{in.}^3/\text{gal}}{\text{in.}^3/\text{ft}^3} = \text{ft}^3/\text{sec}$$

$$= \frac{2250}{60} \times \frac{231}{1728} = 5 \text{ ft}^3/\text{sec flowing}$$

Since $Q = AV$,

$$V = \frac{Q}{A} = \frac{5}{b^2/2}$$

where V is in feet per second and A is in square feet. The velocity is also given by equation (8.7). Thus,

$$V = \frac{1.486}{n} R^{2/3} S^{1/2} = \frac{10}{b^2}$$

Substituting the data for this problem and using n of 0.014 as being the average value from Table 8.1 for smooth and ordinary concrete, we obtain

$$\frac{1.486}{0.014}\left(\frac{b}{4}\right)^{2/3}(0.01)^{1/2} = \frac{10}{b^2}$$

Solving yields

$$b = 1.38 \text{ ft}$$

and

$$\frac{b}{2} = 0.69 \text{ ft}$$

The flow area will be equal to (1.38)(0.69) = 0.952 ft^2 and the velocity is $10/(1.38)^2 = 5.26$ ft/sec.

ILLUSTRATIVE PROBLEM 8.3

If the channel in Illustrative Problem 8.2 is designed to flow five-eighths full, determine the dimensions, the velocity of the water, and the area of the channel.

Solution

Proceeding as in Illustrative Problem 8.2 (see Figure 8.5), we obtain

$$\text{hydraulic radius} = R = \frac{A}{P} = \frac{\frac{5}{8}b^2}{\frac{5}{8}b + \frac{5}{8}b + b} = \frac{5}{18}b = 0.278b$$

$$\text{flow rate} = Q = 5 \text{ ft}^3/\text{sec} = AV$$

$$\text{velocity} = V = \frac{5}{\frac{5}{8}b^2} = \frac{8}{b^2}$$

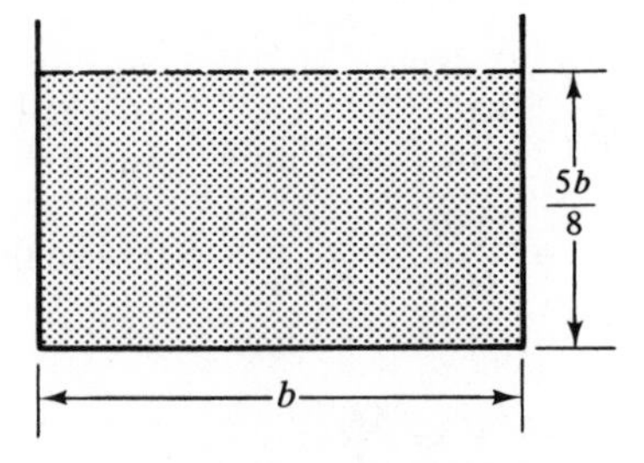

FIGURE 8.5 Illustrative Problem 8.3.

Therefore,

$$V = \frac{1.486}{n} R^{2/3} S^{1/2} = \frac{1.486}{0.014} \times (0.278b)^{2/3}(0.01)^{1/2} = \frac{8}{b^2}$$

Solving, we obtain

$$b = 1.238 \text{ ft}$$
$$\tfrac{5}{8}b = 0.772 \text{ ft}$$
$$\text{area} = 0.955 \text{ ft}^2$$

and

$$\text{velocity} = \frac{8}{(1.238)^2} = 5.24 \text{ ft/sec}$$

Thus, we see that with the same volume flow and the same slope there is a different set of dimensions that yields the same flow area and the same flow velocity.

ILLUSTRATIVE PROBLEM 8.4

If the channel designed in Illustrative Problem 8.1 is flowing with a depth equal to five-eighths of the width, determine the flow and compare with the results of Illustrative Problem 8.2.

Solution

The dimensions of the channel are shown in Figure 8.6. The flow area is $(1.38)(0.863) = 1.19 \text{ ft}^2$. Proceeding as before, we obtain

$$\text{hydraulic radius} = R = \frac{A}{P} = \frac{1.19}{0.863 + 0.863 + 1.38} = 0.383 \text{ ft}$$

$$\text{velocity} = V = \frac{1.486}{n} R^{2/3} S^{1/2}$$

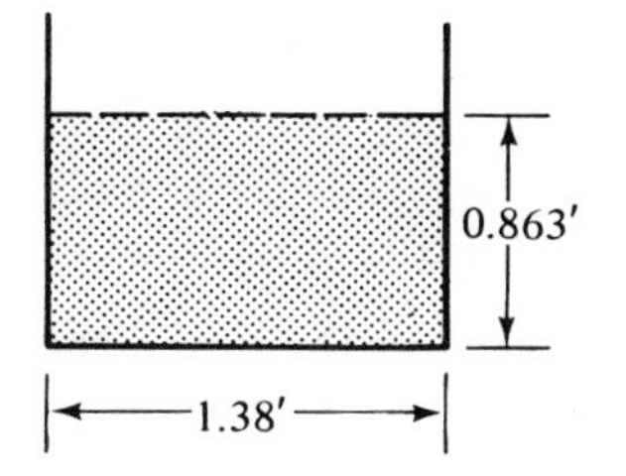

FIGURE 8.6 Illustrative Problem 8.4.

All items are the same as in Illustrative Problem 8.2 except that the hydraulic radius has changed due to the change in dimensions. Therefore, we may form the ratio between the results of Illustrative Problem 8.2 and this problem as follows:

$$\frac{V_2}{V_4} = \left(\frac{R_2}{R_4}\right)^{2/3} = \left(\frac{1.38/4}{0.383}\right)^{2/3} = 0.932$$

where the subscripts 2 and 4 denote, respectively, Illustrative Problems 8.2 and 8.4. Therefore,

$$V_4 = 1.07V_2 = 1.07 \times 5.25 = 5.62 \text{ ft/sec}$$

$$Q = AV = 1.92 \times 5.62 = 6.7 \text{ ft}^3\text{/sec}$$

Thus, increasing the area by 25% leads to a flow increase of 34% when Illustrative Problem 8.4 is compared to Illustrative Problem 8.2.

8.3 OPTIMUM CHANNEL OF RECTANGULAR CROSS SECTION

In many cases it is desired to design a channel to have the maximum quantity of fluid flow for a given type of construction and with a specified slope of channel. For a fixed geometry and flow area, the maximum quantity of flow will occur when the velocity is maximum, and under these conditions equation (8.9) indicates that this occurs when the hydraulic radius is a maximum. Let us now consider a rectangular channel having a width b and a depth of water d, as shown in Figure 8.2a. Since $R = A/P$ and A is given as a fixed quantity, R is maximum when the wetted perimeter P is a minimum. Thus, we have

$$b + 2d = P \tag{8.10a}$$

$$bd = A \tag{8.10b}$$

Therefore,

$$\frac{A}{d} + 2d = P \tag{8.11a}$$

If for convenience we now let $bd = A = 1$, equation (8.11a) becomes

$$\frac{1}{d} + 2d = P \tag{8.11b}$$

Equation (8.11b) can be solved by the methods of calculus or numerically, as is shown in Table 8.2.

TABLE 8.2

Assume d	b	d	$1/d$	$2d$	P
$0.1b$	3.16	0.316	3.16	0.632	3.792
$0.2b$	2.24	0.448	2.24	0.896	3.136
$0.3b$	1.83	0.549	1.83	1.098	2.928
$0.4b$	1.58	0.632	1.58	1.264	2.844
$0.5b$	1.414	0.707	1.414	1.414	2.828
$0.6b$	1.29	0.774	1.29	1.548	2.838
$0.7b$	1.195	0.837	1.195	1.674	2.869
$0.8b$	1.119	0.895	1.119	1.790	2.909
$0.9b$	1.052	0.947	1.052	1.894	2.946
$1.0b$	1.0	1.0	1.0	2.0	3.0

By inspection of Table 8.2 (or by plotting P vs. d) it will be seen that the minimum value of P occurs when $d = \frac{1}{2}b$. In other words, the maximum velocity (or flow) for a fixed area rectangular channel of constant slope occurs when the depth of the water flowing equals half the channel width. In addition, since the perimeter is a minimum with these dimensions, the quantity of material required to construct and line the channel will also be a minimum. These conclusions do not depend on the choice of $A = 1$ in our calculations.

ILLUSTRATIVE PROBLEM 8.5

A rectangular smooth concrete channel having a slope of 1 m in 1000 m is designed to carry 6.88 m^3/s. Determine the dimensions of the channel.

Solution

For optimum channel dimensions the depth should be half the width of the channel. This corresponds to the sketch for Illustrative Problem 8.2.

Proceeding and using the same notation as in Illustrative Problem 8.2,

$$\text{hydraulic radius} = R = \frac{A}{P} = \frac{b^2/2}{2b} = \frac{b}{4}$$

Since $Q = 6.88\ \text{m}^3/\text{s}$,

$$V = \frac{Q}{A}$$

$$= \frac{6.88}{b^2/2} = \frac{13.76}{b^2}$$

Equating this to equation (8.8) yields

$$\frac{13.76}{b^2} = \frac{1.00}{n} R^{2/3} S^{1/2}$$

For the data of this problem, this becomes

$$\frac{13.76}{b^2} = \frac{1.00}{0.014}\left(\frac{b}{4}\right)^{2/3} (0.001)^{1/2}$$

Solving, we obtain $b = 2.78$ m and $b/2 = 1.39$ m.

If this problem is compared to Illustrative Problem 8.1, it will be noted that both channels carry the same discharge, 6.88 m³/s. The perimeter in Illustrative Problem 8.1 (and consequently, the amount of material needed) is 6 m; in this problem it is 5.57 m This agrees with our conclusion that the rectangular channel flowing half-full is the optimum rectangular channel.

ILLUSTRATIVE PROBLEM 8.6

The combined discharge from two 18-in.-diameter concrete storm sewer pipes, each flowing one-half full on a 1.25% grade, is to be carried by a cement lined open rectangular channel on a 1.0% grade (Figure 8.7). Design

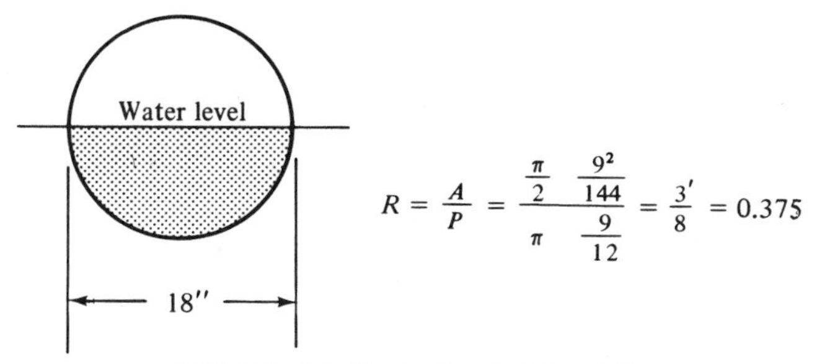

FIGURE 8.7 Illustrative Problem 8.6.

the rectangular channel so that it will have a depth of flow equal to one-half of its width when carrying this discharge.

Solution

Consider the 18-in.-diameter storm sewer pipes:

$$V = \frac{1.486}{n} R^{2/3} S^{1/2} \qquad \text{from Table 8.1, } n = 0.015$$

$$= \frac{1.486}{0.015} (0.375)^{2/3} (0.0125)^{1/2} = 5.71 \text{ ft/sec}$$

Since there are two pipes,

$$\text{total flow} = 2AV = 2\left[\frac{\pi \times (1.5)^2}{8}\right] \times 5.71 = 10.1 \text{ ft}^3\text{/sec}$$

The rectangular channel has an area of $\frac{1}{2}b^2$; therefore,

$$\frac{b^2 V}{2} = 10.1$$

and

$$b^2 V = 20.2$$

or

$$V = \frac{20.2}{b^2}$$

But

$$V = \frac{1.486}{n} R^{2/3} S^{1/2} \qquad n = 0.015 \text{ (Table 8.1)}$$

and

$$R = \frac{b^2/2}{2b} = \frac{b}{4}$$

Therefore,

$$\frac{20.2}{b^2} = \frac{1.486}{0.015} \left(\frac{b}{4}\right)^{2/3} (0.01)^{1/2}$$

$$b = 1.85 \text{ ft} \qquad \text{and} \qquad d = 0.925 \text{ ft}$$

ILLUSTRATIVE PROBLEM 8.7

How much additional depth of channel in Illustrative Problem 8.6 will be required for the channel to carry the combined discharge when the two pipes are flowing just full?

Solution

A circular pipe flowing just full or half-full has the same hydraulic radius R and therefore the same velocity V. The total Q is therefore

$$5.71 \times 2 \times \frac{\pi \times (1.5)^2}{4} = 20.2 \text{ ft}^3/\text{sec}$$

Denoting the cement-lined channel's depth by h,

$$Q = AV = A\left(\frac{1.486}{0.015} R^{2/3} S^{1/2}\right) = 20.2 \text{ ft/sec}$$

Since $R = A/P$,

$$A = \frac{1.486}{0.015}\left(\frac{A}{P}\right)^{2/3} (0.01)^{1/2} = 20.2$$

$$\left(\frac{A}{P}\right)^{2/3} A = \frac{20.2 \times 0.015}{1.486 \times 0.1} = 2.04$$

and

$$\left[2.04 = \left(\frac{A}{P}\right)^{2/3} A\right]^{3/2} = 2.92 = \frac{A}{P}(A)^{3/2}$$

Therefore,

$$\frac{A}{P} = \frac{2.92}{A^{3/2}}$$

From Figure 8.8, we also have

$$\frac{A}{P} = \frac{1.85h}{2h + 1.85}$$

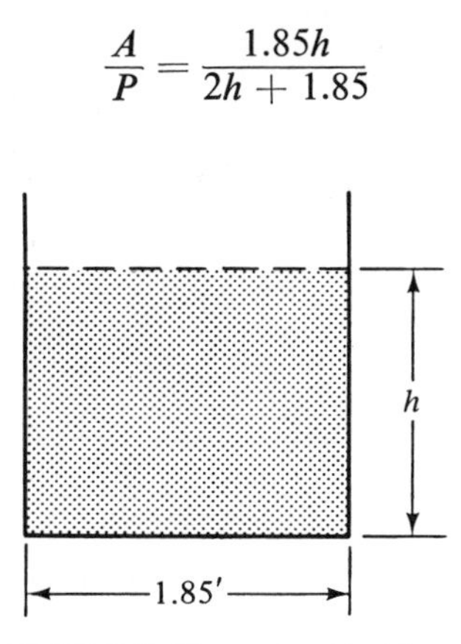

FIGURE 8.8 Illustrative Problem 8.7.

Therefore,

$$\frac{1.85h}{2h + 1.85} = \frac{2.92}{(1.85h)^{3/2}}$$

To solve for h, let us assume various values of h and evaluate each side of the equation. Try

	Left	*Right*
$h = 1$	0.48	1.16
$h = 1.5$	0.57	0.63
$h = 2$	0.63	0.41

Plotting these data (Figure 8.9) yields $h = 1.6$. As a check,

$$R = \frac{1.6 \times 1.85}{3.2 + 1.85} = 0.59$$

$$V = \frac{1.486}{0.015}(0.59)^{2/3} \times (0.01)^{1/2}$$

$$= 7.0 \text{ ft/sec}$$

$$Q = AV = 1.6 \times 1.85 \times 7.0 = 20.7 \text{ ft}^3\text{/sec vs. ft}^3\text{/sec}$$

The difference in these solutions is only 2.5%, which is satisfactory.

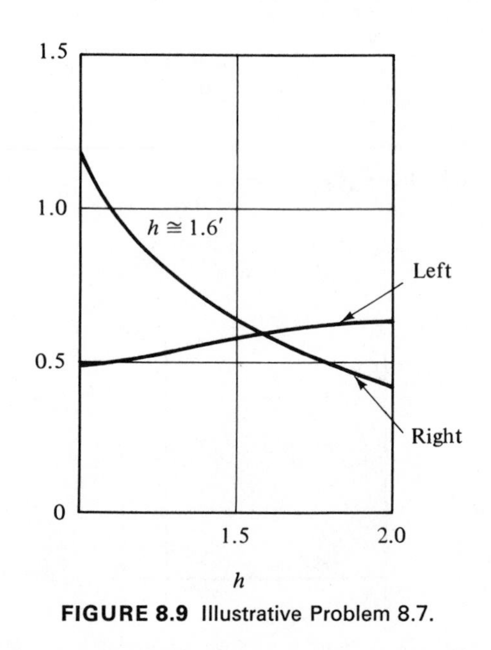

FIGURE 8.9 Illustrative Problem 8.7.

8.4 PIPES FLOWING PARTIALLY FULL

Quite often a large circular pipe will be used as a storm drain or sewer. In this case the pipe may not be flowing full and can be treated as an open channel with uniform flow, as was done in Illustrative Problem 8.6. The Chezy formula was applied to this specific situation to obtain the desired solution, but a more general treatment of this problem will be given in the following paragraphs for pipes of circular cross section.

Consider the circular pipe shown in Figure 8.10, with the liquid flowing above the horizontal center line. The diameter of the pipe will be denoted as

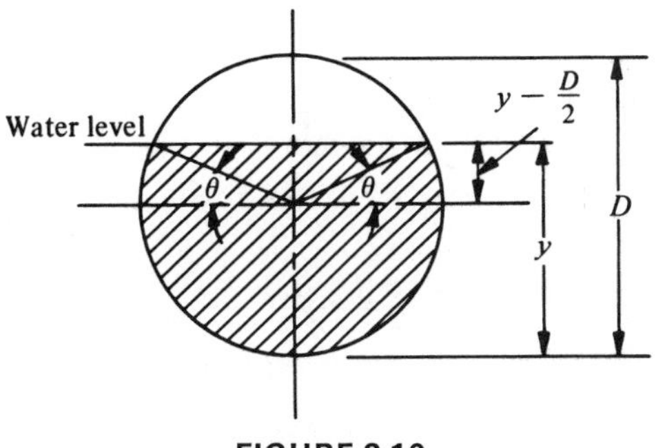

FIGURE 8.10

D and the depth of fluid will be denoted as y. To generalize the solution of this problem, it will be convenient to determine the ratio of the quantity of fluid flowing to the quantity when the pipe flows full. From equation (8.7), for a given pipe with a given slope,

$$\frac{Q}{Q_f} = \frac{AV}{A_f V_f} = \frac{A[(1.486/n)R^{2/3}S^{1/2}]}{A_f[(1.486/n)R_f^{2/3}S^{1/2}]} = \frac{A}{A_f}\left(\frac{R}{R_f}\right)^{2/3} \tag{8.12a}$$

where the subscript f denotes conditions where the pipe is flowing full. Since $R = A/P$,

$$\frac{Q}{Q_f} = \left[\left(\frac{A}{A_f}\right)^{5/3}\right]\left(\frac{P_f}{P}\right)^{2/3} \tag{8.12b}$$

From Figure 8.10, the flow area consists of the area of the sector of the circle plus the area of the triangle. In terms of the angle θ (in radians),

$$A = \left(\frac{\pi D^2}{4}\right)\frac{2[(\pi/2) + \theta]}{2\pi} + \left(y - \frac{D}{2}\right)\left(\frac{D}{2}\cos\theta\right) \tag{8.13}$$

The wetted perimeter P is given by

$$P = \frac{\pi D[(\pi/2) + \theta]}{2\pi} \tag{8.14}$$

The angle θ is given in terms of y and D as the angle whose sine is $(y - D/2)/(D/2)$, or

$$\theta = \arcsin\left(\frac{y - D/2}{D/2}\right) \tag{8.15}$$

For the pipe flowing full,

$$A_f = \frac{\pi D^2}{4} \qquad \text{and} \qquad P_f = \pi D \tag{8.16}$$

Substituting equation (8.13) through (8.16) into equation (8.12b) and simplifying, we obtain

$$\frac{Q}{Q_f} = \left\{\frac{(\pi/2) + \theta}{\pi} + \frac{\cos[(2y/D) - 1]}{\pi}\right\}^{5/3}\left[\frac{\pi}{(\pi/2) + \theta}\right]^{2/3} \tag{8.17}$$

Similarly,

$$\begin{aligned}\frac{V}{V_f} &= \left\{\frac{(\pi/2) + \theta}{\pi} + \frac{\cos\theta\,[(2y/D) - 1]}{\pi}\right\}^{2/3}\left[\frac{\pi}{(\pi/2) + \theta}\right]^{2/3} \\ &= \left\{\frac{\cos\theta\,[(2y/D) - 1]}{(\pi/2) + \theta}\right\}^{2/3}\end{aligned} \tag{8.18}$$

It will be noted from equations (8.15), (8.17), and (8.18) that Q/Q_f and V/V_f are functions only of the ratio y/D. A similar analysis also shows that Q/Q_f and V/V_f are functions of the ratio y/D when the liquid level is below the horizontal center line. These solutions are plotted in Figure 8.11, where it is

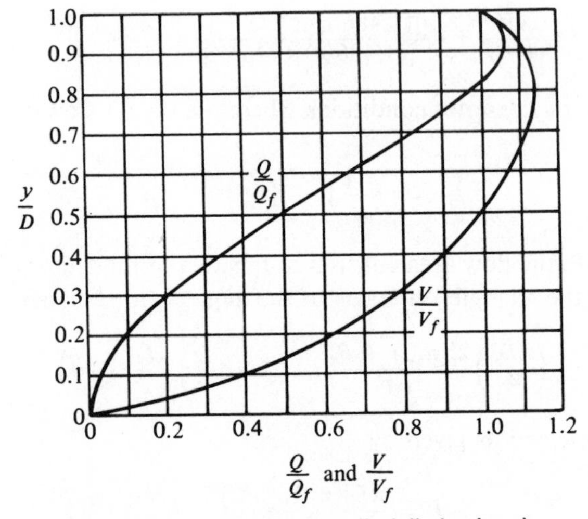

FIGURE 8.11 Flow in partially full circular pipe.

seen that Q/Q_f is maximum at $y = 0.94D$ and V/V_f is a maximum at $0.8D$. Figure 8.11 simplifies the work considerably and is sufficiently accurate for most purposes. Since all functions are given as dimensionless ratios, it is applicable to both the English and SI systems of units.

ILLUSTRATIVE PROBLEM 8.8

Water flows at the rate of 0.3 m³/s in a 1-m-diameter brick lined pipe ($n = 0.015$). If the slope is 0.002, calculate the depth of flow.

Solution

If the drain flows full,

$$R = \frac{\pi D^2/4}{\pi D} = \frac{D}{4} = 0.25$$

and

$$Q_f = AV = \left[\frac{\pi(1)^2}{4}\right]\frac{1.00}{0.018}(0.25)^{2/3}(0.002)^{1/2}$$

$$= 0.774 \text{ m}^3/\text{s}$$

Therefore

$$\frac{Q}{Q_f} = \frac{0.3}{0.774} = 0.387$$

From Figure 8.4, $y/D \simeq 0.43$. Thus,

$$y = 0.43 \times 1 = 0.43 \text{ m}$$

ILLUSTRATIVE PROBLEM 8.9

Water flows at the rate of 1 ft³/sec in a 24-in.-diameter storm drain having a slope of 0.0015 and n of 0.020. Calculate the depth of flow.

Solution

If the drain is flowing full,

$$Q_f = AV = \left[\frac{\pi(2)^2}{4}\right]\frac{1.486}{0.020}\left(\frac{2}{4}\right)^{2/3}(0.0015)^{1/2} = 5.71 \text{ ft}^3/\text{sec}$$

Therefore,

$$\frac{Q}{Q_f} = \frac{1}{5.71} = 0.175$$

From Figure 8.11,

$$\frac{y}{D} = 0.28$$

Thus,

$$y = 0.28 \times 2 = 0.56 \text{ ft}$$

8.5 NONUNIFORM STEADY FLOW

When a fluid flows in an open channel, the velocity is not constant over the cross section of the channel. In general, there are velocity differences from side to side as well as from top to bottom of the channel. If it is desired to express the kinetic energy in terms of the mean (area-weighted) velocity, it is convenient to multiply $V^2/2g$ by a factor α, where V is the average or mean velocity. If the velocity is constant over the area, α equals unity, but depending on the character of the channel, it can equal or exceed 2. Since α cannot be predicted in advance, it is usual to assume its value to be unity: if enough information is available, it should be taken into account. In the following development α will be assumed to be unity.

Let us again consider the case shown in Figure 8.1 and now write an equation for the energy per unit weight, with the reference datum taken as the bottom of the channel. This energy quantity is known as the specific energy and is simply given by

$$E = y + \frac{V^2}{2g} \tag{8.19}$$

In the case of uniform flow, E is a constant, and for nonuniform flow, E may increase or decrease. Since $Q = AV$,

$$V = \frac{Q}{A}$$

and

$$E = y + \frac{Q^2}{2gA^2} \tag{8.20}$$

Per unit width of the channel, $A = y$, and q denotes the volume flow per unit width. Therefore,

$$E = y + \frac{q^2}{2gy^2} \tag{8.21}$$

Figure 8.12 shows a curve of E as a function of y for a fixed value of q. The 45° diagonal line is the potential energy for various values of y, and the horizontal distance between the 45° diagonal and the curve is the kinetic energy. By inspection of Figure 8.12 it can be seen that there is a minimum value of E (at a fixed value of q) that occurs at a single value of the depth, y.

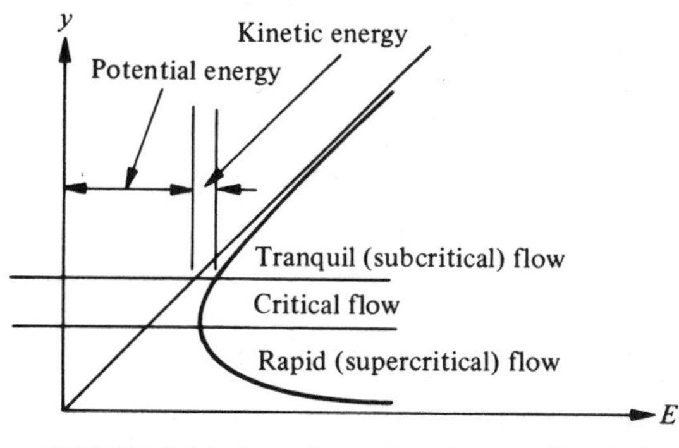

FIGURE 8.12 Specific energy diagram (fixed q).

This depth is called the *critical depth*, and the flow at the point is called the *critical flow*. For any other value of E greater than this minimum, there can physically exist two values of y. If the depth is less than the critical depth, continuity considerations require the velocity to be greater than the critical velocity, and the flow is called either *supercritical* or *rapid*. If y is greater than the critical depth, the velocity will be less than the critical velocity, and the flow is called *subcritical* or *tranquil*. For a rectangular channel it is not difficult to establish a relation between the minimum value of E and the value of the y for the critical depth. Using the methods of calculus, it is found that

$$E_{\min} = \tfrac{3}{2}y_c \tag{8.22}$$

where $E_{\min}$ is the minimum energy and y_c the critical depth.

The foregoing considerations were for the case of constant q. If we now consider E to be constant and y and q to be variable, q is determined by solving equation (8.21). Thus,

$$q^2 = (E - y)2gy^2 \tag{8.23a}$$

and for constant E, q is a function only of y. Plotting this function yields Figure 8.13. Once again it is possible to solve for the value of maximum q as a function of y, and the critical depth is found to occur at

$$y_c = \tfrac{2}{3}E \tag{8.23b}$$

which is also the point of maximum q. Also, for a given q, the critical depth is given by

$$y_c = \left(\frac{q^2}{g}\right)^{1/3} \tag{8.24}$$

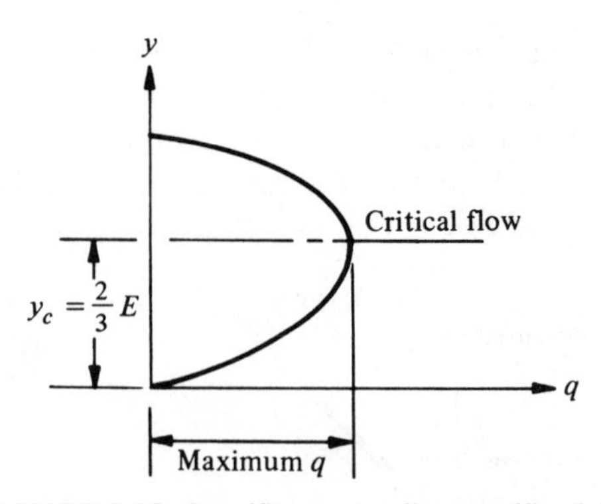

FIGURE 8.13 Specific energy diagram (fixed E).

In addition to these relations, it is possible to express the character of the flow in terms of the velocity and the depth. Thus, from equation (8.23),

$$\frac{q^2}{y_c^3 g} = 1 \tag{8.25}$$

Since $V^2 = q^2/y^2$ or $V^2/y = q^2/y^3$,

$$\frac{q^2}{y_c^3 g} = \frac{V^2}{y_c g} = 1 \tag{8.26}$$

Therefore, if $V^2/gy > 1$, the flow is rapid or supercritical, while $V^2/gy < 1$ is the criteria that the flow is tranquil or subcritical.

ILLUSTRATIVE PROBLEM 8.10

A channel is 10 ft wide and 5 ft deep. The rate of flow is 500 ft³/sec. Determine whether the flow is subcritical or supercritical. Also determine the second depth of flow that is possible for the same specific energy.

Solution

The quantity of flow is $AV = 500$ ft³/sec. Therefore,

$$V = \frac{500}{50} = 10 \text{ ft/sec}$$

$$\frac{V^2}{gy} = \frac{(10)^2}{32.2 \times 5} < 1$$

The flow is thus subcritical.

From equation (8.21) (for constant E),

$$y_1 + \frac{q^2}{2gy_1^2} = y_2 + \frac{q^2}{2gy_2^2}$$

Since q is the flow per unit width,

$$q = \frac{500}{10} = 50 \text{ ft}^3\text{/sec/ft of width}$$

Therefore,

$$5 + \frac{(50)^2}{2g(5)^2} = y_2 + \frac{(50)^2}{2gy_2^2}$$

$$6.55 = y_2 + \frac{38.9}{y_2^2}$$

Simplifying yields

$$y_2^3 - 6.55y_2^2 + 38.9 = 0$$

Solving numerically, we obtain

$$y_2 \simeq 3.65 \text{ ft}$$

ILLUSTRATIVE PROBLEM 8.11

A rectangular channel is to carry 12 m³/s. Calculate the critical depth and critical velocity if the channel is 6 m wide.

Solution

$$q = \frac{12}{6} = 2 \text{ m}^3\text{/s per metre of width}$$

From equation (8.24),

$$y_c = \left(\frac{q^2}{g}\right)^{1/3} = \left(\frac{2^2}{9.81}\right)^{1/3} = 0.742 \text{ m}$$

and

$$\text{critical velocity} = V_c = \frac{q}{y_c} = \frac{2}{0.742} = 2.7 \text{ m/s}$$

ILLUSTRATIVE PROBLEM 8.12

A rectangular channel is to carry 300 ft³ of water/sec. Calculate the critical depth and critical velocity for a channel width of 14 ft.

Solution

Per unit width,

$$q = \frac{300}{14} = 21.4 \text{ ft}^3/\text{sec per foot of width}$$

From equation (8.24),

$$y_c = \left[\frac{q^2}{g}\right]^{1/3} = \left[\frac{(21.4)^2}{32.2}\right]^{1/3} = 2.42 \text{ ft}$$

The critical velocity

$$V_c = \frac{q}{y_c} = \frac{21.4}{2.42} = 8.85 \text{ ft/sec}$$

8.6 HYDRAULIC JUMP

It has already been noted in Section 8.5 that at a fixed value of q there are two equally possible stable depths of flow for a given specific energy. However, it is possible under certain conditions for the character of this flow to undergo an abrupt change from one of these stable flow states to the other. Thus, a channel in which the depth of flow is less than the critical depth can undergo a sudden change to a depth greater than the critical depth. In the course of this change the velocity is decreased, there is a loss in energy as the depth of flow is increased, and the flow passes through the critical state. These phenomena are known as the *hydraulic jump* and can be a desirable method of decreasing the velocity in a channel and converting part of the kinetic energy of the flow to potential energy beyond the jump.

To compute the height of this jump (Figure 8.14), certain simplifying assumptions will be made:

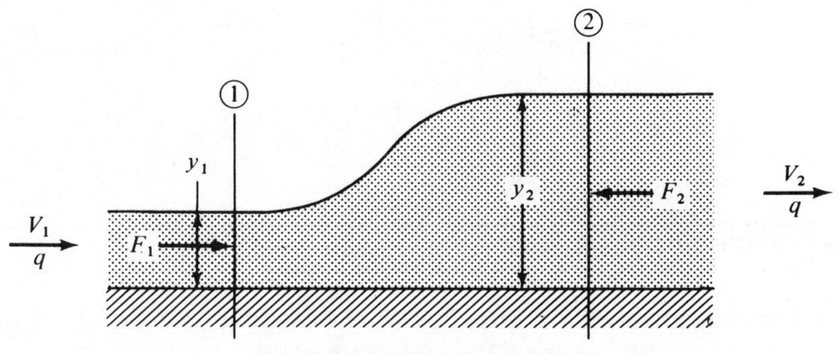

FIGURE 8.14 Hydraulic pump.

1. The frictional forces on the sides and bottom of the channel are negligibly small for the relatively short length of the jump when compared to the other energy terms.
2. The slope of the channel is assumed to be zero (i.e., the channel is horizontal).
3. The hydrostatic pressure forces are assumed to act horizontally.
4. The flow into and out of the jump is steady.
5. For this analysis we shall restrict ourselves to a rectangular channel.

The forces F_1 and F_2 are hydrostatic forces and are simply given as the product of the respective areas multiplied by the hydrostatic pressure exerted by the fluid at the centroid of the area. Mathematically, F_1 and F_2 are given per unit width as

$$F_1 = \frac{\gamma y_1^2}{2} \qquad \text{and} \qquad F_2 = \frac{\gamma y_2^2}{2} \tag{8.27}$$

We can now write the equation of impulse and momentum for this problem in the direction of flow as

$$F(\Delta t) = m(\Delta V) \tag{8.28a}$$

where F is the sum of the forces in the direction of flow, m the mass undergoing acceleration per unit width of channel, and ΔV the change in velocity of the fluid in the direction of flow. Rewriting equation (8.28a), we obtain

$$F_1 - F_2 = \frac{m}{\Delta t}(V_2 - V_1) = \dot{m}(V_2 - V_1) \tag{8.28b}$$

Substituting for F_1 and F_2 and noting that $\dot{m} = \gamma q/g$ yields

$$\frac{\gamma}{2}(y_1^2 - y_2^2) = \frac{\gamma q}{g}(V_2 - V_1) \tag{8.29}$$

But V is simply q/y; therefore,

$$\frac{q^2}{gy_1^2} + \frac{y_1^2}{2} = \frac{q^2}{gy_2^2} + \frac{y_2^2}{2} \tag{8.30}$$

Solving equation (8.30), we obtain

$$y_1 = \frac{y_2}{2}\left(-1 + \sqrt{1 + \frac{8q^2}{gy_1^3}}\right) \tag{8.31a}$$

and

$$y_2 = \frac{y_1}{2}\left(-1 + \sqrt{1 + \frac{8q^2}{gy_1^3}}\right) \tag{8.31b}$$

From equations (8.31a) and (8.31b), y_2, the depth after the jump for a rectangular channel, is related to y_1, the depth before the jump, and the initial velocity V_1 by

$$y_2 = -\frac{y_1}{2} + \sqrt{\frac{2V_1^2 y_1}{g} + \frac{y_1^2}{4}} \tag{8.32}$$

The loss of energy $E_1 - E_2$ is given per unit mass as

$$E_1 - E_2 = \left(y_1 + \frac{V_1^2}{2g}\right) - \left(y_2 + \frac{V_2^2}{2g}\right) \tag{8.33}$$

Expressed per unit width of channel, the loss of energy in the hydraulic jump is

$$E_1 - E_2 = \left(y_1 + \frac{q^2}{2gy_1^2}\right) - \left(y_2 + \frac{q^2}{2gy_2^2}\right) \tag{8.34}$$

ILLUSTRATIVE PROBLEM 8.13

A channel is 2 ft wide and the flow is 15 in. above the base of the channel. If 15 million gal/day is flowing, what is the depth of water downstream in the channel? What is the energy loss in this process?

Solution

It is first necessary to determine whether the flow is greater or less than the critical flow:

$$y_c = \left(\frac{q^2}{g}\right)^{1/3}$$

Since there are 231 in.^{3} in 1 gal,

$$q = \frac{\text{gal/day}}{\text{hr/day} \times \text{sec/hr}} \times \frac{\text{in.}^3\text{/gal}}{\text{in.}^3\text{/ft}^3} \times \frac{1}{\text{channel width (ft)}} = \frac{\text{ft}^3\text{/sec}}{\text{ft width of channel}}$$

$$= \frac{15{,}000{,}000}{24 \times 3600} \times \frac{231}{1728} \times \frac{1}{2} = 11.6 \text{ ft}^3\text{/sec/ft of channel width}$$

Therefore,

$$y_c = \left[\frac{(11.6)^2}{32.2}\right]^{1/3} = 1.61 \text{ ft}$$

The depth is initially less than the critical depth and will jump to a greater depth. From equation (8.32),

$$y_2 = -\frac{y_1}{2} + \sqrt{\frac{2V_1y_1}{g} + \frac{y_1^2}{4}}, \qquad V_1 = \frac{11.6}{15/12} = 9.27 \text{ ft/sec}$$

$$= -\frac{1.25}{2} + \sqrt{\frac{2 \times (9.27)^2 \times 1.25}{32.2} + \frac{(1.25)^2}{4}}$$

$$= -\frac{1.25}{2} + \sqrt{6.68 + 0.39}$$

$$= -0.625 + 2.65 = 2.03 \text{ ft}$$

The energy lost is

$$\left(y_2 + \frac{V_1^2}{2g}\right) - \left(y_2 + \frac{V_2^2}{2g}\right) \text{ per lb of fluid flowing}$$

But $V_2 = 11.6/2.03 = 5.71$ ft/sec:

$$\left[1.25 + \frac{(9.27)^2}{2g}\right] - \left[2.03 + \frac{(5.71)^2}{2g}\right] = 2.59 - 2.54 = 0.05 \text{ ft lb/lb}$$

The total energy lost is the total mass flowing multiplied by 0.05 ft lb/lb. Thus,

$$\text{total energy lost} = (11.6 \times 62.4 \times 0.05)(24 \times 3600)$$
$$= 3{,}130{,}000 \text{ ft lb/day}$$
$$= 0.066 \text{ hp}$$

8.7 WEIRS

The quantity of fluid flowing in an open channel is usually measured by introducing a measuring device into the stream or channel. The simplest and most widely used device is a *weir*, which consists of an obstruction placed in the stream at right angles to the flow. As shown in Figure 8.15, the weir causes the stream to back up and either to flow over it or if the weir is constructed as a notch to flow through it. The notch is usually rectangular, triangular, or trapezoidal in shape and can be installed in the stream in almost any desirable manner. Some usual installations are shown in Figure 8.16. In Figure 8.16a the notch does not extend across the entire channel and is said to have two end contractions; Figure 8.16b shows a weir with one end contraction; and Figure 8.16c shows a weir with the notch extending the full width of the channel and having no end contractions. Since the end contrac-

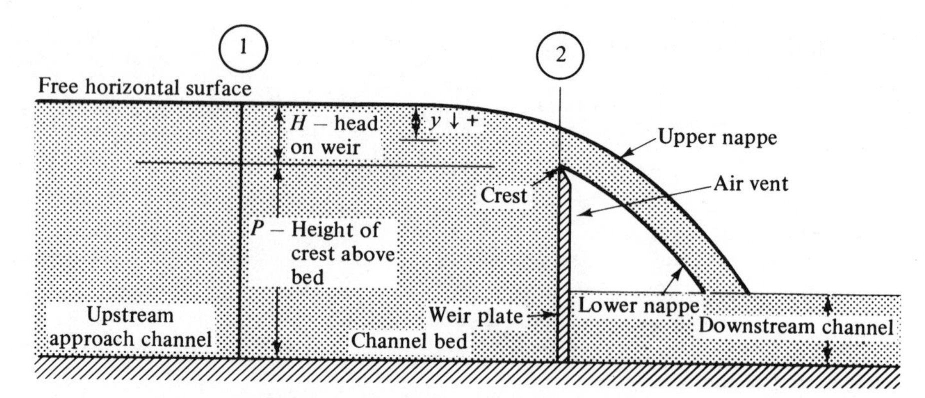

FIGURE 8.15 Simplified flow over a weir.

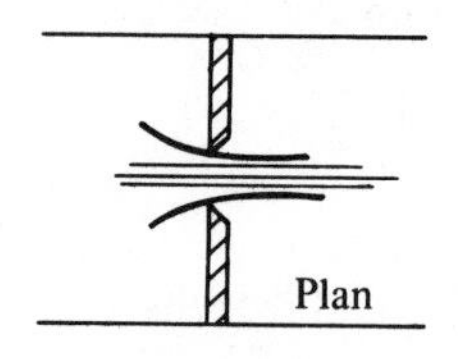

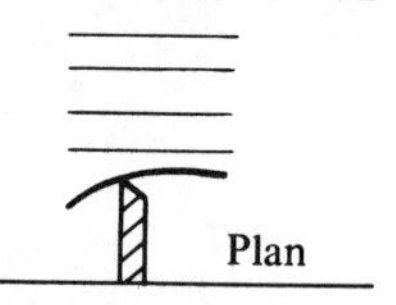

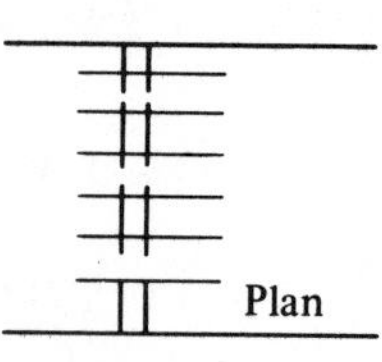

(a) Two end contractions (b) One end contraction (c) End contractions suppressed

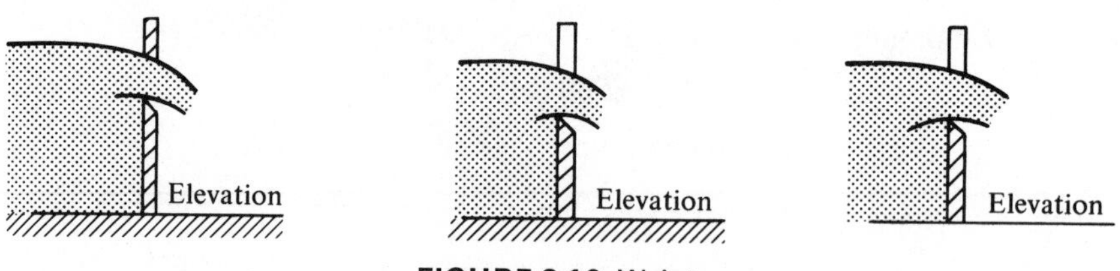

FIGURE 8.16 Weir types.

tions in Figure 8.16c have been suppressed, it is commonly called a *suppressed weir*.

To develop a simplified analysis of a weir, the following assumptions will be made:

1. The pressure on the upper nappe and lower nappe is atmospheric. Unless this lower nappe is vented to the atmosphere, the water will tend to adhere to the lower edge of the weir.
2. The weir plate is vertical with a smooth upstream face and the flow is normal to the plate.
3. The crest is sharp and horizontal and the flow is normal to the crest.
4. Pressure losses are negligible due to the flow over the weir.

5. The channel is uniform with smooth sides upstream and downstream of the weir.
6. The approach velocity to the weir is uniform and there are no surface waves.

It is obvious that the mathematical model postulated by the foregoing assumptions does not represent the actual flow conditions in weirs. However, it does permit a rational approach to the problem of computing the flow over a weir, and the results thus obtained can be modified to conform to the experimentally determined flow.

Let us consider a rectangular suppressed weir based upon the assumptions given above, and for the present let us also impose the condition that the velocity of approach to the weir is negligibly small. Writing an energy equation between sections 1 and 2 shown in Figure 8.15 and recalling the conditions imposed by the assumptions made yields

$$V = \sqrt{2gy} \tag{8.35}$$

where V is the velocity at a depth y in the weir. The velocity is seen to increase as $\sqrt{y}$. Let us now consider the quantity of fluid flowing over the weir. As shown in Figure 8.17, the volume of fluid flowing through the small area

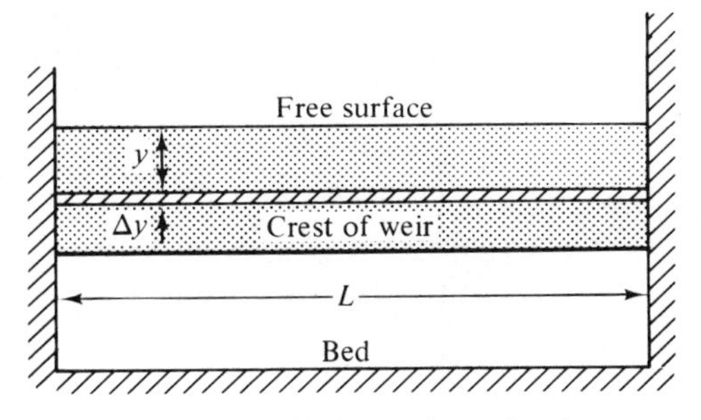

FIGURE 8.17 Front view of weir.

$(\Delta y)L$ per unit of time is $(\Delta y)LV$. Since $V = \sqrt{2gy}$, the quantity of fluid flowing through this small area is $\sqrt{y}\ \Delta y L \sqrt{2g}$. To obtain the total flow through the weir it is necessary to sum up all of these volumes as y goes from zero to H. The complete solution is given by equation (8.36) as

$$Q = \tfrac{2}{3}\sqrt{2g}LH^{3/2} \tag{8.36}$$

Equation (8.36) cannot be expected to yield accurate results when applied to the flow pattern that actually exists in weirs. To account for the assumptions made in the analysis, this equation is usually multiplied by an experimentally determined coefficient C. Thus,

$$Q = C\tfrac{2}{3}\sqrt{2g}\,LH^{3/2} \tag{8.37}$$

The most common form of equation (8.37) with an empirical value for C incorporated in it is

$$Q = 3.33LH^{3/2} \tag{8.38a}$$

Equation (8.38a) is in the English system of units and with L and H in feet, Q will be given in ft³/sec. In SI units, equation (8.38a) becomes,

$$Q = 1.84LH^{3/2} \tag{8.38b}$$

Equations (8.38a) and (8.38b) are the well-known *Francis formula*. Since many weir equations are empirical, the conversion factor of 1 ft³/sec = 0.02832 m³/s is useful in converting from the English system to the SI system.

ILLUSTRATIVE PROBLEM 8.14

Water is flowing through a rectangular sharp-edged weir notch 2 ft wide, 6 in. high, and set 2 ft above the bottom of an approach channel 2 ft wide and 2 ft, 6 in. deep. If the water just fills the approach channel, how much is flowing over the weir notch?

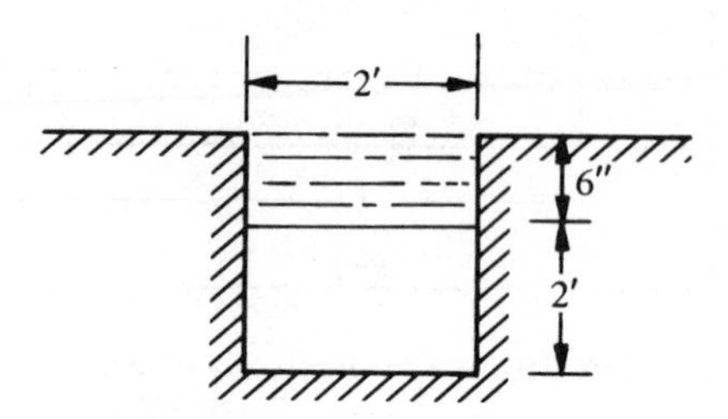

FIGURE 8.18 Illustrative Problem 8.14.

Solution

Figure 8.18 depicts this suppressed weir. As a first approximation we will neglect the approach velocity. Applying the Francis formula,

$$\begin{aligned} Q &= 3.33LH^{3/2} \\ &= 3.33 \times 2 \times (\tfrac{1}{2})^{3/2} = 2.36 \text{ ft}^3/\text{sec} \end{aligned}$$

If the velocity of approach, V_1, is not negligible, the Francis formula is modified to account for this by adding a term incorporating V_1. Thus,

$$Q = 3.33L\left[\left(H + \frac{V_1^2}{2g}\right)^{3/2} - \left(\frac{V_1^2}{2g}\right)^{3/2}\right] \tag{8.39}$$

ILLUSTRATIVE PROBLEM 8.15

How does the approach velocity affect the flow in Illustrative Problem 8.14?

Solution

Using the result of Illustrative Problem 8.14 as a first approximation, we obtain

$$V_1 = \frac{Q}{A} = \frac{2.36}{2.5 \times 2} = 0.472 \text{ ft/sec}$$

$$Q = 3.33 \times 2\left\{\left[\left(\frac{1}{2} + \frac{(0.472)^2}{2g}\right)\right]^{3/2} - \left[\frac{(0.472)^2}{2g}\right]^{3/2}\right\}$$

$$= 3.33 \times 2\left[\left(\frac{1}{2} + 0.0023\right)^{3/2} - (0.0023)^{3/2}\right]$$

$$= 2.36 \text{ ft}^3\text{/sec}$$

It is apparent that for this case the effect of the approach velocity is negligible on the quantity of fluid flowing.

For small discharge quantities, the V-notch weir or triangular weir is widely used. Referring to Figure 8.19 and denoting the half-angle of the notch by α, the theoretical formula for the discharge through a V notch is

$$Q = \frac{8}{15}\sqrt{2g}H^{5/2} \tan \alpha \tag{8.40}$$

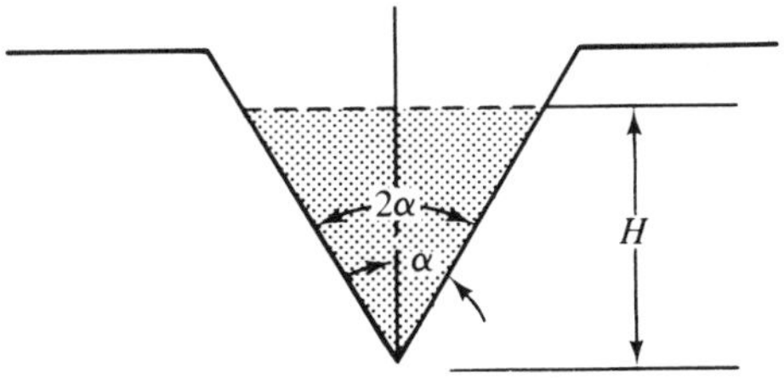

FIGURE 8.19 V-notch (triangular) weir.

The actual flow quantity through *V*-notch weirs has been found to be closely 60% of the value given by equation (8.40). For 90° triangular weirs, the flow can be calculated from

$$Q = 2.5H^{2.5} \qquad \text{ft}^3/\text{sec} \tag{8.41a}$$

In SI Units, equation (8.41a) becomes

$$Q = 1.38H^{2.5} \qquad \text{m}^3/\text{s} \tag{8.41b}$$

There are several reasons the actual discharge over a weir differs from the theoretical discharge given by equation (8.36). Among the reasons for this difference are (1) the assumption of a uniform approach velocity is not found in actual weirs; (2) the weir is not perfectly sharp and smooth; (3) viscous effects along the weir face have been neglected; and (4) the channel of approach is nonuniform. Other effects also tend to cause the actual discharge over a weir to deviate from the theoretical discharge based upon the idealized flow model that we have assumed.

Quite often it is found that the use of a weir as a measuring device is either not practical or desirable. The introduction of a weir into a channel with a very small grade may cause backing up upstream of the weir for an undesirable distance. Excessive sedimentation may occur due to the presence of the weir and the loss of head at the weir may be too great.

To overcome this problem, an instrument called the flume has been developed and is widely used throughout the world. One type of venturi flume, called the *Parshall flume*, has certain outstanding advantages. Among these are:

1. Installation is simple.
2. The loss in energy through the flume is very small.
3. Sedimentation in the flume is eliminated.
4. It is easily operated.

As shown in Figure 8.20, the Parshall flume provides a smooth transition from the standard section to a reduced section and another smooth transition back to the standard section. The Parshall flume contracts the flow from the sides and bears the same relation to a sharp-edged weir that a venturi bears to an orifice in a pipe. Calibrations have been made of this flume, yielding the following equations:

$$Q = 4WH_A^{1.522W^{00.26}} \qquad \text{(for } W \text{ from 1 to 8 ft)} \tag{8.42}$$

and

$$Q = (3.688W + 2.5)H_A^{1.6} \qquad \text{(for } W \text{ from 8 to 50 ft)} \tag{8.43}$$

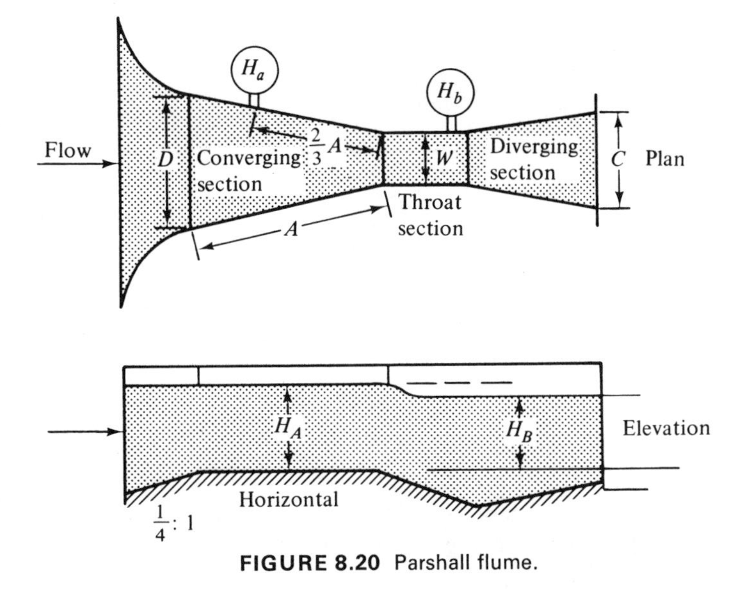

FIGURE 8.20 Parshall flume.

In equations (8.42) and (8.43), Q is in cubic feet per second, W is in feet, and H_A is measured as shown in Figure 8.20 in feet.

8.8 CLOSURE

The fluid mechanics of open channel flow has been treated in a simplified manner due to the difficulty of analyzing this problem from a completely theoretical viewpoint. Our approach has been to set up a simplified model that can be treated mathematically and then to use the resulting formulation with empirically determined coefficients to make these results conform with test data. This procedure is often used in many fields and represents a workable compromise when other procedures prove inadequate. As a result of this approach we have derived formulas that apply to the flow in open channels and weirs. However, the results are limited to water as the working fluid and for a limited range of the stream conditions. In those cases where there is a need for greater accuracy then these formulas can give, it is necessary to calibrate the weir or channel.

PROBLEMS

8.1 Determine the hydraulic radius for the shapes in Figure P8.1. Assume each to be flowing full.

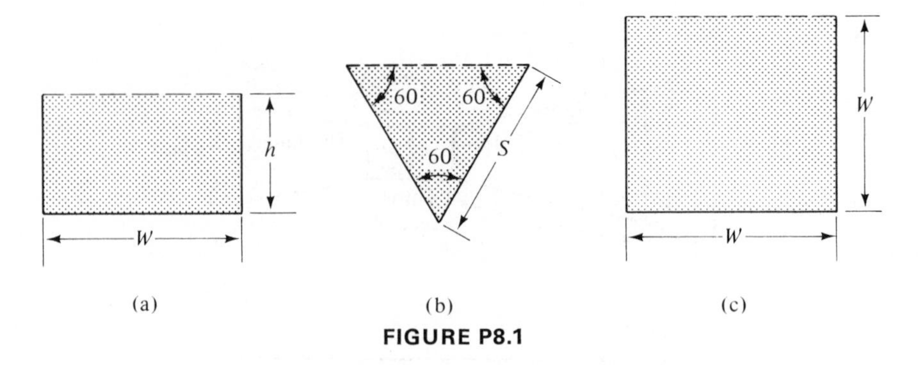

FIGURE P8.1

8.2 For each of the shapes shown in Problem 8.1, water is flowing half-full. Determine the hydraulic radius in each case.

8.3 A square culvert 6 ft on each side has a slope of 1 ft in 1000 ft and is carrying water to a depth of 4 ft (Figure P8.3). If $n = 0.013$, compute the quantity of water being discharged.

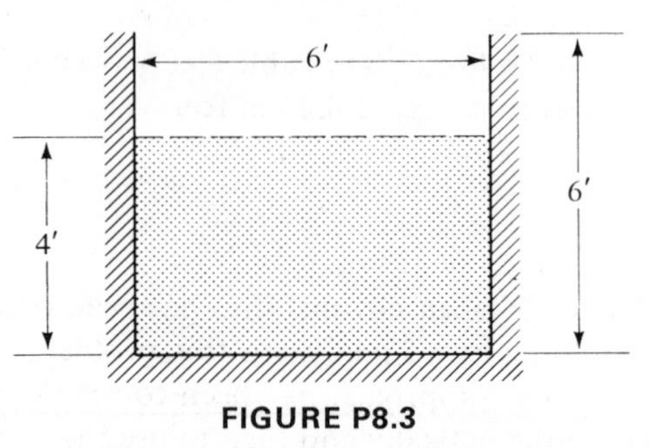

FIGURE P8.3

8.4 A rectangular channel 2 m wide × 1 m high flows full. If the channel is steel with a slope of 1 m in 1500 m, what is the discharge?

8.5 A square channel is used for irrigation. The original channel is 3 m × 2 m. If there is a lack of water and the channel is flowing half-full, what is the discharge? Assume that a brick lining and a slope of and a slope of 1 m in 5000 m.

8.6 A circular steel culvert is flowing half-full. If the diameter of the culvert is 6 ft and the slope is 1 ft in 1000 ft, what is the discharge?

8.7 If the culvert in Problem 8.6 flows with a depth of water of 4 ft, what will the discharge be?

8.8 Solve Problem 8.7 using Figure 8.11. Compare your results.

8.9 Calculate the uniform flow in a brick-lined trapezoidal canal ($n = 0.018$) whose slope is 1 ft in 10,000 ft (Figure P8.9). Assume the width

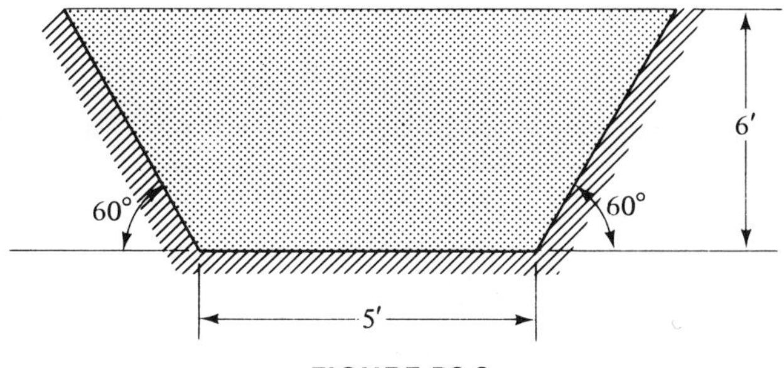

FIGURE P8.9

at the bottom to be 5 ft and the sides to slope from 60° to the horizontal. The depth of flow in the canal is 6 ft.

8.10 A rectangular irrigation channel is 6 ft wide, has a slope of 1 in 1000, and $n = 0.011$. If Q is 102 ft^3/sec, what is the depth of water in the channel?

8.11 A wooden flume carries 1.0 m^3/s. If the flume is rectangular, with a base of 1 m, what should its slope be if it is made to the best dimensions?

8.12 What are the best dimensions of a rectangular channel whose flow cross-sectional area is 10 m^2?

8.13 What will the flow be in Problem 8.12 if the channel has a slope of 1 : 500 and is made of smooth concrete?

8.14 What are the best dimensions of a rectangular channel whose flow cross-sectional area is 150 ft^2?

8.15 What will the flow be in Problem 8.14 if the channel slopes 1 ft in 750 ft and is made of brick?

8.16 An irrigation channel has been constructed with a bottom 20 ft wide and side slopes at 45°, all lined with concrete (Figure P8.16). If the canal is 10 ft deep and delivers 1050 ft^3/sec of water, what is the necessary drop of level per mile? Assume that the canal flows full.

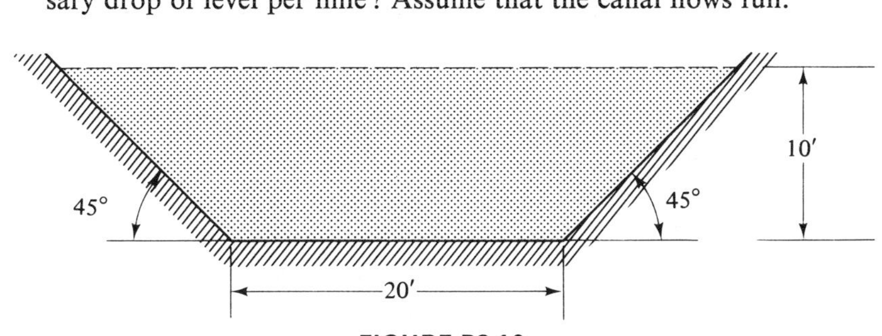

FIGURE P8.16

8.17 A steel flume is shaped as shown in Figure P8.17. If it is flowing full and is required to carry 15 ft³/sec, what slope is required?

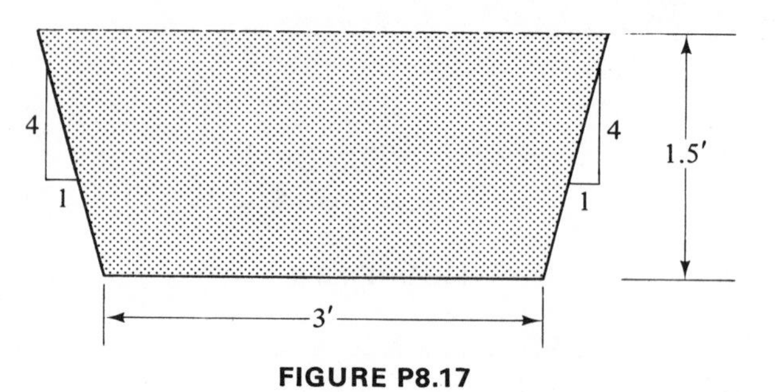

FIGURE P8.17

8.18 A rectangular irrigation channel is 4 m wide, has a slope of 1 in 900, and $n = 0.01$. If the flow is 4 m³/s and is uniform, what is the depth of water in the channel?

8.19 A rectangular channel is to carry 80 ft³/sec with a slope of 1.5 ft in 1000 ft. If $n = 0.015$, what are the best dimensions of the channel? Assume that the channel is flowing full.

8.20 If $n = 0.020$ in Problem 8.19, determine the best dimensions of the channel. Compare the results of these problems.

8.21 A rectangular channel is made 10 ft wide. What is the critical depth of flow if 500 ft³/sec is being carried?

8.22 A rectangular channel is made 4 m wide. What is the critical depth of flow if 20 m³/s flows?

8.23 A rectangular channel is constructed to be 15 ft wide. At some time 2000 ft³/sec is to be carried in the channel with a velocity of 10 ft/sec. Determine whether the flow is rapid or tranquil and also its specific energy.

8.24 A rectangular channel is made 6 m wide. If it is to carry 70 m³/s at a velocity of 4 m/s, will the flow be tranquil or rapid? What is its specific energy?

8.25 The specific energy in a rectangular channel is 5 ft. If the channel is 4 ft wide and the flow is 25 ft³/sec, determine the possible flow depths in the channel.

8.26 If a rectangular channel is 10 ft wide, determine the maximum flow for a specific energy of 8 ft.

8.27 A rectangular channel carries a flow of 400 ft³/sec. If the channel is 8 ft wide and the water flows at a depth of 2 ft, calculate the height of the flow after a hydraulic jump.

8.28 What is the specific energy loss due to the jump in Problem 8.27?

8.29 A sharp-crested weir is constructed at the end of a concrete canal 10 ft wide (Figure P8.29). The weir is 8 ft high and the channel walls extend beyond the top of the weir to a height of 9.44 ft above the bottom of the canal. The nappes are completely ventilated below the weir. How much water is flowing over the weir?

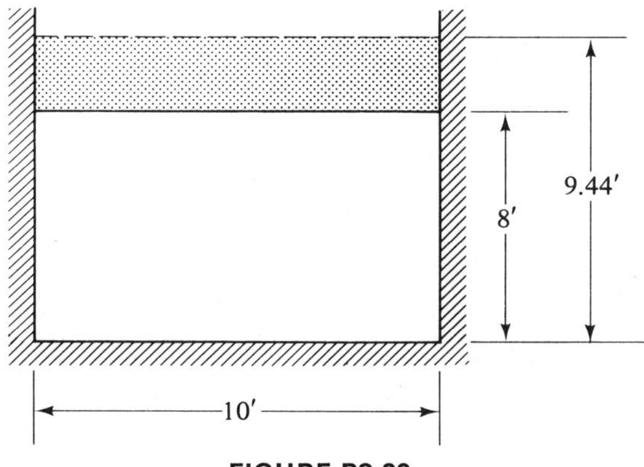

FIGURE P8.29

8.30 If the weir described in Problem 8.29 is made 3 m wide and 2.5 m high and if the walls of the weir extend to a height of 3.5 m above the bottom of the canal and the weir is flowing full, how much water is flowing over the weir?

8.31 A 90° V-notch weir is to discharge 20 ft^3/sec. Determine the head on the weir.

8.32 A 90° V-notch weir has a head of 1 m on it. Calculate the discharge through the weir.

8.33 A 90° V-notch weir is to discharge 1 m^3/s. Determine the head on the weir.

8.34 A 90° V-notch weir has a head of 2 ft on it. What is the discharge through the weir?

8.35 If the overall notch angle, 2α, is 60° in a V-notch weir, determine the head for a discharge of 20 ft^3/sec. Compare the result with Problem 8.31.

8.36 If the overall notch angle 2α is 60° in a V-notch weir and the weir has a head of 1 m on it, determine the discharge through the weir. Compare your result with Problem 8.32.

8.37 If 2α is 60° in Problem 8.33, what will the head on the weir be?

8.38 Solve Problem 8.34 for $\alpha = 60°$.

8.39 In connection with a water turbine test, the discharge water goes into a flume 3 ft wide. At the end of this flume the flow is measured by a sharp-crested weir 3 ft high with no end contractions. If the head on the weir is 3.5 ft, what is the flow in cubic feet per second?

REFERENCES

1. *Fluid Mechanics for Engineers* by P. S. BARNA, Butterworth & Co. (Publishers) Ltd., London, 1957.
2. *Fluid Mechanics for Hydraulic Engineers* by H. ROUSE, McGraw-Hill Book Company, New York, 1938.
3. *Basic Fluid Mechanics* by J. L. ROBINSON, McGraw-Hill Book Company, New York, 1963.
4. *Fluid Mechanics* by R. C. BINDER, 4th ed., Prentice-Hall, Inc., Englewood Cliffs, N.J., 1962.
5. *Elementary Fluid Mechanics* by J. K. VENNARD, John Wiley & Sons, Inc., New York, 1961.
6. *Mechanics of Fluids* by G. MURPHY, International Textbook Company, Scranton, Pa., 1942.
7. *Fluid Mechanics for Engineers* by M. L. ALBERTSON, J. R. BARTON, and D. B. SIMONS, Prentice-Hall, Inc., Englewood Cliffs, N.J., 1960.
8. *Applied Fluid Mechanics* by M. P. O'BRIEN and G. H. HICKOX, McGraw-Hill Book Company, New York, 1937.
9. *Applied Fluid Mechanics* by R. L. MOTT, 2nd ed., Charles E. Merrill Publishing Co., Columbus, Ohio, 1979.
10. *Fundamentals of Fluid Mechanics*, by I. A. SULLIVAN, Reston Publishing Co., Reston, Va., 1978.
11. *Fluid Mechanics* by V. L. STREETER and E. B. WYLIE, 7th ed., McGraw-Hill Book Company, New York, 1979.

Pumps

9.1 INTRODUCTION

In order to force a fluid to flow against a resistance, to develop pressure, to transmit power, and so on, some form of machine must be provided. In the pump system mechanical energy is converted to fluid energy. For most modern pumps we find that electric motors provide the mechanical driving force to the device known as the *pump*, which converts the mechanical energy to fluid power. Pumps are made in all sizes, from small pumps used to supply control signals in fluidic systems to huge pumps used in conjunction with earth-moving equipment. In the selection of a pump for a given application, it is necessary to know many factors, such as the capacity required, the properties of the fluid being pumped, conditions at inlet and outlet of the pump, the power source for driving the pump, and so on. The purpose of this chapter is to describe the types of pumps in general use and the factors that determine their performance. The student should note that a good source of performance data for pumps is the literature published by manufacturers.

9.2 CLASSIFICATION OF PUMPS

Although there are many types of pumps, it is possible to classify pumps into three different categories: *displacement*, *delivery*, and *motion*. Table 9.1 shows the various major categories and subcategories under each major division.

TABLE 9.1

Pump Classification
I. Displacement
a. Positive
b. Nonpositive
II. Delivery
a. Constant volume
b. Variable volume
III. Motion
a. Rotary
b. Reciprocating

Figure 9.1 illustrates four *positive displacement* types of pumps: the external gear pump, the internal gear pump, the vane pump, and the lobe pump. As the name "positive displacement pumps" implies, these pumps are designed to provide a given amount of fluid to a system for each revolution of the pump. The positive displacement pump is made with very close clearances between the rotating and stationary parts to minimize flow back through the pump or *slip*. In general, these pumps will pump against high pressures but their volumetric capacity is low. Because of the positive displacement per revolution feature, it is necessary to protect positive displacement pumps with relief valves to prevent damage by overpressurization.

The *external gear pump* shown in Figure 9.1a consists of a drive gear and a driven gear enclosed within a precision housing. As the gears rotate, fluid enters the space at the inlet to the pump. The fluid is then trapped between the gear teeth and the casing and is transported around the periphery of both gears to the outlet. The remeshing of the teeth at the outlet forces the fluid out of the outlet of the pump. Care must be taken in operating external gear pumps to ensure the correct rotation of the pump. If the pump is operated in the wrong direction, it can be severely damaged or destroyed. This type of gear pump can be used in fluid power applications for pressures up to 3000 psi (21 MPa) and capacities up to 150 gpm (0.01 m^3/s or 0.6 m^3/min). This type of pump has a low initial cost, a long operating life, and relatively high efficiencies.

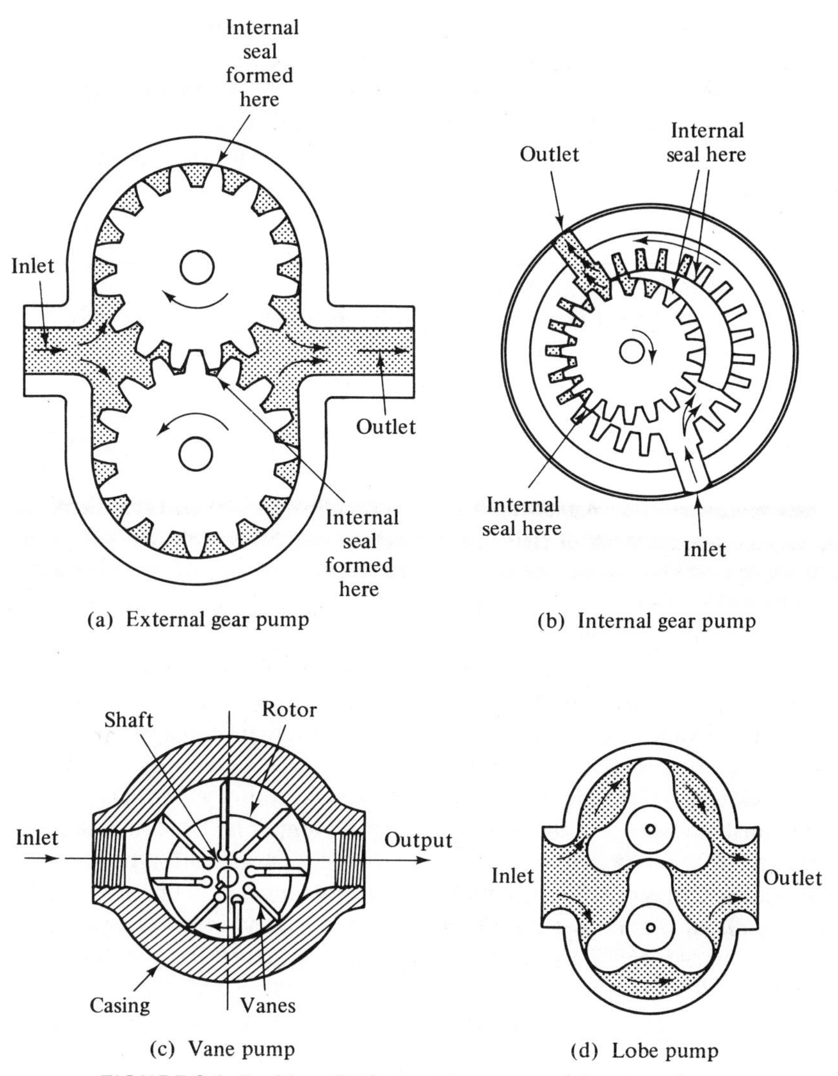

FIGURE 9.1 Positive displacement pumps: (a) external gear pump; (b) internal gear pump; (c) vane pump; (d) lobe pump.

Figure 9.1b shows an *internal gear pump* where the basic action is similar to that of the external gear pump. The unit has an outer ring gear and an off-center gear with fewer teeth. There is a crescent-shaped spacer around which the fluid is carried.

The *vane pump* is shown in Figure 9.1c. As can be seen, it consists of a series of vanes that are almost radially located in a relatively large diameter rotor. These units use spring force, fluid pressure, and centrifugal force to cause the vanes to stay in contact with the outer casing. This action traps fluid in the space between the vanes and casing and the fluid is transported to the outlet. These pumps show a very good reliability and are not costly. As the vanes wear, the wear is compensated for by the outward motion of the vanes. Vane pumps have been widely used on machine tools and have been used for pressures up to 1500 psi (10.5 MPa).

The pump shown in Figure 9.1d is the *lobe pump*. Although there are similarities between the lobe pump and the external gear pump, there are certain significant differences. In the lobe pump both rotors are externally driven and neither element contacts the other. The effect of the rotary motion of the lobes is to transfer fluid around their outer peripheries to the outlet port. This type of pump has a higher volume delivery rate per revolution than does the gear pump and it is also quieter. Thus, we find it used as a supercharger on engines.

Nonpositive displacement pumps are generally used to transfer large volumes of fluids at pressures that are relatively low. Among the many types of nonpositive displacement pumps that can be found, the principal ones are the *centrifugal type* of unit and the *axial flow* type of unit. Figure 9.2 shows these units schematically.

Figure 9.2a shows the type of centrifugal pump known as the *diffuser* type and Figure 9.2b shows the *volute* type. In the diffuser-type pump there is a series of fixed vanes surrounding the impeller. In the diffuser there is a reduction in velocity and an increase in static pressure which tends to make the static pressure distribution around the impeller uniform. In Figure 9.2b, which illustrates the volute-type centrifugal pump, the fluid is discharged directly from the impeller into the volute. Notice that in both types of centrifugal pump there is only a single moving part, the impeller. This class of pump (the centrifugal type) takes fluid in at the center of the impeller and ejects it to the discharge. Since there must be large clearances for the centrifugal action to be effective, these pumps have high slip rates. Centrifugal pumps are relatively low in cost, are highly reliable, and can handle almost all types of fluids.

The *propeller-type* pump shown in Figure 9.2c is also a nonpositive displacement pump that is sometimes called an *axial flow* pump. These units are built in a large range of sizes where a large discharge rate is required under a relatively low head. Since the flow is straight-through, such pumps can handle solids in suspension quite readily. For this reason they are used in such applications as sewage disposal, irrigation, and drainage.

Figures 9.3 and 9.4 show two large commercial types of centrifugal fans having different types of blading. Figure 9.5 shows three general types of

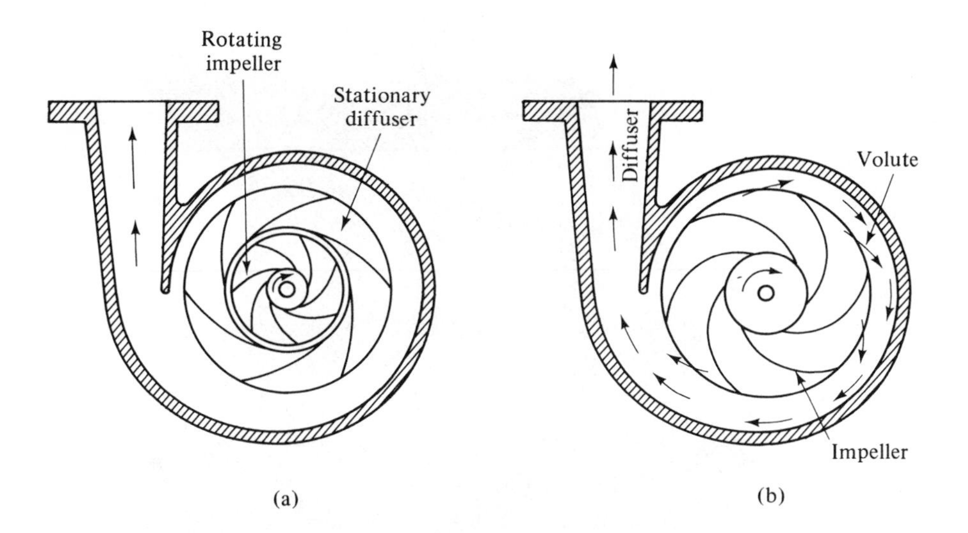

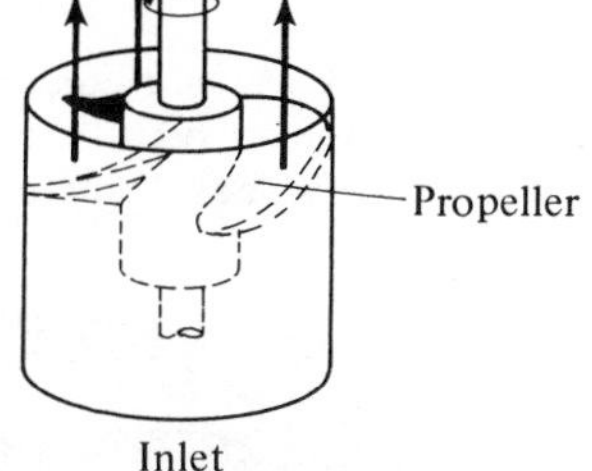

(c) Propeller

FIGURE 9.2 Nonpositive displacement pumps: (a) diffuser; (b) volute; (c) propeller.

centrifugal fans and different blading configurations. Figure 9.6 indicates a simple type of axial flow fan.

The performance of a fan is usually given graphically. Figure 9.7 shows a typical performance curve for a centrifugal fan. Efficiency, horsepower, and pressure developed are given as a function of capacity (discharge rate).

One advantage of the centrifugal pump is that the discharge valve can be closed without damaging the pump. Thus, if for some reason the discharge becomes blocked, the unit will not be damaged. It is unnecessary to protect this pump with a pressure relief valve. The only effect of a blockage is a churning of the fluid with a subsequent rise in the fluid temperature.

FIGURE 9.3 Fan with backward-curved blades and vane-control inlet, suitable for forced draft service, handling room air. (Courtesy of Babcock & Wilcox Corporation).

FIGURE 9.4 Fan with radial-tip blades and double inlet, arranged for induced draft service. (Courtesy of Babcock & Wilcox Corporation).

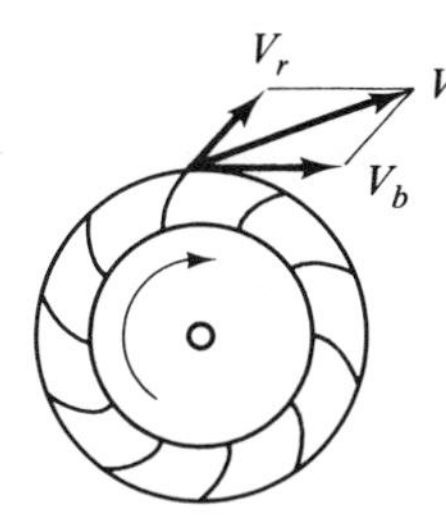

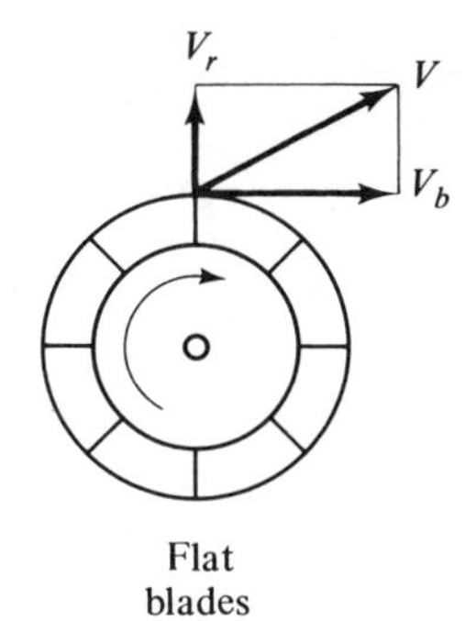

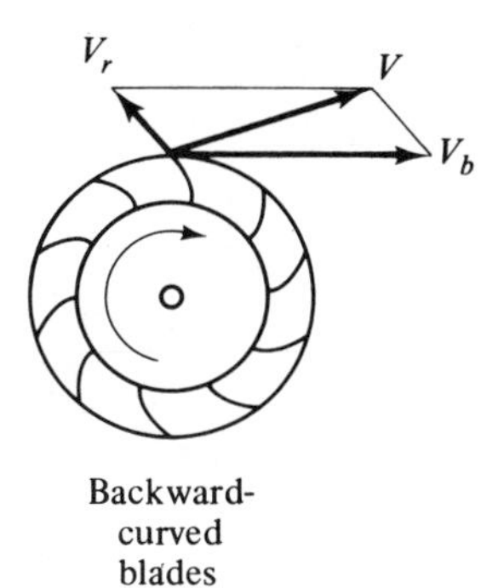

Forward-curved blades

Flat blades

Backward-curved blades

V = Absolute velocity of air leaving blade (shown equal for all three blade types)

V_r = Velocity of air leaving blade relative to blade

V_b = Velocity of blade tip

FIGURE 9.5 Three general types of centrifugal fans with vectors showing relative tip speed required for equal velocity leaving blades.

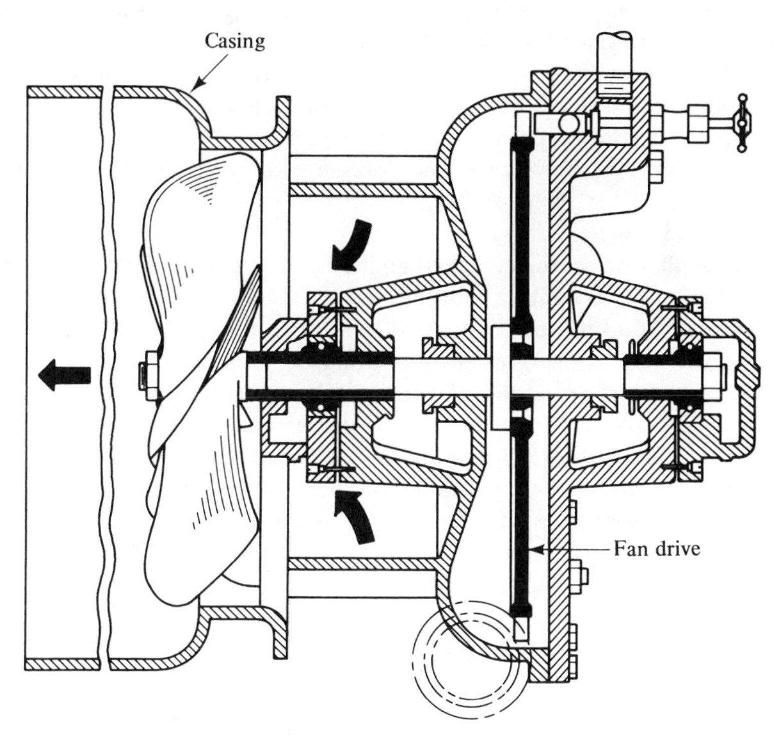

FIGURE 9.6 Simple type of axial-flow fan.

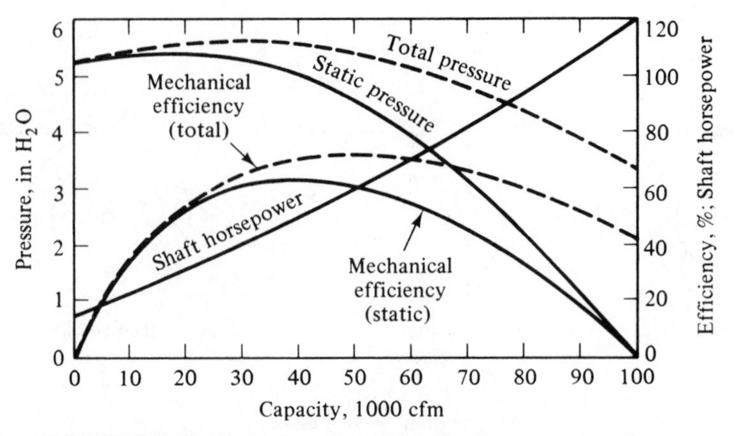

FIGURE 9.7 Typical characteristic performance curves for a centrifugal pump.

9.3 PUMP PERFORMANCE—CAVITATION

Pumps are often operated with the inlet pressure reduced to the point that bubbles of vapor may be formed due to the generation of vapor in the fluid. For any fluid there is a pressure caused by molecules of the fluid that escape from the fluid and exist just above the fluid surface. The pressure of these molecules has been termed the *vapor pressure* of the fluid. For any given fluid, there is a definite relation between the vapor pressure and the temperature of the fluid. Figure 9.8 shows the vapor pressure for water as a function of temperature.

Vapor bubbles generated at the inlet to the pump are carried along until a region of higher pressure is reached and then they suddenly collapse. If the vapor bubbles come in contact with the walls at the time they collapse, pitting of the surface can occur due to the high local pressures generated. The entire process of the formation, growth, and collapse of a bubble can occur in milliseconds in a turbomachine. Experimental data indicate pressures of the order of 10^9 Pa (143,000 psi) occurring from the collapse of vapor bubbles, which is consistent with the damage observed due to cavitation. The ultimate failure of the surface of the material is usually a fatigue failure.

The basic measure to protect the pump against cavitation is the proper design of the system to avoid pressures that approach the vapor pressure of the fluid. In order to have an index of performance to use to specify the minimum suction conditions for a turbomachine, we will have to define certain terms. This can best be done by referring to Figure 9.9, which shows the sketch of a pump system connected between two reservoirs. A term is now

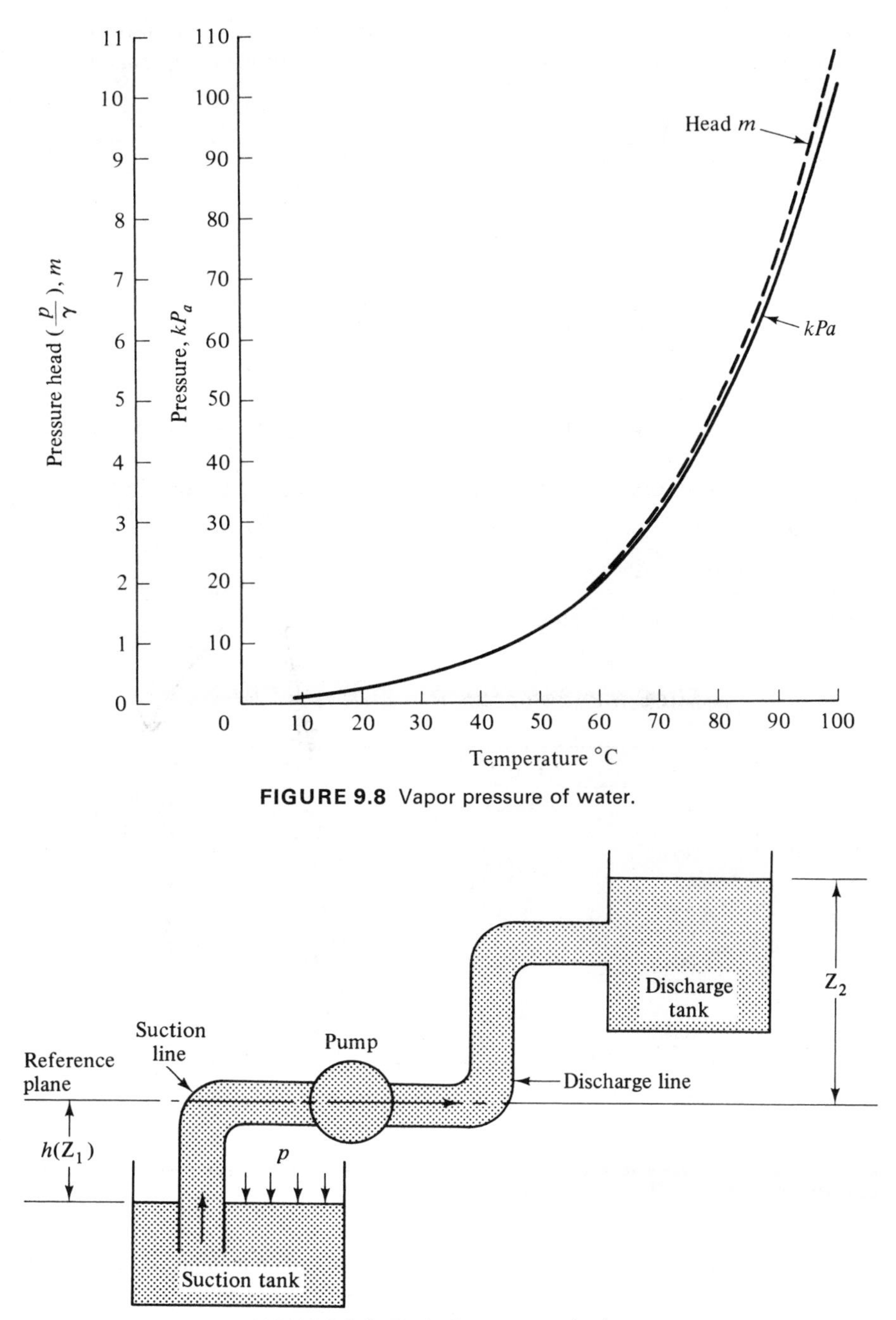

FIGURE 9.8 Vapor pressure of water.

FIGURE 9.9 Typical pump terminology.

defined, called the *net positive suction head* (NPSH), to aid in evaluation of the possibility of cavitation in the pump. Using the pump center line as a reference,

$$\text{NPSH} = \frac{p}{\gamma} + \frac{V^2}{2g} + (-h) - h_L - h_p \tag{9.1}$$

where p = static pressure on the fluid surface (m or ft)
γ = specific weight of the fluid (N/M^3 or lb/ft^3)
h = distance from center line of pump to fluid level (m or ft); note that if the reservoir is *above* the center line of the pump, this term will be positive while a reservoir below the pump causes this term to be negative
h_L = head lost due to friction and other factors in the suction line (m or ft)
h_p = vapor pressure of the liquid corresponding to the liquid temperature (m or ft)

Since water is such a widely used fluid, its vapor pressure is given either from Figure 9.8 or from Table 9.2.

TABLE 9.2 Selected Properties of Water†

Temperature (°C)	γ (N/m^3)	p (Pa)	h_p (m)
10	9802.08	1227.6	0.125
20	9788.38	2339	0.239
30	9764.01	4246	0.435
40	9730.11	7384	0.759
50	9688.77	12 349	1.275
60	9640.19	19 940	2.068
70	9587.41	31 190	3.253
80	9528.71	47 390	4.973
90	9465.25	70 140	7.410
100	9397.22	101 350	10.785

†Data from Ref. 1.

ILLUSTRATIVE PROBLEM 9.1

A pump is placed so that its center line is 1.5 m above the level in a reservoir that is open to the atmosphere. If water is being pumped at the rate of 1 m/s in the suction line and friction losses are estimated at 1.2 m, determine the NPSH if the water is at 40°C.

Solution

Refer to Figure 9.9. Thus, for 40°C (see Table 9.2),

$$\begin{aligned}
\text{NPSH} &= \frac{p}{\gamma} + \frac{V^2}{2g} - h - h_L - h_p \\
&= \frac{101\,325}{9730.11} + \frac{(1)^2}{2 \times 9.81} - 1.5 - 1.2 - 0.759 \\
&= 10.41 + 0.05 - 1.5 - 1.2 - 0.76 \\
&= 7.00 \text{ m}
\end{aligned}$$

The NPSH can be interpreted as the total suction head of liquid above the vapor-pressure head. It is usual for a manufacturer to run a test to determine the NPSH for the machine that will permit it to operate efficiently and without noise or damage. As long as the machine is operated at a NPSH *above* this value, the operation will be satisfactory.

ILLUSTRATIVE PROBLEM 9.2

Solve Illustrative Problem 9.1 if the pump is placed so that its center line is 1.5 m below the level in the reservoir. Assume all other data to be the same.

Solution

The only term that changes is h, which will now enter equation (9.1) as a positive term. Thus,

$$\begin{aligned}
\text{NPSH} &= \frac{p}{\gamma} + \frac{V^2}{2g} + h - h_L - h_p \\
&= 10.41 + 0.05 + 1.5 - 1.2 - 0.76 \\
&= 10.0 \text{ m}
\end{aligned}$$

Note that, in general, the higher NPSH is desirable.

9.4 PUMP PERFORMANCE CALCULATIONS

As in any technology, certain terms are used which are by common usage generally understood. The tank from which the fluid is being pumped is called the *suction tank*, and the tank that the fluid is pumped to is called the *discharge tank*. The distance from the pump center line to the level in the suction tank (Z_1 in Figure 9.9) is known as the *static discharge head*. If we now apply a Bernoulli equation across the pump,

$$\frac{p_1}{\gamma} + \frac{V_1^2}{2g} + Z_1 + E_p = \frac{p_2}{\gamma} + \frac{V_2^2}{2g} + Z_2 \tag{9.2}$$

or

$$E_p = \frac{p_2 - p_1}{\gamma} + \frac{V_2^2 - V_1^2}{2g} + (Z_2 - Z_1) \tag{9.3}$$

Notice that we have written this equation across the pump and that sections 1 and 2 refer to the pump inlet and outlet, respectively. The term *total head* or *dynamic head* is used to designate the energy added to the pump per unit weight and is given by E_p in equation (9.3). If we multiply E_p by the weight flow rate ($\dot{w}$), we obtain the power added to the fluid by the pump as

$$\text{hp} = \frac{\dot{w}E_p}{550} = \frac{\gamma Q E_p}{550} \tag{9.4}$$

or

$$\text{hp} = \frac{\dot{w}E_p}{746} = \frac{\gamma Q E_p}{550} \tag{9.5}$$

in English and SI units, respectively.

ILLUSTRATIVE PROBLEM 9.3

A pump steadily pumps water from one reservoir to another. If the pump delivers 0.1 m^3/s and has a 150-mm inlet diameter and a 75-mm outlet diameter, determine the power added to the water. Assume the pump outlet to be 1 m above the inlet and that static taps indicate an outlet pressure 70 kpa greater than the inlet pressure. Use $\gamma = 9810\ N/m^3$.

Solution

Since $Q = AV$, $V = Q/A$. At the inlet,

$$V_1 = \frac{0.1}{(\pi/4)(0.150)^2} = 5.7\ \text{m/s}$$

at the outlet,

$$V_2 = \frac{0.1}{(\pi/4)(0.075)^2} = 22.6 \text{ m/s}$$

Using equation (9.3), we obtain

$$\begin{aligned} E_p &= \frac{p_2 - p_1}{\gamma} + \frac{V_2^2 - V_1^2}{2g} + Z_2 - Z_1 \\ &= \frac{70 \times 10^3}{9810} + \frac{(22.6)^2 - (5.7)^2}{2 \times 9.81} + 1 \\ &= 7.14 + 24.4 + 1 = 32.5 \text{ m} \end{aligned}$$

The power delivered to the fluid is

$$\begin{aligned} \text{hp} &= \frac{\gamma Q E_p}{746} \\ &= \frac{9810 \times 0.1 \times 32.5}{746} \\ &= 42.7 \text{ hp} \end{aligned}$$

In Illustrative Problem 9.3 we considered the pump only and not the system that it is part of. If it is desired to know the operating point of a pump in a given system, we must consider that the pump will be required to deliver the fluid against a static discharge head and to overcome all the flow losses in the system.

Let us assume that we have a pump which has the characteristics shown by the performance curves of Figure 9.1 connected in the system shown in Figure 9.9 (repeated). For each assumed value of flow in the system, we will calculate E_p for the system, including all flow losses. If the system is to flow at the assumed rate, the pump is required to provide this head. We will therefore call this the *required head* and the pump head the *available head*. Plotting both curves on the same coordinates gives us Figure 9.10, where we have indicated the head available from the pump and the head required by the system. The head h is the static head that the system must overcome since it is the value at zero flow. The losses in the system will be the difference between the head required by the system and h. The operating point of the system will be at point ②, where both the head available and the head required curves intersect and both have the same value. By varying the flow resistances in the system (by such methods as opening or closing valves) we can shift the head required curve to give us operating points to the left or right of point ② in Figure 9.10.

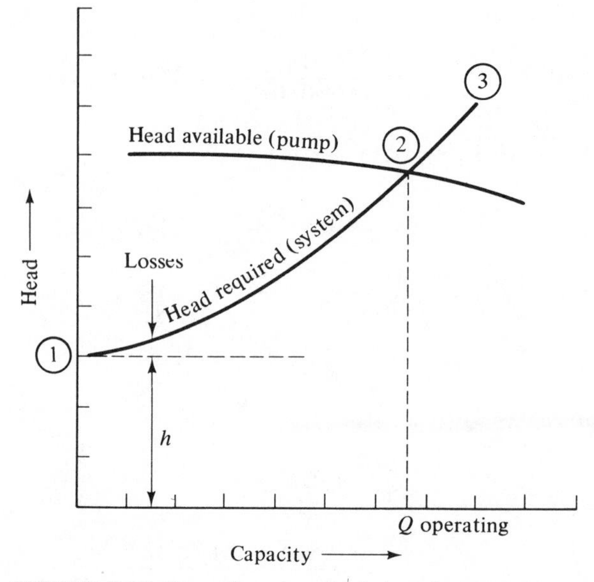

FIGURE 9.10 Operating characteristics of a system.

9.5 THE AFFINITY LAWS

Data on an original or prototype pump are usually obtained from tests of a model pump. To ensure that the data from the model are applicable to the prototype, it is important that certain similarity conditions be met. In general, we can state that the two pumps must be geometrically similar and that also the flow patterns in the two pumps must be similar. Two units that are geometrically similar and have similar vector diagrams are said to be *homologous*.

The discharge $Q = AV$ and since the area is proportional to the impeller diameter D (or any other characteristic dimension), we can say that Q is proportional to VD^2. By noting that the velocity V is proportional to the peripheral speed of the impeller, ND, where N is the rpm of the impeller, we can write the following relation:

$$\frac{Q}{ND^3} = \text{constant} \tag{9.6}$$

which expresses the necessary condition for homologous units (similar geometry and similar flow patterns).

Let us now consider the head developed by a pump. If we start with $Q = AV$ again and note that V is proportional to $\sqrt{H}$, Q will be proportional to

$D^2 \sqrt{H}$, or

$$\frac{Q}{D^2 \sqrt{H}} = \text{constant} \tag{9.7}$$

Eliminating Q in equations (9.6) and (9.7) yields

$$\frac{H}{N^2 D^2} = \text{constant} \tag{9.8}$$

The last consideration that we need concerns power, P. Power will be proportional to γQH. If we use equations (9.6) and (9.8) for Q and H, respectively, we have P proportional to $N^3 D^5$, or

$$\frac{P}{N^3 D^5} = \text{constant} \tag{9.9}$$

Note that equations (9.6), (9.8), and (9.9) have assumed constant (the same) efficiency.

The utility of equations (9.6), (9.8), and (9.9) lies in the fact that most centrifugal pumps are either run at various speeds to obtain the desired capacity and head or are run at constant speed using different impellers in the same casing to achieve the same results. Let us consider the case of a machine having a given diameter impeller operating at different rotational speeds. From equations (9.6), (9.8), and (9.9), *for constant diameter*,

$$\frac{Q_1}{N_1} = \frac{Q_2}{N_2}$$
$$\frac{H_1}{N_1^2} = \frac{H_2}{N_2^2} \tag{9.10}$$
$$\frac{P_1}{N_1^3} = \frac{H_2}{N_2^3}$$

and *for constant speed*,

$$\frac{Q_1}{D_1^3} = \frac{Q_2}{D_2^3}$$
$$\frac{H_1}{D_1^2} = \frac{H_2}{D_2^2} \tag{9.11}$$
$$\frac{P_1}{D_1^5} = \frac{P_2}{D_2^5}$$

Equations (9.6), (9.8), and (9.9) or their equivalent forms, equations (9.10) and (9.11), are known as the *affinity laws* for pumps and are extremely useful. They permit tests at a given speed or given impeller diameter to be converted to yield data at another speed or diameter in homologous units.

ILLUSTRATIVE PROBLEM 9.4

A test is conducted on a centrifugal pump and the pump is found to develop a total head of 40 ft while delivering 100 cfm. The pump was operated at 1100 rpm. What head and capacity would be expected if the pump were operated at 1200 rpm?

Solution

Since this is the same pump operating at different speeds, equations (9.10) are applicable. Thus,

$$\frac{Q_1}{N_1} = \frac{Q_2}{N_2} \quad \text{or} \quad Q_2 = Q_1\frac{N_2}{N_1} = 100 \times \frac{1200}{1100} = 109.1 \text{ cfm}$$

and

$$\frac{H_1}{N_1^2} = \frac{H_2}{N_2^2} \quad \text{or} \quad H_2 = H_1\left(\frac{N_2}{N_1}\right)^2 = 40 \times \left(\frac{1200}{1100}\right)^2 = 47.6 \text{ ft}$$

ILLUSTRATIVE PROBLEM 9.5

A pump has an impeller diameter of 1.5 m and operates at 1200 rpm. If the speed is increased to 1400 rpm, what diameter impeller should be used to maintain a constant power input to the pump?

Solution

We can either use equations (9.10) and then (9.11) or equation (9.9) directly to solve this problem. If we use equation (9.9), we can write it as

$$\frac{P_1}{N_1^3 D_1^5} = \frac{P_2}{N_2^3 D_2^5}$$

Since $P_1 = P_2$, $N_1^3 D_1^5 = N_2^3 D_2^5$, or

$$D_2 = D_1\left(\frac{N_1}{N_2}\right)^{3/5}$$

$$D_2 = 1.5\left(\frac{1200}{1400}\right)^{3/5} = 1.367 \text{ m}$$

9.6 SPECIFIC SPEED

We can use equations (9.6) and (9.8) to develop another performance factor which is widely used for both preliminary design and selection of pumps. If we eliminate D from these equations, we obtain

$$\frac{N^{2/3}Q^{1/3}}{H^{1/2}} = \text{constant} \tag{9.12}$$

If we raise both sides of (9.12) to the $\frac{3}{2}$ power,

$$\frac{N\sqrt{Q}}{H^{3/4}} = \text{constant} \tag{9.13}$$

At this point we define the term *specific speed*, N_s, as the value of N in equation (9.13) when $H = 1$ and $Q = 1$; in other words, it is equal to the constant in this equation. Therefore, we can write equation (9.13) as

$$N_s = \frac{N\sqrt{Q}}{H^{3/4}} \tag{9.14}$$

In the conventional English system, N is usually given in rpm, Q is in gal/min, and H is in feet. These values are those at the point of maximum efficiency for the shaft speed used. Unfortunately, the units gal/min and feet do not yield a dimensionless form for N_s. In order to write equation (9.14) in consistent units, such as SI, we can rewrite it as

$$N_s = \frac{N\sqrt{Q}}{gH^{3/4}} \tag{9.15}$$

Since most published data have been obtained prior to the use of SI units, the published values use equation (9.14) for the definition of specific speed. To convert from equation (9.14) to (9.15), it is necessary to divide the values obtained from equation (9.14) by 17 200 to convert them to SI values consistent with equation (9.15).

The utility of the specific-speed concept is that certain combinations of head, speed, and capacity are typical of a given type of pump. Thus, specific speed can tell us the combinations of these factors that are both possible and desirable. It has been found that the specific speed of centrifugal pumps is low while axial-flow pumps have high specific speeds. Pumps with a combination of these types (mixed-flow) have intermediate values of specific speed. For single-stage centrifugal pumps, N_s ranges from 500 to 5000 while single-stage axial flow pumps have specific speeds that lie in the range 5000 and 10,000 in English conventional units. Figure 9.11 shows how impeller design

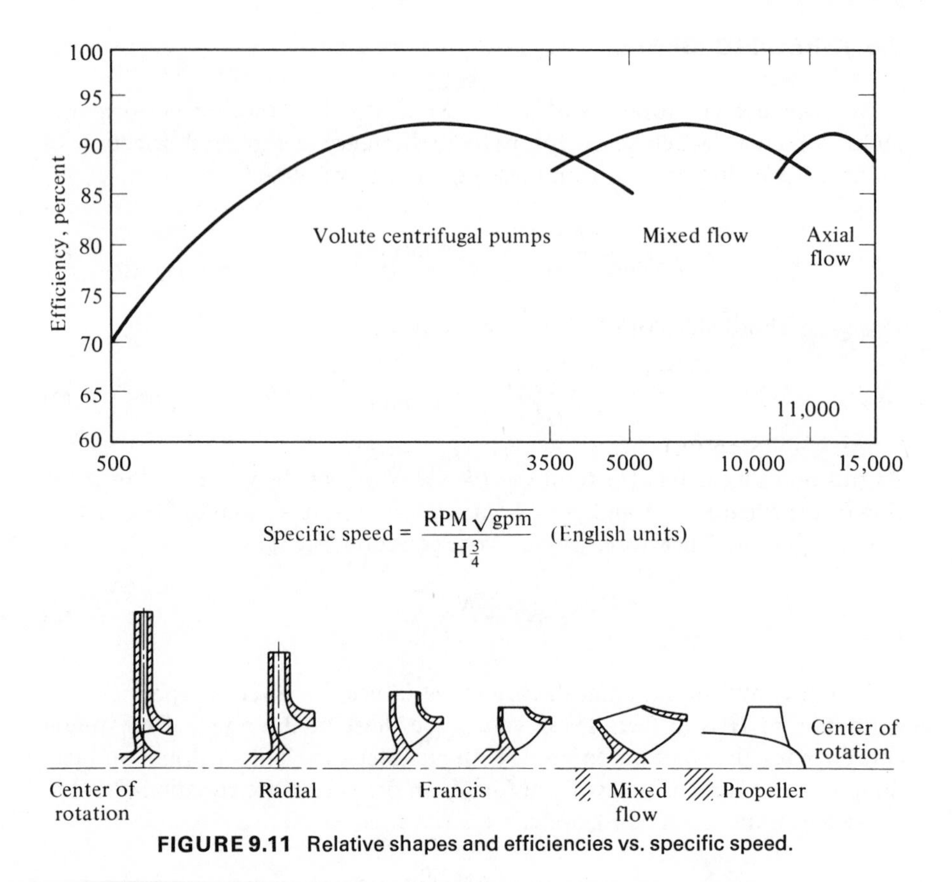

FIGURE 9.11 Relative shapes and efficiencies vs. specific speed.

and optimum efficiency vary as a function of specific speed for centrifugal, mixed-flow, and axial-flow pumps.

ILLUSTRATIVE PROBLEM 9.6

A centrifugal pump operates at 3600 rpm, developing a head of 100 ft while pumping 2000 gal/min. What is its specific speed in both English and SI units?

Solution

From equation (9.14),

$$N_s = \frac{N\sqrt{Q}}{H^{3/4}} = \frac{3600\sqrt{2000}}{100^{3/4}}$$

$$= 5091$$

In SI units,

$$N_s = \frac{5091}{17\,200} = 0.296$$

ILLUSTRATIVE PROBLEM 9.7

What type of pump should be selected that will deliver 2000 gal/min against a head of 25 ft while operating at 3600 rpm?

Solution

N_s from equation (9.14) is

$$N_s = \frac{N\sqrt{Q}}{H^{3/4}} = \frac{3600\sqrt{2000}}{(25)^{3/4}} = 14\,400$$

$$= 0.837 \qquad \text{in SI units}$$

From our discussion or Figure 9.11, this should be an axial flow pump.

9.7 CLOSURE

A single short chapter in a book on fluid mechanics really cannot do justice to a topic as vast as this one. However, most technical personnel get involved in the selection of pumps rather than in their design. To make a meaningful selection, it is necessary to know the types of units that are available, to be able to evaluate the operating conditions using a Bernoulli equation and pump operating characteristic curves, to know how to use test data taken under one set of conditions when these conditions are changed, and to be able to select the pump type best suited for a given application. We have covered each of these topics in sufficient detail for most selection purposes. For specific details, and the design of pumps, reference should be made to the vast literature on this subject as well as the extensive data that pump manufacturers will quite willingly supply.

PROBLEMS

9.1 A pump is located with its center line 2 ft above the level of a reservoir that is open to the atmosphere. Water is being pumped at the rate of

15 ft/sec in the suction line and friction losses are estimated to be 3 ft. Determine the NPSH if the water is at 60°C. ($\gamma = 61.3$ lb/ft³.)

9.2 Solve Problem 9.1 if the pump is located with its center line below the level in the reservoir.

9.3 A pump delivers water at 68°F from a suction tank to a discharge tank at the rate of 10 litres/s. If the suction line has a 75-mm inside diameter and the pump center line is located 2 m above the level of the suction reservoir, determine the NPSH. Suction-line losses are estimated to be 1.0 m and the reservoir is open to the atmosphere.

9.4 Solve Problem 9.3 if the pump is located below the level in the reservoir.

9.5 A pump delivers water from a lower reservoir to an upper reservoir. If the flow rate is 0.05 m³/s and the pump has a 100-mm inlet diameter and a 50-mm outlet diameter, calculate the power added to the water. The pump outlet is 2 m above the inlet, and static pressure taps indicate an outlet pressure that is 100 kPa greater than the inlet pressure. $\gamma =$ 9810 N/m³.

9.6 A pump delivers water from a lower to a higher reservoir. If the pump delivers 100 gal/min and the outlet is 3 ft above the inlet, determine the pump power. The inlet pipe has a 3-in. inside diameter and the outlet pipe has a 1-in. inside diameter. Pressure taps indicate the outlet pressure to be 6 psi higher than the inlet pressure. Use $\gamma = 62.4$ lb/ft³.

9.7 A test is conducted on a pump and test data show the pump capable of developing a total head of 15 m while delivering 20 litres/s when operated at 1000 rpm. Estimate the head and capacity when the pump is operated at 800 rpm.

9.8 A pump delivers 500 gal/min at 1000 rpm against a total head of 45 ft. Determine its performance at 1100 rpm.

9.9 A pump has an impeller that is 5 ft in diameter and is operated at 1450 rpm. If the speed is changed to 1150 rpm, what impeller diameter should be used to maintain a constant power input to the pump?

9.10 The impeller of a pump is 1.25 m in diameter and the pump is operated at 1050 rpm. If the speed is increased to 1175 rpm, what impeller diameter should be used to maintain a constant power input to the pump!

9.11 Determine the initial specific speed of the pump in Problem 9.7.

9.12 Determine the initial specific speed of the pump in Problem 9.8.

9.13 A pump is to deliver 500 gal/min against a head of 40 ft. If it is operated at 1150 rpm, what type of pump should be selected?

9.14 A pump is to deliver 4000 litres/min against a head of 20 m. If the pump is operated at 3600 rpm, select the type of pump to be used.

REFERENCES

1. *Steam Tables*, International Edition, by J. H. KEENAN, G. F. KEYES, P. G. HILL, and J. G. MOORE, John Wiley & Sons, Inc., New York, 1969.
2. *Engineering Fluid Mechanics* by J. A. ROBERSON and C. T. CROWE, Houghton Mifflin Company, Boston, 1975.
3. *Fluid Mechanics* by R. C. BINDER, 4th ed., Prentice-Hall, Inc., Englewood Cliffs, N.J., 1962.
4. *Applied Fluid Mechanics* by R. L. MOTT, 2nd ed., Charles E. Merrill Publishing Co., Columbus, Ohio, 1979.
5. *Engineering Applications of Fluid Mechanics* by J. C. HUNSAKER and B. G. RIGHTMIRE, McGraw-Hill Book Company, New York, 1947.
6. *Fluid Mechanics* by V. L. STREETER and E. B. WYLIE, 7th ed., McGraw-Hill Book Company, New York, 1979.
7. *Basic Fluid Power* by D. A. PEASE, Prentice-Hall, Inc., Englewood Cliffs, N.J., 1967.
8. *Fundamentals of Fluid Mechanics* by J. A. SULLIVAN, Reston Publishing Co., Reston, Va., 1978.
9. *Steam—Its Generation and Use*, 38th ed., Babcock & Wilcox, New York 1972.

Flow about Immersed Bodies

10.1 INTRODUCTION

It is a trite truism that we on earth live at the bottom of an ocean of air. This fact must be accounted for in the design of all vehicles that move in this environment (i.e., aircraft, trains, automobiles, etc.). Buildings, bridges, and other structures must also be designed to withstand the dynamic forces that this environment imposes on them. Recent advances in oceanography have opened up an entirely new field, aptly called hydrospace, a relatively viscous, dense, and hostile environment in which man and his vehicles are totally immersed. The flow about an object may be due to the motion either of the object relative to the fluid or of the fluid relative to the object. In this chapter we shall consider the incompressible steady flow of a fluid relative to an object. It should also be noted that the principles developed in this chapter can also be applied to the study of turbines, fans, pumps, propellers, and many other flow systems.

10.2 GENERAL CONSIDERATIONS

When a body is fully submerged in a viscous, incompressible fluid and moves with a velocity V relative to the fluid, the body will experience a resistance to its motion due to the fluid. In Section 2.7 we discussed this type of physical interaction when we considered the motion of a spherical object falling freely in a still viscous fluid. In general, we can assume that the resultant force on any such body can be represented by two perpendicular components. One of these components is usually taken in a direction parallel to the direction of flow and is known as *drag*. The other component is taken perpendicular to the direction of flow and is known as *lift*. The body (airfoil) shown in Figure 10.1 illustrates these concepts. Quantitatively, we can ex-

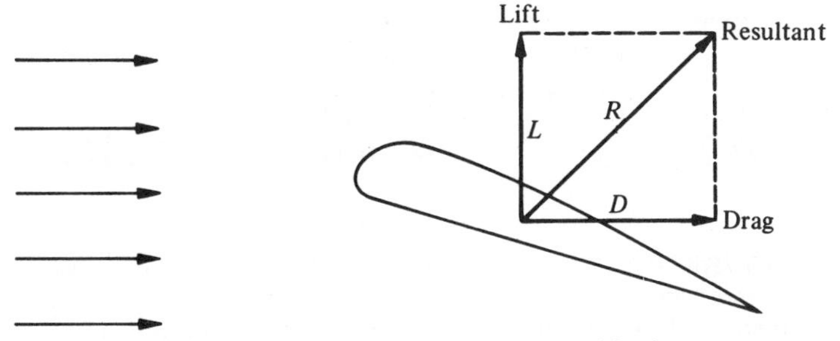

FIGURE 10.1 Forces on an immersed body.

press the lift and drag on a body in terms of a lift coefficient (C_L) and a drag coefficient (C_D). Equations (10.1) and (10.2) are the defining equations;

$$D = C_D \frac{\rho V^2}{2} A \tag{10.1}$$

and

$$L = C_L \frac{\rho V^2}{2} A \tag{10.2}$$

where D and L are the lift and drag forces, ρ the density of the fluid, V the relative velocity of the fluid to the body, and A a characteristic area that we will discuss in detail later. The term $\rho V^2/2$ is called the *dynamic* or *impact pressure*. The coefficients C_D and C_L are dimensionless and do not depend upon the system of units used. Both C_D and C_L can be written as function of the Reynolds number for incompressible flow,

$$C_D = \varphi_1(\text{Re}) \tag{10.3a}$$

$$C_L = \varphi_2(\text{Re}) \tag{10.3b}$$

Figures 10.2 and 10.3 show the drag coefficients for spheres, plates and cylinders.

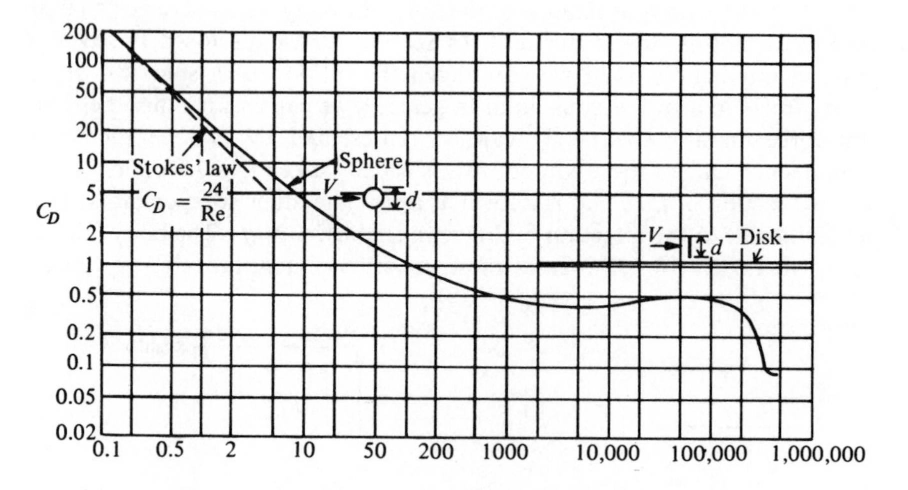

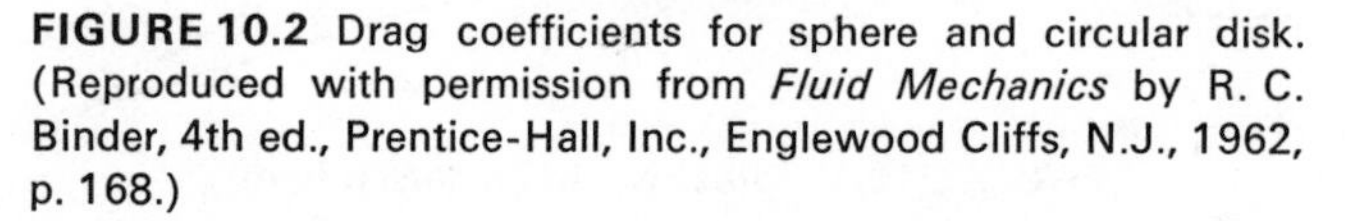

FIGURE 10.2 Drag coefficients for sphere and circular disk. (Reproduced with permission from *Fluid Mechanics* by R. C. Binder, 4th ed., Prentice-Hall, Inc., Englewood Cliffs, N.J., 1962, p. 168.)

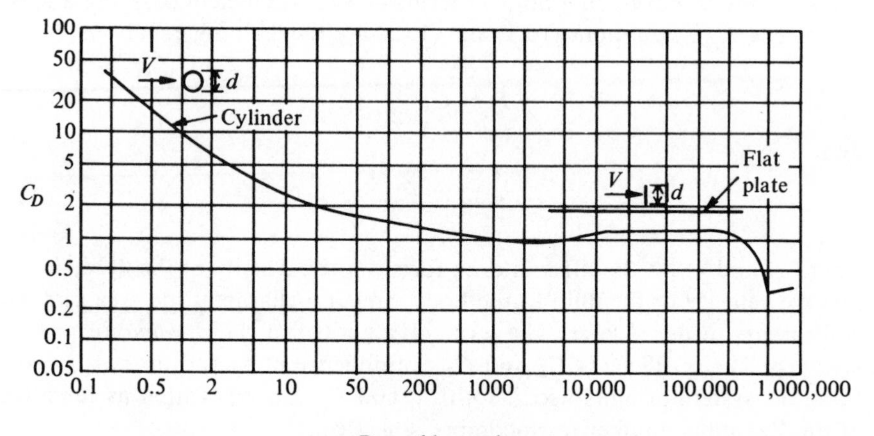

FIGURE 10.3 Drag coefficients for cylinder and flat plate. (Reproduced with permission from *Fluid Mechanics* by R. C. Binder, 4th ed., Prentice-Hall, Inc., Englewood Cliffs, N.J., 1962, p. 170.)

ILLUSTRATIVE PROBLEM 10.1

In Chapter 2, Stokes' law was discussed in relation to a sphere falling at a constant velocity in a viscous fluid. Derive a relation from Stokes' law for C_D of a sphere for these conditions.

Solution

From Chapter 2, Stokes' law is

$$F = 6\pi r_o \mu V$$

where r_o is the outside radius of the sphere, μ the viscosity of the fluid, and V the relative velocity between the sphere and the undisturbed fluid. Equating this to equation (10.1),

$$D = F = 6\pi r_o \mu V = C_D \frac{\rho V^2}{2} A$$

but the projected area of the sphere is πr_o^2:

$$6\pi r_o \mu V = C_D \left(\frac{\rho V^2}{2}\right)(\pi r_o^2)$$

Rearranging and simplifying, we obtain

$$C_D = \frac{2(6\pi r_o \mu V)}{\pi r_o^2 \rho V^2} = \frac{24}{D_o V \rho / \mu} = \frac{24}{\text{Re}}$$

The conclusion from this illustrative problem is that Stokes' law yields a drag coefficient for spheres at low Reynolds numbers that is solely a function of Reynolds number. This equation is plotted in Figure 10.2, showing the Stokes' law line and experimental data for sphere at various Reynolds numbers.

When a flat plate is placed parallel to the direction of fluid motion, as shown in Figure 10.4a, it experiences a drag force due to fluid friction only on both faces. As the velocity of the fluid relative to the plate is increased (with a concurrent increase in Reynolds number), the drag force decreases. At some value of Reynolds number (approximately 10^6) a sudden increase in drag occurs while a still further increase in Reynolds number yields a continual decrease in drag force. If the drag coefficient is plotted as a function of Reynolds number for smooth flat plates where the linear dimension in Reynolds number

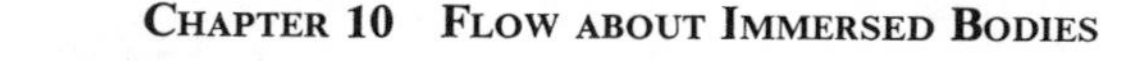

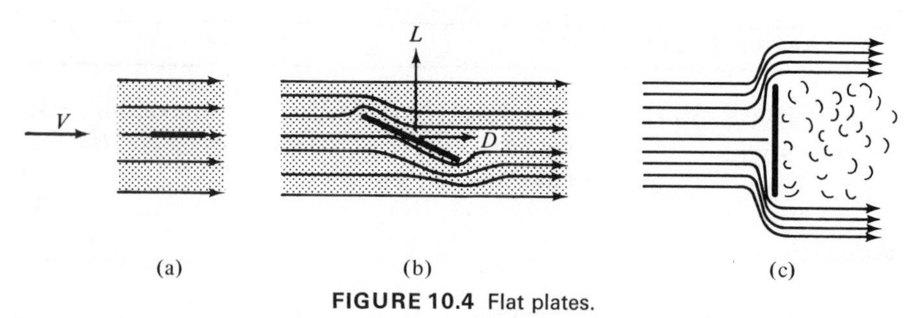

FIGURE 10.4 Flat plates.

is taken to be the length of the plate parallel to the flow, the curves in Figure 10.5 are determined. These two curves appear to be similar to the curves for smooth tubes in the Moody diagram and basically are found to depend on whether the boundary layer is found to be laminar or turbulent.

When the plate is tilted at an angle to the stream, as shown in Figure 10.4b, the resultant force on the plate can be resolved into two component forces, lift (L) and drag (D). The data shown in Figure 10.6 are based upon tests on small rectangular plates whose longer side was six times the shorter side, with air striking the longer side first. Small corrections may be needed in applying these data to longer plates or plates of different shapes. It should be noted that these curves show no dependence on Reynolds number and that at an angle of zero degrees when the plate is parallel to the flow the drag coefficient is given as a single value. Since the conditions of the test are not specifically given and the Reynolds number is also not given, this curve should be used

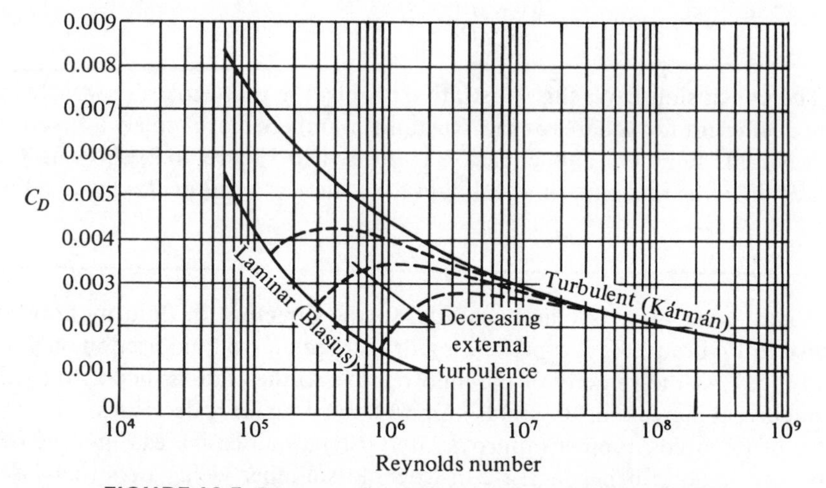

FIGURE 10.5 Drag on flat plate parallel to flow direction. (Reproduced with permission from *Fluid Mechanics* by R. C. Binder, 4th ed., Prentice-Hall, Inc., Englewood Cliffs, N.J., 1962, p. 173.)

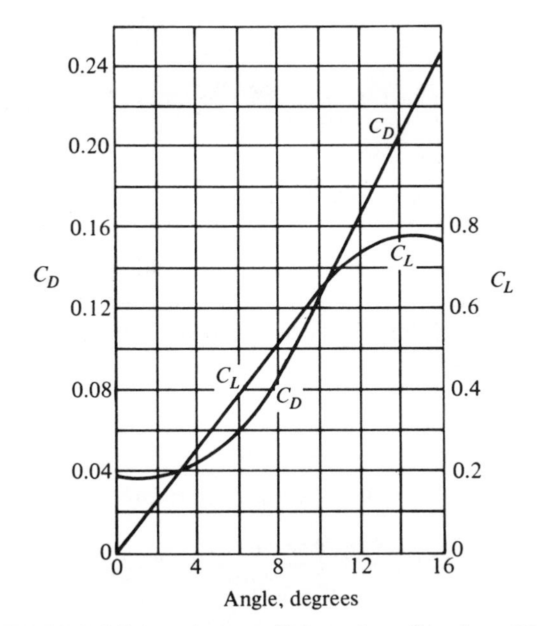

FIGURE 10.6 Lift anu drag coefficients for a flat plate. (Reproduced with permission from *Elements of Practical Aerodynamics* by B. Jones, 3rd ed., John Wiley & Sons, Inc., New York, 1942, p. 21.)

for design with caution. However, the qualitative trend shows an increase in both lift and drag coefficients as the angle the plate makes with the stream increases. At some finite angle, approximately 15°, C_L starts to decrease, but the drag coefficient C_D continues to increase.

ILLUSTRATIVE PROBLEM 10.2

Using the data shown in Figure 10.6, determine the lift and drag forces on a plate 6 ft long and 1 ft wide if the plate is set at an angle of 8° to an airstream having a velocity of 35 ft/sec. What is the resultant force on the plate? The specific weight of air can be taken as 0.075 lb/ft³.

Solution

Using the data from Figure 10.6 at 8°, $C_L = 0.51$ and $C_D = 0.086$. Therefore,

$$L = C_L \frac{\rho}{2} V^2 A = 0.51 \times \frac{0.075}{32.17 \times 2}(35)^2 \times 6 \times 1 = 4.37 \text{ lb}$$

$$D = C_D \frac{\rho}{2} V^2 A = 0.086 \times \frac{0.075}{32.17 \times 2}(35)^2 \times 6 \times 1 = 0.74 \text{ lb}$$

The resultant force equals the vector sum of L plus D. Thus,

$$R = \sqrt{(L)^2 + (D)^2} = \sqrt{(0.74)^2 + (4.37)^2} = 4.43 \text{ lb}$$

ILLUSTRATIVE PROBLEM 10.3

Using SI data, Illustrative Problem 10.2 would have the plate 1.83 m long and 0.305 m wide, the velocity would be 10.67 m/s and the specific weight of air would be 11.78 N/m.[3] What is the resultant force on the plate using the data of Figure 10.6?

Solution

$C_L = 0.51$ and $C_D = 0.086$ as before. Therefore,

$$L = C_L \frac{\rho V^2}{2} A = 0.51 \times \frac{11.78 \times (10.67)^2}{2 \times 9.81} \times 1.83 \times 0.305 = 19.46 \text{ N}$$

$$D = C_D \frac{\rho V^2}{2} A = 0.086 \times \frac{11.78 \times (10.67)^2}{2 \times 9.81} \times 1.83 \times 0.305 = 3.28 \text{ N}$$

The resultant force is

$$R = \sqrt{(19.46)^2 + (3.28)^2} = 19.73 \text{ N}$$

which is in agreement with Illustrative Problem 10.2 since 19.73 N $\times$ 0.2248 = 4.44 lb.

Increasing the plate angle still further until the plate is perpendicular to the direction of flow yields the flow pattern shown in Figure 10.4c. For this condition the drag force is due to the pressure difference on both sides of the plate. Figure 10.7 shows the drag coefficient for finite flat plates perpendicular

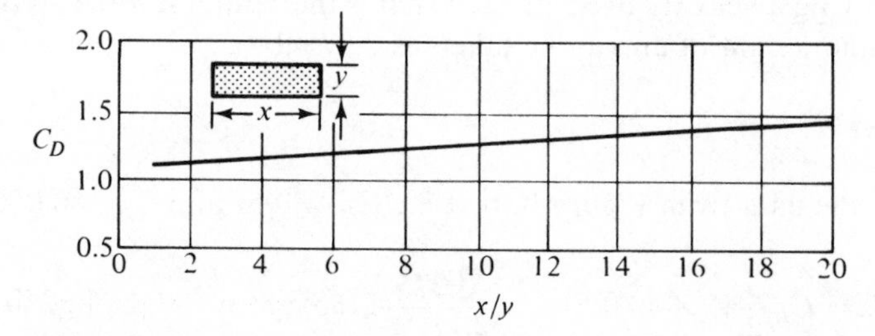

FIGURE 10.7 Drag coefficients; flat plate perpendicular to the flow. (Reproduced with permission from *Fluid Mechanics* by R. C. Binder, 4th ed., Prentice-Hall, Inc., Englewood Cliffs, N.J., 1962, p. 171.)

to the direction of flow. Comparison of the data for a square plate ($x = y$) with the data for the disk shown in Figure 10.2 shows good agreement. As the value of x/y increases, the value of C_D approaches the value given for flat plates in Figure 10.3.

In Illustrative Problem 10.1, C_D for a sphere was determined to be 24/Re, based upon Stokes' law. Stokes' law is based upon laminar flow, and it will be seen from Figure 10.2 that for Reynolds numbers greater than 0.5, the discrepancy between values of C_D equal to 24/Re, and the experimentally determined value of C_D increases. For small Reynolds numbers, the value of C_D for various bodies is given in Table 10.1. Note that for a flat plate oriented

TABLE 10.1†

Object	Re	C_D
Sphere	<0.5	24/Re
Disk flow	<0.5	20.4/Re
Disk flow	<0.1	13.6/Re
Circular cylinder	<0.1	$8\pi/\text{Re}[2.0 - \ln \text{Re}]$
Flat plate perpendicular to flow	<0.1	$8\pi/\text{Re}[2.2 - \ln \text{Re}]$
Flat plate parallel to flow	<0.01	4.12/Re

†Data for this table are based upon *Fluid Mechanics for Engineers* by M. L. Albertson, J. R. Barton, and D. B. Simons, Prentice-Hall, Inc., Englewood Cliffs, N. J., 1960, p. 395.

parallel to the direction of flow, the proper area to use in the drag equation is the area of both sides of the plate. It will be noted that C_D in every case is much greater than the values shown for turbulent flow on Figures 10.2 and 10.3. This was also the case for the flow of fluids in pipes, where it will be recalled that the largest values of friction factor occur when the flow is laminar for small values of Reynolds number.

ILLUSTRATIVE PROBLEM 10.4

A 6-in. by 1-in. flat plate is placed parallel to the flow with the 6-in. edge facing into the flow. If the velocity of the fluid is 0.01 ft/sec, evaluate the drag on the plate. Assume that the viscosity of the fluid is 1×10^{-4} lb sec/ft^2 and that the flow is laminar.

Solution

Using $C_D = 4.12/\text{Re}$ from Table 10.1,

$$\text{drag} = D = C_D \frac{\rho V^2}{2}(A)$$

Therefore,

$$D = \frac{4.12}{DV\rho/\mu}\left(\frac{\rho V^2}{2}\right)A$$

and

$$D = \frac{4.12\mu}{(1/12)V\rho}V\rho\frac{V}{2}\left[\frac{1}{12}\left(\frac{6}{12}\right)\right] \times 2$$

Note that twice the area is used for flow parallel to the plate since the fluid travels over both sides of the plate. Rearranging yields

$$D = 4.12\mu V\frac{6}{12} = 2.06\mu V$$

For this problem,

$$D = 2.06 \times 1 \times 10^{-4} \times 0.01 = 2.06 \times 10^{-6} \text{ lb}$$

It should be noted that whenever C_D can be written as a constant/Re, the resulting equation for the drag will be a function of only the product of viscosity, the velocity, and a characteristic length [i.e., D = constant/($L\mu V$)].

The previous discussion has been principally concerned with laminar flow about an immersed object. As the velocity of the fluid relative to the body is increased (with the subsequent increase in Reynolds number) we have already noted that the drag coefficient decreases. At a Reynolds number of approximately 2×10^5 it will be seen from Figures 10.2 and 10.3 that there is a sharp discontinuity, indicating a marked decrease in drag coefficient for both the cylinder and sphere. As a matter of fact, the drag coefficients decrease to approximately one-third of their value just prior to this occurrence. It has been found experimentally that the Reynolds number at which this abrupt decrease in drag coefficient occurs is dependent on the turbulence in the undisturbed fluid and the roughness of the body. Thus, if the turbulence in the undisturbed fluid is large and/or the surface of the body is made very rough (relatively), the noted decrease in drag coefficient will occur at Reynolds numbers less than 2×10^5. This behavior is again found to be similar to the effects discussed in Chapter 6 on the incompressible flow of fluids in pipes. In both instances (flow inside pipes and flow around immersed bodies) the underlying behavior is found to reside in the boundary layer adjacent to either the pipe wall or adjacent to the immersed object. When studying the flow around an immersed body, the abrupt transition in the drag coefficient is found to be caused by the boundary layer changing from a laminar boundary layer to a turbulent boundary layer on the fore part of the body. This same phenomenon has been discussed earlier in this section during our discussion on the flow over flat plates.

Drag coefficients for cylinders and flat plates are given in Table 10.2. In

TABLE 10.2 Drag Coefficients for Cylinders and Flat Plates†

Object (flow from L to R)		L/d	Re	C_D
1. Circular cylinder, axis perpendiculer to the flow		1	10^5	0.63
		5		0.74
		20		0.90
		∞		1.20
		5	$>5 \times 10^5$	0.35
		∞		0.33
2. Circular cylinder, axis parallel to the flow		0	$>10^3$	1.12
		1		0.91
		2		0.85
		4		0.87
		7		0.99
3. Elliptical cylinder	(2 : 1)		4×10^4	0.6
			10^5	0.46
	(4 : 1)		2.5×10^4 to 10^5	0.32
	(8 : 1)		2.5×10^4	0.29
			2×10^5	0.20
4. Airfoil	(1 : 3)	∞	$>4 \times 10^4$	0.07
5. Rectangular plate for which L = length and d = width		1	$>10^3$	1.16
		5		1.02
		20		1.50
		∞		1.90
6. Square cylinder	■		3.5×10^4	2.0
	◆		$10^4 \times 10^5$	1.6
7. Triangular cylinder				
	120°		$>10^4$	2.0
				1.72
	60°			2.20
				1.39
	30°		$>10^5$	1.80
				1.0
8. Hemispherical shell			$>10^3$	1.33
			10^3 to 10^5	0.4
9. Circular disk, normal to the flow			$>10^3$	1.12
10. Tandem disks; spacing is L		0	$>10^3$	1.12
		1		0.93
		2		1.04
		3		1.54

†Reproduced with permission from *Fluid Mechanics for Engineers* by M. L. Albertson, J. L. Barton, and D. B. Simons, Prentice-Hall Inc., Englewood Cliffs, N. J., 1960, p. 407.

the turbulent range the drag coefficient for these bodies decreases with increased Reynolds number.

10.3 LIFT AND DRAG ON AIRFOILS

10.3a GENERAL

In Section 10.2 the discussion was directed to the general problem of forces on an immersed body. At this point the discussion will be directed to the forces (lift and drag) that occur when an airplane moves through the air at Mach numbers $< < 1$. Prior to to our study of specific subtopics under this heading, it will be useful to define and illustrate our terminology.

Airplane. An airplane is a mechanically driven, fixed-wing aircraft, heavier than air, that is supported by the dynamic reaction of the air against its wings. This definition can be amplified by considering an airplane in steady, level flight. For this airplane to be in equilibrium (it is not accelerating) the lift forces on it must equal its weight, and the drag forces must be countered by an equal but oppositely directed thrust from the airplane's power plant. Figure 10.8 illustrates the forces on an airplane in level flight.

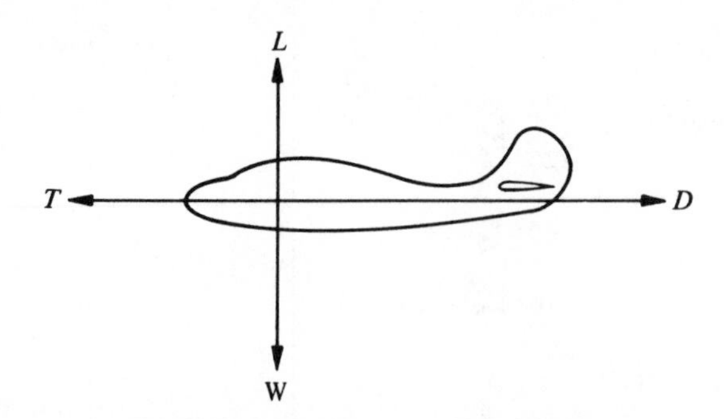

FIGURE 10.8 Airplane in level flight.

Airfoil. An airfoil is any surface, such as the airplane wing, aileron, or rudder, designed to obtain reaction from the air through which it moves. Figure 10.9 shows an airfoil and the pressure distribution on this airfoil. Along the upper surface there is a reduced pressure (negative pressure), while underneath the airfoil the pressure is greater than ambient (positive). This pressure distribution occurs from the acceleration of the air over the upper surface of the airfoil and the deceleration of the air over the lower surface of the airfoil. Significantly, the greater portion of the lift is obtained from the upper airfoil surface.

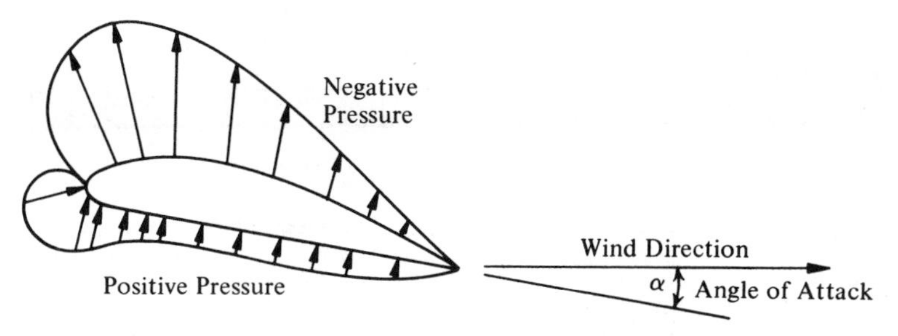

FIGURE 10.9 Pressure distribution on an airfoil section.

Angle of Attack. The angle of attack is the acute angle between a reference line in a body and the line of the relative wind direction projected on a plane containing the reference line and parallel to the plane of symmetry. The relative wind is the velocity of the air with reference to the body and it is measured at a distance from the body to minimize the disturbing effect of the body. Figure 10.9 also illustrates the angle of attack on an airfoil section.

Angle, Zero Lift. The zero lift angle is that angle of attack of an airfoil when its lift is zero. Figure 10.10 shows the characteristics of a Clark Y airfoil. With an airfoil that is completely symmetrical the lift coefficient (and consequently the lift) would be zero at a zero angle of attack. However, this airfoil can be seen to be asymmetric, and it is found that C_L is zero at an angle of attack of $-5°$ (i.e., that angle where the sum of the positive lift forces equals the sum of the negative lift forces). For angles of attack greater than the angle of zero lift, it will be seen that the lift coefficient is directly proportional to the angle of attack when this angle is measured relative to the zero lift angle; that is,

$$C_L = K(\alpha - \alpha_{0L}) \tag{10.4}$$

where K is the slope of the curve of C_L against the angle of attack, α the angle of attack, and α_{0L} the angle of zero lift.

ILLUSTRATIVE PROBLEM 10.5

Using the data for the Clark Y airfoil shown in Figure 10.10, it is found that C_L is 0.5 at an angle of attack of 2°. Estimate C_L at an angle of attack of 6°.

Solution

From Figure 10.10, the angle of zero lift is $-5°$. From the stated condition at 2°,

$$0.5 = K[2 - (-5)]$$

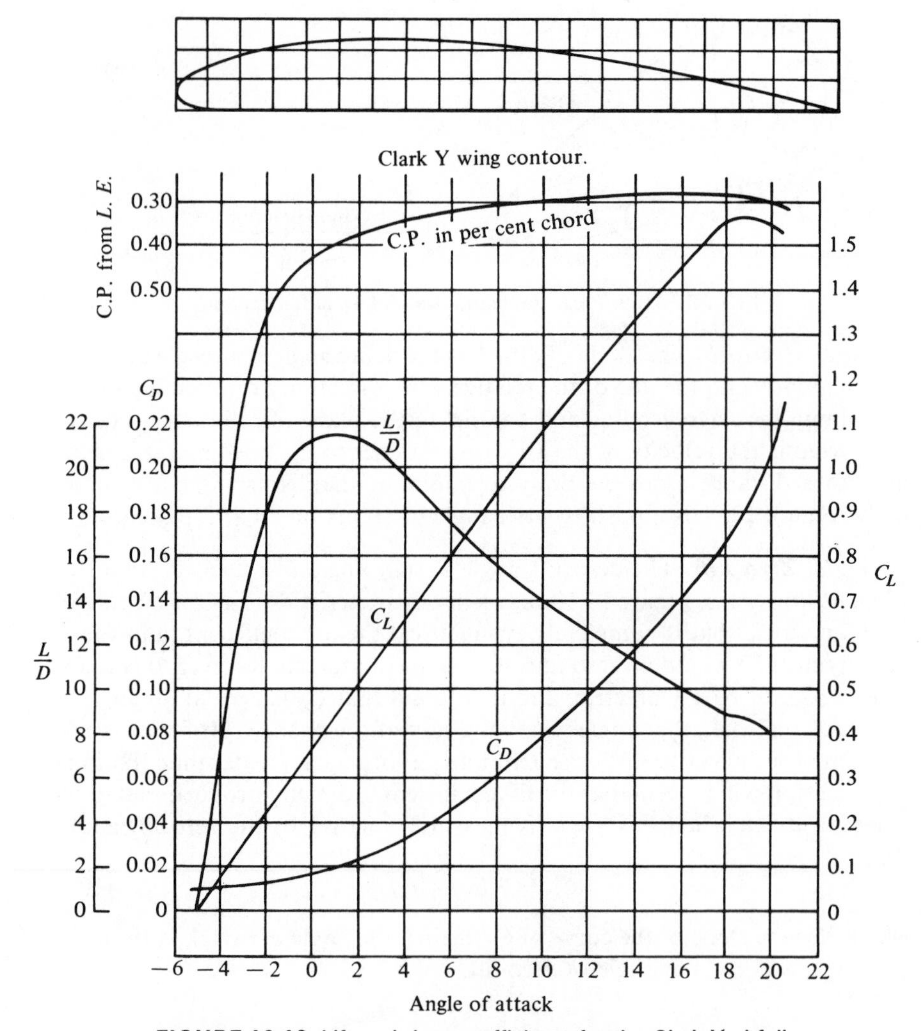

FIGURE 10.10 Lift and drag coefficients for the Clark Y airfoil. (Reproduced with permission from *Elements of Practical Aerodynamics* by B. Jones, John Wiley & Sons, Inc., New York, 1942, p. 39.)

Therefore,

$$K = \frac{0.5}{2 - (-5)} = 0.0714$$

For 6°,

$$C_L = 0.0714[6 - (-5)] = 0.785$$

From Figure 10.10, C_L for 6° is approximately 0.79, which is in good agreement with the calculated value.

As the angle of attack is increased the lift coefficient deviates from a linear relationship with respect to angle of attack. For the Clark Y airfoil the maximum value of C_L occurs at $18\frac{1}{2}°$; above this angle (known as the *burble point* or *stall angle*) the lift decreases with increasing angle of attack.

Chord. The chord of an airfoil is an arbitrary datum line from which the ordinates and angles of an airfoil are measured. It is usually the straight line tangent to the lower airfoil surface at two points of the straight line joining the leading and trailing edge of the airfoil. This definition is illustrated for the two airfoils shown in Figure 10.11. The chord length (customarily given the symbol c) is the length of the projection of the airfoil profile on its chord.

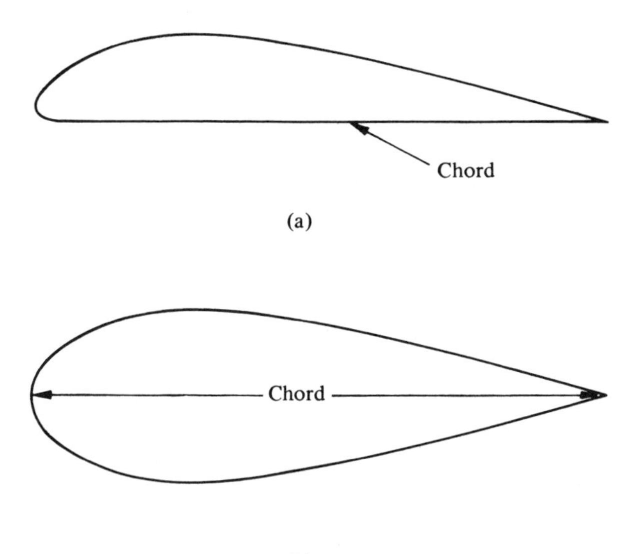

FIGURE 10.11 Chord for unsymmetric and symmetric airfoils.

Wing Area. To calculate both the lift and drag on an airplane, it is necessary to know the area term in equation (10.2). For purposes of standardization the wing area is measured from the projection of the actual outline on the plane of the chords without deduction for the area that may be blanketed by the fuselage or nacelles. The wing is thus considered to extend without interruption through the fuselage and nacelles, and the wing area always includes flaps and ailerons.

Aspect Ratio (A.R.). The aspect ratio of a wing is defined as the ratio of the span of the wing to its mean chord. In general, wings are not rectangular when viewed from above, and the aspect ratio in this case is defined as the ratio of the square of the span (the distance from wing tip to wing tip is denoted as b) to the area; that is,

$$\text{A.R.} = \frac{b^2}{A} \tag{10.5}$$

Camber. The curvature of an airfoil is known as its camber, with upper camber referring to the upper surface, lower camber referring to the lower surface, and mean camber referring to the mean surface. It is usual to express camber as the ratio of the departure of the curved surface from a straight line joining the extremeties of the curve to the length of this straight line.

Center of Pressure (C.P.). The location of the resultant aerodynamic force on an airfoil is not fixed relative to the airfoil. The center of pressure is the point in the chord of airfoil (prolonged if necessary), which is the intersection of the chord and the line of action of the resultant air force. It is practically expressed in terms of the ratio of the distance of the center of pressure from the leading edge to the chord length and is usually stated as percent of chord. Values of C.P. in percent of chord for the Clark Y airfoil at various angles of attack are shown in Figure 10.10.

Stall. When an airfoil is operated at an angle of attack greater than the angle of attack at maximum lift (for the Clark Y airfoil this is $18\frac{1}{2}°$), the airfoil is said to be operating in the stall condition. At stall the air separates from the airfoil or wing and forms an eddying wake with severe turbulence, and as a consequence it is found that as the stall condition is approached there is a pronounced increase in the drag coefficient. Inspection of Figure 10.10 shows the drag coefficient increasing continuously as the angle of attack is increased; the percentage increase becoming larger as the stall condition is approached.

10.3b Steady Level Flight

Let us now return to the basic lift and drag equations and apply them to the airplane in level flight. In the following material the data given in Figure 10.10 for the Clark Y airfoil will be used as illustrative of airfoil data in general. This is not meant to infer that all airfoil data are the same or even close to that shown for the Clark Y airfoil. For other airfoils it is necessary to obtain data for the specific airfoil in question. In many instances such data will be found in the extensive compilations from wind tunnel tests conducted by NASA. From equation (10.2) and Figure 10.8 we can conclude the following:

1. For an airplane in steady flight the lift will just equal the weight of the airplane. Therefore,

$$W = C_L \frac{\rho}{2} V^2 A \tag{10.6}$$

Rearranging, we obtain

$$V = \sqrt{\left(\frac{W}{A}\right)\frac{2}{\rho C_L}} \tag{10.7a}$$

The term W/A is the weight loading of the wing in lb/ft² or N/m², and therefore the velocity at which a plane must fly in steady level flight is proportional to the square root of the wing loading.

Using the specific weight of air to be 0.0765 lb/ft³ at sea level, V in mi/hr, W in lb, and A in ft², we can write equation (10.7a) as

$$V = 19.77\sqrt{\frac{1}{C_L}\left(\frac{W}{A}\right)} \tag{10.7b}$$

In terms of SI units with the specific weight of air taken to be 12.02 N/m³, V in km/h, W in newtons, and A in m², we can write equation (10.7a) as

$$V = 4.6\sqrt{\frac{1}{C_L}\left(\frac{W}{A}\right)} \tag{10.7c}$$

ILLUSTRATIVE PROBLEM 10.6

An airplane is to fly at 300 mi/hr at sea level. If the airfoil is a Clark Y airfoil having an angle of 6°, determine the wing loading in pounds per square foot of wing.

Solution

For the units of this problem we will use equation (10.7b),

$$V = 19.77\sqrt{\frac{1}{C_L}\left(\frac{W}{A}\right)} \tag{10.7b}$$

Using the data of this problem and $C_L = 0.79$ for an angle of attack of 6°, we obtain

$$300 = 19.77\sqrt{\frac{1}{0.79}\left(\frac{W}{A}\right)}$$

Solving for W/A yields

$$\frac{W}{A} = 181.9 \text{ lb/ft}^2$$

ILLUSTRATIVE PROBLEM 10.7

An airplane is to fly at 483 km/h at sea level. If the airfoil is a Clark Y airfoil having an angle of 6°, determine the wing loading in N/m² of wing.

Solution

This problem is the same as Illustrative Problem 10.6 with the problem given in SI units. Therefore,

$$V = 4.6\sqrt{\frac{1}{C_L}\left(\frac{W}{A}\right)}$$

and using $C_L = 0.79$ as before, we obtain

$$483 = 4.6\sqrt{\frac{1}{0.79}\left(\frac{W}{A}\right)}$$

Solving yields

$$\frac{W}{A} = 8710 \text{ N/m}^2$$

As a check:

$$181 \text{ lb/ft}^2 \times 4.448\,\frac{\text{N}}{\text{lb}} \times \left(3.281\,\frac{\text{m}}{\text{ft}}\right)^2 = 8667 \text{ N/m}^2$$

The agreement is very good since 483 is closely 300 mi/hr.

2. For an airplane in steady level flight the velocity of flight is inversely proportional to the square root of the lift coefficient. Since the lift coefficient exhibits a maximum value at the stall point, the maximum velocity (stall speed) that the airplane can have in steady level flight must occur at an angle of attack coinciding with the maximum value of C_L. For the Clark Y airfoil we have already noted this angle to be $18\frac{1}{2}$°. This minimum velocity will also be (very closely) the lowest landing and/or takeoff speed of the airplane.

ILLUSTRATIVE PROBLEM 10.8

Using the data given in Illustrative Problem 10.7 and the calculated wing loading determined in this problem, evaluate the landing and takeoff speed of the airplane.

Solution

At $18\frac{1}{2}°$, C_L (max) for the Clark Y airfoil is found to be closely 1.56. Therefore,

$$V_{\text{landing or takeoff}} = 4.6\sqrt{\frac{1}{C_L}\left(\frac{W}{A}\right)}$$

$$= 4.6\sqrt{\frac{1}{1.56}(8710)}$$

$$= 343.7 \text{ km/h}$$

3. When the airplane is in steady level flight the propulsion system must provide a thrust equal and opposite to the drag force on the plane. An increase in thrust while in level flight will cause the velocity of the plane to increase until the drag force once again equals the applied thrust. Conversely, a decrease in thrust will cause the airplane to decrease its velocity until the drag equals the thrust. The power input by the propulsion system is the product of the thrust multiplied by the velocity of the airplane. However, in steady level flight thrust necessarily equals drag and the power required becomes

$$P = TV = DV \quad (10.8a)$$

In terms of horsepower with D in pounds and V in feet per second,

$$\text{hp} = \frac{DV}{550} = \frac{C_D(\rho/2)V^3 A}{550} \quad (10.8b)$$

where 1 hp = 550 ft lb/sec. In terms of miles per hour and a specific weight of 0.0765, we can write

$$\text{hp} = 6.8 \times 10^{-6}\, C_D V^3 A \quad (10.8c)$$

In SI units, 1 hp = 746 W = 746 N·m/s. In terms of km/h, a specific weight of 12.02 N/m³ and area in m²,

$$\text{hp} = 1.76 \times 10^{-5}\, C_D V^3 A \quad (10.8d)$$

ILLUSTRATIVE PROBLEM 10.9

What horsepower is required for the condition given in Illustrative Problem 10.6? Assume that the wing has an area of 550 ft².

Solution

At an angle of attack of 6°, the Clark Y airfoil has $C_D = 0.045$, and applying equation (10.8c), we have

$$\text{hp} = 6.8 \times 10^{-6} \times 0.045 \times (300)^3 \times 550 = 4540 \text{ hp}$$

ILLUSTRATIVE PROBLEM 10.10

What horsepower is required for the condition given in Illustrative Problem 10.7 if the wing area is 51 m²?

Solution

The angle of attack is again 6° and $C_D = 0.045$. Using equation (10.8d), we obtain

$$\text{hp} = 1.76 \times 10^{-5} \times 0.045 \times (483)^3 \times 51 = 4550 \text{ hp}$$

As noted before this is the SI equivalent of Illustrative Problem 10.9 and agreement is very good.

---

4. The general conclusions reached in the earlier parts of this section are valid even when an airplane is operated at different altitudes. We have already concluded that steady level flight requires the lift to equal the weight of the airplane, and this must be true at all altitudes. However, if the angle of attack is constant, C_L is constant, and the velocity must increase to maintain level flight at altitudes since the density decreases with increasing altitude. Again, if the angle of attack is constant, C_D is constant, and for steady level flight it is necessary to evaluate the drag at increasing altitude since the velocity has increased but the density has decreased with increasing altitude. Let the subscript o denote sea level (zero altitude) and x any altitude; C_L and C_D are constant at all altitudes for a given angle of attack. Consider the lift equation

$$W = C_L \frac{\rho_o}{2} V_o^2 A = C_L \frac{\rho_o}{2} V_x^2 A \tag{10.9}$$

Therefore,

$$V_x^2 = \frac{\rho_o}{\rho_x} V_o^2 \tag{10.10}$$

From the drag equation,

$$D_o = C_D \frac{\rho_o}{2} V_o^2 A \tag{10.11}$$

and

$$D_x = C_D \frac{\rho_x}{2} V_x^2 A \tag{10.12}$$

Substituting (10.10) into (10.12), we obtain

$$D_x = C_D \frac{\rho_x}{2}\left(\frac{\rho_o}{\rho_x}\right) V_o^2 A = C_D \frac{\rho_o}{2} V_o^2 A \tag{10.13}$$

The somewhat surprising conclusion from equation (10.13) is that for a given angle of attack the drag is the same regardless of altitude. However, the power required does not stay constant, as can be shown by the following reasoning. From equation (10.8b) we can write

$$(\text{hp})_o = \frac{D_o V_o}{550} \tag{10.14}$$

and

$$(\text{hp})_x = \frac{D_x V_x}{550} = \frac{D_o V_x}{550} \tag{10.15}$$

since $D_o = D_x$.

From equation (10.10),

$$V_x = V_o \sqrt{\frac{\rho_o}{\rho_x}} \tag{10.16}$$

Substitution of (10.16) into (10.15) yields

$$(\text{hp})_x = \frac{D_o V_o}{550} \sqrt{\frac{\rho_o}{\rho_x}} = (\text{hp})_o \sqrt{\frac{\rho_o}{\rho_x}} \tag{10.17}$$

Since ρ_x is always less than ρ_o, the horsepower required at any altitude will be greater than the horsepower required at sea level for a fixed angle of attack.

ILLUSTRATIVE PROBLEM 10.11

A Clark Y airfoil having an area of 450 ft² is operated at a 4° angle of attack. If the airplane is designed to fly at 200 mi/hr at sea level, determine its velocity at an altitude where the density is eight-tenths of the density at sea level. Also determine the horsepower required at sea level and at altitude. Use $\rho = 0.002378$ slug/ft³ at sea level.

Solution

At 4°,

$$C_L = 0.65$$

$$C_D = 0.034$$

From equation (10.10),

$$V_x = V_o\sqrt{\frac{\rho_o}{\rho_x}} = 200\sqrt{\frac{1}{0.8}} = 224 \text{ mi/hr}$$

At sea level,

$$D_o = C_D\frac{\rho_o}{2}V_o^2 A \qquad 200 \text{ mi/hr} = 294 \text{ ft/sec}$$

Therefore,

$$D_o = 0.034\frac{0.002378}{2}(294)^2 \times 450 = 1560 \text{ lb}$$

$$\frac{D_oV_o}{550} = \frac{1560 \times 294}{550} = 835 \text{ hp}$$

At altitude,

$$(\text{hp})_x = 835\sqrt{\frac{1}{0.8}} = 935 \text{ hp}$$

---

5. The ability of an airfoil to perform its function of providing lift with the minimum amount of drag is expressed in the ratio L/D. Aerodynamically, the wing having the largest value of this figure of merit would perform in the most desirable manner. The ratio of L/D can also be interpreted by referring to Figure 10.1, where it will be seen that the ratio of lift to drag is the tangent of the angle that the resultant force on the airfoil makes with respect to the line of the relative wind direction. Mathematically, we can write

$$\frac{L}{D} = \frac{C_L\rho/2V^2A}{C_D\rho/2V^2A} = \frac{C_L}{C_D} \tag{10.18}$$

The ratio of L/D has been plotted in Figure 10.10 for the Clark Y airfoil. For this airfoil section it will be seen that the maximum value of L/D occurs at an angle of attack of approximately +1°. For a given angle of attack, an airfoil having the maximum value of L/D will require less thrust from its propulsion system.

ILLUSTRATIVE PROBLEM 10.12

At $18\frac{1}{2}°$ the Clark Y airfoil has its maximum lift and is said to be at the stall point. If an airplane using this airfoil weighs 10,000 lb, determine its drag at the stall point. Also determine the minimum drag for this airplane. Neglect all effects except those due to the airfoil. Discuss the values obtained.

Solution

At $18\frac{1}{2}°$,

$$\frac{L}{D} = 8.8 \qquad \text{(Figure 10.10)}$$

However, $L = W$; therefore,

$$\frac{W}{D} = 8.8$$

$$D = \frac{10{,}000}{8.8} = 1138 \text{ lb}$$

At $+1°$, L/D is maximum and equal to 21.5.

$$\frac{W}{D} = 21.5$$

$$D = \frac{10{,}000}{21.5} = 466 \text{ lb}$$

At the higher angle of attack the required thrust is 1135 lb, which is almost $2\frac{1}{2}$ times greater than the value at the lower angle of attack. However, it must be noted that the velocity of the airplane is not the same for both cases. Since velocity is inversely proportional to the square root of the lift coefficient, we can write

$$\frac{V_{18.5}}{V_1} = \sqrt{\frac{(C_L)_1}{(C_L)_{18.5}}} = \sqrt{\frac{0.45}{1.56}} = 0.54$$

The velocity of the airplane at $18\frac{1}{2}°$ is therefore approximately one-half the velocity that the airplane would have a 1°. At takeoff, where it is desired that the airplane should become airborne at the minimum possible velocity (using the least length of runway), the high angle of attack is achieved using flaps and other auxiliary devices, but this is achieved only at the expense of requiring more thrust from the airplane's propulsion system.

10.3c Polar Diagram

The curves for C_L, C_D, and L/D as a function of angle of attack are satisfactory for evaluating the performance of a given airfoil, and this type of plot can also be useful when comparing airfoils. However, this requires that three curves must be plotted and evaluated in each case. It is much more convenient to plot these data on a single diagram, called a *polar diagram* (in which C_L is the ordinate and C_D the abscissa), as shown in Figure 10.12 for the Clark Y airfoil. If we first consider the plot to have the same scales for C_L and C_D, the ratio of lift to drag (L/D) would be the slope of the straight line drawn from the origin to the curve. The maximum value of L/D would be the slope of the line drawn from the origin tangent to the curve. In every case a line from the origin to the curve would give the direction of the resultant force and be pro-

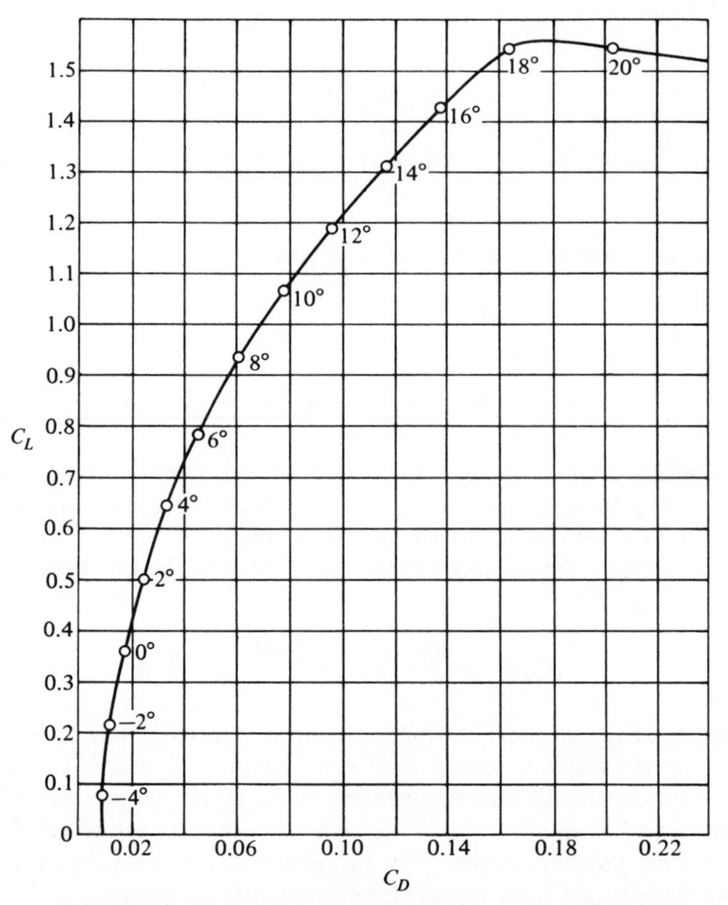

FIGURE 10.12 Polar curve for Clark Y airfoil. (Reproduced with permission from *Elements of Practical Aerodynamics* by B. Jones, John Wiley & Sons, Inc., New York, 1942, p. 61.)

portional to it in magnitude. The angle of zero lift is readily obtained, and the angle of minimum drag can be found by drawing a vertical tangent to the curve. In addition, the stall point and the maximum value of C_L are also readily determined from this curve.

Since values of C_D are approximately one order of magnitude less than the value of C_L at a given angle of attack, it is usual to use scales that are unequal for ease of reading, as done in Figure 10.12. When the scales are unequal, the line drawn from the origin to the curve no longer gives the direction, nor is it proportional to the magnitude of the resultant force on the airfoil. However, the tangent to the curve drawn from the origin always locates the angle of attack and the values of C_L and C_D for the maximum value of L/D.

ILLUSTRATIVE PROBLEM 10.13

From Figure 10.12 determine the stall angle and C_L, C_D and L/D at this angle.

Solution

From Figure 10.12, the stall angle is approximately $18\frac{1}{2}°$, $C_L = 1.56$, $C_D = 0.175$, and

$$\frac{L}{D} = \frac{C_L}{C_D} = \frac{1.56}{0.175} = 8.9$$

ILLUSTRATIVE PROBLEM 10.14

An airplane operating at sea level utilizes a Clark Y airfoil. If the airplane weighs 10,000 lb and is operated at 120 mi/hr, determine the power required to keep it in level flight. The wing area is to be taken as 500 ft². $\rho = 0.002378$ slug/ft³.

Solution

It is first necessary to determine the angle of attack at which the plane is operating. For this C_L is evaluated, noting that 120 mi/hr = 176 ft/sec:

$$L = W = C_L \frac{\rho}{2} V^2 A$$

$$10{,}000 = C_L \left(\frac{0.002378}{2}\right)(176)^2 500$$

and

$$C_L = 0.544$$

From Figure 10.12, the angle of attack is approximately 2.6° and $C_D =$ 0.027. Therefore, from equation (10.8c),

$$\text{hp} = 6.8 \times 10^{-6} \times 0.027 \times (120)^3 \times 500 = 159 \text{ hp}$$

10.3d Drag

If the span of an airfoil were infinite and the airfoil had an infinite aspect ratio (A.R.), it would be found to have a uniform lift across the span. For airfoils of finite length the distribution of the lift across the span is not uniform, decreasing in magnitude near the wing tips. As shown in Figure 10.9, the pressure below the airfoil is positive, while the pressure above the airfoil is negative. This pressure difference causes air on the bottom of the wing near the wing tips to flow outward and upward toward the top surface of the wing; on top the incoming air causes the airflow to move inward toward the center of the airfoil. The result of this flow pattern is to produce vortices at the wing tips, which can often be seen as an airplane passes through moist air due to condensation caused by the reduced pressure and temperature within these vortices.

As a consequence of the foregoing there is a downward flow of air at the wing tips causing changes in both the lift and drag on the wing. The magnitude of this effect is dependent on the span and aspect ratio of the airfoil; the greater the aspect ratio, the smaller is the effect on lift and drag.

Figure 10.13 shows an airfoil of finite span and the velocity of the relative wind in the undisturbed stream. For an infinite span and infinite aspect ratio there is no induced downward motion of the air so that the relative wind in the undisturbed air forward of the wing is the same as the relative wind at the

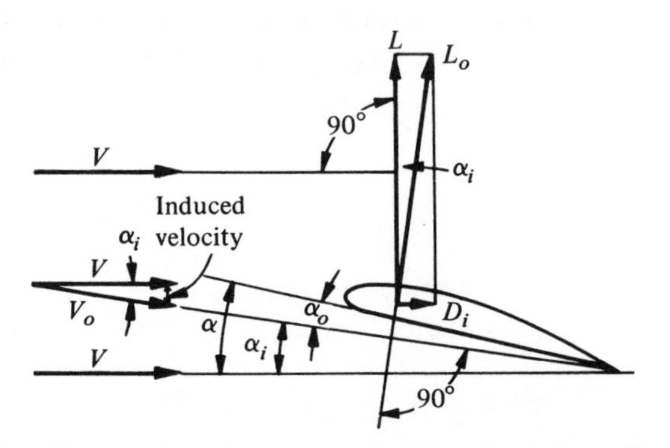

FIGURE 10.13 Finite airfoil showing induced drag. (Reproduced with permission from *Elementary Fluid Mechanics* by J. K. Vennard, 4th ed., John Wiley & Sons, Inc., New York, 1961, p. p. 522.)

wing. Due to the downward induced velocity of the finite wing, the relative wind at the airfoil does not equal the relative wind forward of the airfoil in the undisturbed air stream. If we denote the angle between the relative wind V and the true relative wind V_o to be α_i, we obtain the geometric angle of attack α from Figure 10.13 to be

$$\alpha = \alpha_o + \alpha_i \tag{10.19}$$

where α_o is the angle of attack for an airfoil of infinite span and infinite aspect ratio. Let us now consider that the airfoil has an infinite span and infinite aspect ratio. If this were the case, α_i, the direction of the relative wind, would coincide with the line of V_o, the effective velocity, and L_o would be the lift normal to the direction of V_o. Because L_o is not vertical, we can consider it to be composed of two components; one of these, L, is the lift directed normal to V, and the other, D_i, is called the induced drag and is directed parallel to V. Note that D_i, the induced drag, is in addition to the drag that we obtain for an airfoil of infinite span and infinite aspect ratio. Denoting the drag of an airfoil of infinite span to be the profile drag D_o, the total drag on a finite wing D can be written as the sum of the induced and profile drags.

$$D = D_i + D_o \tag{10.20}$$

Dividing all terms in equation (10.20) by $(\rho/2)V^2A$, we obtain

$$C_D = C_{D_i} + C_{D_o} \tag{10.21}$$

The importance and usefulness of the foregoing lies in the fact that airfoils of finite span that are tested in wind tunnels do not necessarily have the aspect ratio of the airfoil under consideration by the designers. In the past the aspect ratio most commonly used was 6, and by custom it can be assumed that published airfoil data are for an aspect ratio of 6 unless otherwise stated. It is presently the custom to furnish airfoil data for an infinite aspect ratio and to correct these data to any desired aspect ratio. The correction procedure consists of converting the data from one aspect ratio to an infinite aspect ratio and then reconverting to the desired aspect ratio. By this process it is necessary to report test data only for one aspect ratio (usually infinite), thereby eliminating the necessity for conducting wind tunnel tests at all aspect ratios of interest.

10.4 AIRCRAFT MANEUVERS

The previous discussion has considered only the airplane in steady level flight. A complete discusison of the mechanics of unsteady flight is beyond the scope of this text, but we have already developed enough information to consider certain steady-state aircraft maneuvers: (1) glide, (2) banked turn, and (3) climb.

When an airplane is said to be in a *glide* we shall assume that the power is off (engine completely throttled or dead) and that the propulsion system does not provide any thrust or drag. The glide angle is defined as the angle below the horizontal that the airplane descends along under these conditions. Denoting this angle as θ, the angle of incidence to be α, and the lift, drag, and weight as L, D, and W, respectively, we obtain the free-body diagram shown in Figure 10.14. In this figure the lift is perpendicular to the flight path, and the drag is parallel to the flight path. For a steady glide, the lift, drag, and weight must be a system of concurrent forces in equilibrium, and from Figure 10.14 and the conditions of equilibrium, we can write

$$L = W \cos \theta \tag{10.22}$$

but

$$L = C_L \frac{\rho}{2} V^2 A \tag{10.23}$$

Therefore,

$$\cos \theta = \frac{L}{W} = \frac{C_L(\rho/2)V^2 A}{W} \tag{10.24}$$

However,

$$D = W \sin \theta \tag{10.25}$$

and

$$W = \frac{D}{\sin \theta} = \frac{C_D(\rho/2)V^2 A}{\sin \theta} \tag{10.26}$$

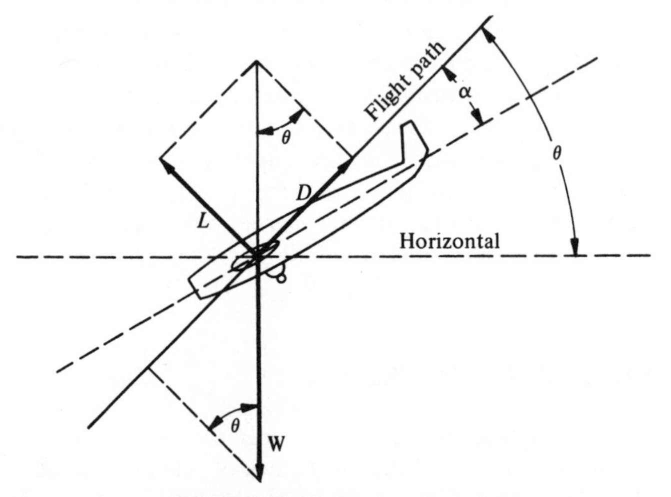

FIGURE 10.14 Airplane in a glide.

Combining equations (10.24) and (10.26), we obtain

$$\tan\theta = \frac{D}{L} = \frac{C_D}{C_L} \tag{10.27}$$

or

$$\cot\theta = \frac{L}{D} = \frac{C_L}{C_D} \tag{10.28}$$

The lift-to-drag ratio, L/D, can be read directly from curves such as Figure 10.10 or 10.12 for a given airfoil. Note that for each angle of incidence there is a single value of L/D, and therefore there is a single value of the glide angle. Also, since the ratio of lift to drag (and consequently θ) is not a function of the air density, the glide angle is the same for all altitudes.

ILLUSTRATIVE PROBLEM 10.15

An airplane using a Clark Y airfoil of A.R. 6 has a power plant failure and must glide to a nearby airport. Assuming the airplane to be at an altitude of 1000 ft, can the plane glide to the airport if the airport is 2 mi away?

Solution

The maximum horizontal distance that the plane can glide occurs when the plane is operated at the minimum glide angle, and this corresponds to a maximum value of L/D. For the airplane in question, Figure 10.10 indicates a maximum value of L/D to be closely 21.5 at an angle of attack of approximately 1°. The $\tan\theta = 1/21.5 = 0.0465$ and $\theta \simeq 2.7°$. Referring to Figure 10.15, the maximum distance the plane can glide is found as follows:

$$\tan = \frac{1000}{x} \qquad x = \frac{1000}{\tan\theta} = \frac{1000}{1/21.5} = 21{,}500 \text{ ft}$$

or 4.08 mi. Therefore, it is possible for the plane to safely glide to the airport.

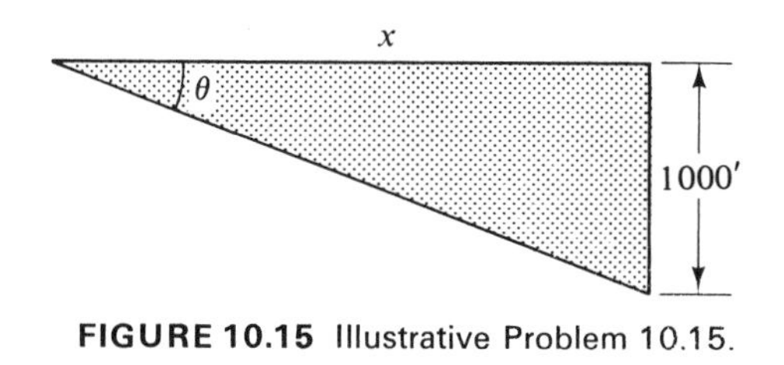

FIGURE 10.15 Illustrative Problem 10.15.

As the angle of the glide is made steeper and steeper, a vertical "dive" occurs when the glide angle is made 90°. Under this condition the lift is zero, the angle of attack corresponds to the angle of zero lift, and the plane will reach a velocity, termed the terminal velocity, V_T, of

$$V_T = \sqrt{\frac{2W}{C_D \rho A}} \tag{10.29}$$

The terminal velocity is seen to decrease as the altitude decreases, since the air density increases with decreasing altitude. If the airplane is placed in a dive (either deliberately or otherwise) with the power on, the thrust of the airplane's power plant must be accounted for. For this condition the drag on the airplane must equal the sum of the weight and thrust; thus,

$$T + W = D = C_D \frac{\rho}{2} V^2 A \tag{10.30}$$

and the terminal velocity with power is given by

$$V_{TP} = \sqrt{\frac{2(W + T)}{C_D \rho A}} \tag{10.31}$$

Depending on the altitude and thrust, it is possible for the plane to continue to accelerate and never reach its terminal velocity. Unless the airplane's controls can be made to bring it out of the dive without structural damage, it will crash and be destroyed.

ILLUSTRATIVE PROBLEM 10.16

As a result of a malfunction in the controls of an airplane, it is placed in a dive. If the thrust is 2000 lb, the weight is 3000 lb, and the airfoil is a Clark Y airfoil of A.R. 6, determine the terminal velocity at sea level if the wing area is 600 ft². Use $\rho = 0.002378$ slug/ft³.

Solution

For the Clark Y airfoil of A.R. 6, the angle of zero lift is obtained from Figure 10.10 as approximately −5°. C_D corresponding to this angle of attack is closely 0.01, and the terminal velocity will be

$$V_{TP} = \sqrt{\frac{2(2000 + 3000)}{0.01(0.002378)600}} = 836 \text{ ft/sec}$$

or

$$V_{TP} = 570 \text{ mi/hr}$$

ILLUSTRATIVE PROBLEM 10.17

The pilot of the plane described in Illustrative Problem 10.16 is able to cut (turn off) his engine. What will his terminal velocity be?

Solution

Since $T = 0$, equation (10.31) becomes

$$V_{TP} = \sqrt{\frac{2W}{C_D \rho A}} = \sqrt{\frac{2(3000)}{0.01 \times 0.002378 \times 600}}$$
$$= 648 \text{ ft/sec}$$

or

$$V_{TP} = 442 \text{ mi/hr}$$

While the reduction in terminal velocity is only 22%, it will give the pilot more time, there will be smaller forces on the control surfaces, and the possibility of structural damage will be greatly reduced.

When an airplane in steady level flight changes its flight direction by turning, it is necessary to consider the centrifugal force acting during the turn. As the airplane is turned in a horizontal plane it is usual to *bank*, that is, to depress the inner wing and elevate the outer wing. Without banking, the airplane tends to move outward; with banking the motion of the airplane depends on the angle of bank, the velocity of the airplane, and the radius of the turn. Consider the free-body diagram of the banked airplane shown in Figure 10.16. For equilibrium, the centrifugal force WV^2/gR must be balanced by the horizontal component of the lift, and the weight must be balanced by the vertical component of the lift. Thus,

$$L \sin \theta = \frac{WV^2}{gR} \tag{10.32}$$

and

$$L \cos \theta = W \tag{10.33}$$

From equations (10.32) and (10.33),

$$\tan \theta = \frac{V^2}{gR} \tag{10.34}$$

The angle of bank is thus seen to be independent of the weight of the airplane, the wing area, and the properties of the airfoil. If the angle of the bank

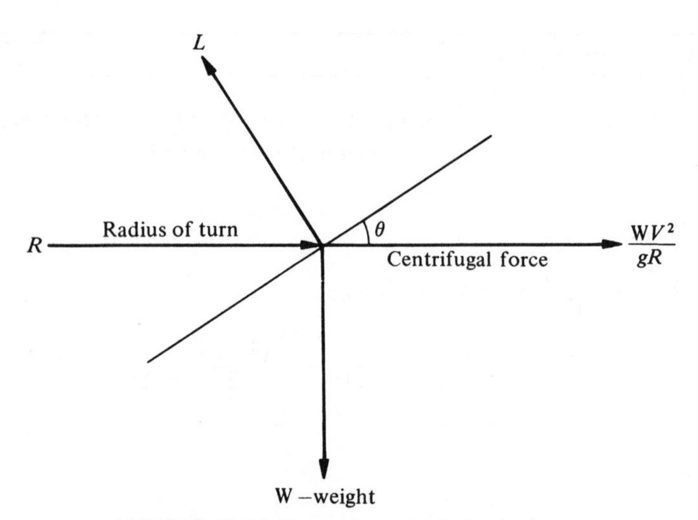

FIGURE 10.16 Forces in a steady banked turn.

is made less than the value given by equation (10.34), the plane will move outward and is said to skid. If the angle of bank exceeds the value given by equation (10.34), the airplane will move inward and downward and is said to slip.

ILLUSTRATIVE PROBLEM 10.18

An airplane in a race over a closed course is required to turn around a pylon at each end of the course. If the plane is traveling at 640 km/h and is restricted to a turn radius that must not exceed 3/4 km, determine the best bank angle.

Solution

For proper banking without skidding or slipping,

$$\tan \theta = \frac{V^2}{gR}$$

$$640 \text{ km/h} = \frac{640 \times 1000}{3600} \text{ m/s} = 178 \text{ m/s} \qquad \text{and} \qquad R = 750 \text{ m}$$

$$\tan \theta = \frac{(178)^2}{9.81 \times 750} = 4.31$$

$$\theta = 76.9^\circ$$

The plane must make a very steep bank if it is to perform within the required limitations.

ILLUSTRATIVE PROBLEM 10.19

If the pilot in Illustrative Problem 10.18 weighs W, what is his apparent weight in the turn?

Solution

The total force experienced by any object in the airplane is the vector sum of its weight and the centrifugal force acting on the body. Therefore,

$$\text{total force} = \text{apparent weight} = \sqrt{\left(\frac{WV^2}{gR}\right)^2 + W^2}$$

Factoring,

$$\text{apparent weight} = W\sqrt{\left(\frac{V^2}{gR}\right)^2 + 1} = W\sqrt{\tan^2\theta + 1} = \frac{W}{\cos\theta}$$

The ratio of the apparent weight to W is the g loading; that is, the force the body would experience in a gravitational field greater than earth's gravitational field by some multiple of the earth's gravitational field. Therefore, the g load is

$$g\text{ load} = \frac{\text{apparent weight}}{W} = \frac{1}{\cos\theta} = \sec\theta$$

For the problem under consideration, $\theta = 76.9°$, $\cos\theta = 0.2267$, and

$$g\text{ load} = \frac{1}{0.2267} = 4.41$$

Therefore, the apparent weight of the pilot is 4.41 W. Notice that the g load is a function only of the bank angle, θ.

10.5 CLOSURE

In this chapter we have discussed the flow about immersed bodies from both general and applied aspects. For a subject as vast as this particular topic, it would indeed try anyone to attempt more than a survey of selected portions of this subject within the limits of a single chapter in an introductory textbook. However, we have been able to discuss and differentiate between the laminar and turbulent flow regions for flow about immersed bodies. We have also been able to study the airfoil in compressible subsonic flow and relate the performance of the airplane to the properties of the airfoil. Wherever possible the development was based upon reasonable mathematical models, but it was found necessary ultimately to resort to experimentally determined characteristics for each of the topics covered.

This chapter is intended to be an introduction to selected areas of aerodynamics and hydrodynamics. Mastery of it will provide a sound basis for further study in these areas.

PROBLEMS

Use

$$\mu = 3.73 \times 10^{-7} \text{ lb sec/ft}^2 \qquad \mu = 1.80 \times 10^{-5} \text{ Pa}\cdot\text{s}$$

$$\rho = 0.002378 \frac{\text{lb sec}^2}{\text{ft}^4} \qquad (\gamma = 0.0765 \text{ lb/ft}^3)$$

$$\rho = 1.225 \text{ kg/m}^3 \qquad (\gamma = 12.02 \text{ N/m}^3)$$

for air in the following problems unless otherwise noted.

10.1 A sphere whose specific weight is 100 lb/ft^3 settles in air at a velocity of 0.3 ft/sec. Evaluate the diameter of the sphere. Neglect the buoyant effect of the air.

10.2 A sphere having an unknown specific weight settles in still air at a velocity of 0.1 m/s. If the diameter of the sphere is 0.05 mm and the buoyant effect of the air is neglected, determine the specific weight of the material of the sphere.

10.3 If a sphere is immersed in a fluid whose relative velocity with respect to the sphere is 0.01 ft/sec, determine the drag force on the sphere. Assume the Reynolds number to be less than 0.5, the viscosity of the fluid to be 3×10^{-5} lb sec/ft^2, and the diameter of the sphere to be 0.001 in.

10.4 Using the data for cylinders in Figure 10.3, derive a relationship for the drag coefficient on a cylinder for Re ≤ 0.5 of the form $C_D =$ constant/Re.

10.5 Using the results of Problem 10.4, solve Problem 10.3 for a cylinder placed normal to the flow and having a length of 1 in. This is essentially a short, fine wire having a large ratio of length to diameter.

10.6 Air flows at 140 km/h perpendicular to a billboard that is 3 m long and 2 m high. What force is exerted on the billboard?

10.7 What force is exerted on a billboard 10 ft long and 5 ft high by a 100-mi/hr wind perpendicular to it?

10.8 Air flows edge-on to a rectangular plate 2×1 ft. Determine the drag on the plate if the air velocity is 30 mi/hr and the stream strikes the 2-ft side first. Assume the flow to be laminar.

10.9 If the plate in Problem 10.8 is to be towed through the air at 30 mi/hr, what horsepower is required?

10.10 What is the drag on a spherical ball 3 in. in diameter on top of a flag-pole when the relative wind velocity is 60 mi/hr?

10.11 A spherical ball, 75 mm in diameter, is placed on top of a flagpole. If the relative wind velocity is 100 km/h, what is the drag force on the sphere?

10.12 Determine the ratio of the drag on a flat plate for angles of the relative wind of 16° and 8°.

10.13 A plate 1 ft long and 8 in. wide is set at an angle of 12 deg to an air stream with a relative velocity with respect to the plate of 75 ft/sec. Determine the lift, drag, and total force on the plate.

10.14 A plate is 0.3 m long and 0.2 m wide. If it is set at an angle of 6° with respect to an airstream with a relative velocity of 20 m/s with respect to the plate, determine the lift, drag, and total force on the plate.

10.15 A hemispherical shell is placed in a fluid so that the relative velocity of the flūid impinges on the circular plane face. Assume the hemisphere to be 4 in. in diameter, the velocity of the fluid to be 10 ft/sec relative to the hemisphere, the Reynolds number to be greater than 1000, and the fluid to have a specific weight of 100 lb/ft^3. For these conditions determine the drag on the hemisphere.

10.16 If the hemisphere in Problem 10.15 is turned through an angle of 180°, what will the drag be?

Unless otherwise stated, assume a Clark Y airfoil with an A.R. of 6 for the following problems.

10.17 What is the lift, drag, and ratio of lift to drag at an angle of attack of 9° on a wing having an area of 400 ft^2 at a relative velocity of 180 mi/hr? How much horsepower is required to keep this wing in level flight?

10.18 Determine the lift, drag, and the ratio of lift to drag for a wing having an area of 60 m^2 when the angle of attack is 9° and the relative wind velocity is 300 km/h.

10.19 Calculate the horsepower required to keep the wing of Problem 10.18 flying on a steady, level course.

10.20 An airplane has a wing loading (W/A) of 100 lb/ft^2 and is operated at 240 mi/hr. What is its angle of attack?

10.21 If the wing loading on an airplane is 4000 N/m^2 and the plane operates at 360 km/h, what is the angle of attack?

10.22 What wing area is required to support an airplane weighing 5000 lb and flying at an angle of attack of 7° with a velocity of 180 mi/hr?

10.23 Determine the wing area required to support an airplane weighing 25 kN at an angle of attack of 7° if the velocity is 360 km/h.

10.24 What wing loading will an airplane have if it operates at an angle of attack of 4° at 200 mi/hr?

10.25 If the plane in Problem 10.24 operates at an altitude where the density of the air is half that at sea level, determine the permissible wing loading.

10.26 Determine the landing speed of the airplane in Problem 10.20.

10.27 An airplane takes off from New York (sea level) and is to land at Denver (altitude 7500 ft). The specific weight of air at Denver is approximately 0.8 times the value at New York. Determine the ratio of the takeoff speed from New York to the landing speed at Denver.

10.28 Using the data given in Problem 10.27, determine the ratio of the horsepower for takeoff from New York to the horsepower required for takeoff from Denver. Assume the same angle of attack for each case.

10.29 It is desired to operate an airplane at the angle of attack for maximum lift. If the plane weighs 6000 lb and has a wing area of 200 ft^2, determine the operating velocity, the drag, and the horsepower required.

10.30 Determine all of the items in Problem 10.29 if the plane is operated at an angle of attack of 14°. Compare your results with Problem 9.29.

10.31 A plane weighs 30 kN and has a wing area of 25 m^2. If the plane is operated at the angle of maximum lift, determine the operating velocity, the drag, and the horsepower required.

10.32 The largest engine an airplane can utilize generates 1000 hp. At what angle of attack should the plane be operated with this engine if the design speed is 300 mph and the wing area is 200 ft^2?

10.33 An airplane operates at a condition such that $C_L = 0.8$ and the drag is 500 lb. What is the lift?

10.34 What is the least glide angle for a wing having an area of 450 ft^2 and weighing 2500 lb?

10.35 Evaluate the terminal velocity of the airplane in Problem 10.34.

10.36 If the airplane in Problem 10.34 has a thrust of 2000 lb, determine its terminal velocity in a power dive

10.37 If the airplane in Problem 10.34 is at 750 ft when the glide is started, determine the farthest distance the plane can glide.

10.38 What angle of bank is required for an airplane traveling at 240 mph in a horizontal turn of radius equal to 1 mi?

10.39 A plane banks in a turn while traveling at 360 km/h. What is the required angle of bank if the turn radius is 3 km?

10.40 Evaluate the *g* force and radius of turn for an airplane weighing 3500 lb that is turning at a bank angle of 45° with a velocity of 180 mi/hr.

10.41 A pilot of a racing plane places it in a 75° banked turn. Determine the *g* force on the pilot.

REFERENCES

1. *Fluid Mechanics* by R. C. BINDER, 4th ed., Prentice-Hall, Inc., Englewood Cliffs, N.J., 1962.
2. *Fluid Mechanics for Engineers* by M. L. ALBERTSON, J. R. BARTON, and D. B. SIMONS, Prentice-Hall, Inc., Englewood Cliffs, N.J., 1960.
3. *Elementary Fluid Mechanics* by J. K. VENNARD, John Wiley & Sons, Inc., New York, 1961.
4. *Elements of Practical Aerodynamics* by B. JONES, John Wiley & Sons, Inc., New York, 1942.
5. *Basic Fluid Mechanics* by J. L. ROBINSON, McGraw-Hill Book Company, New York, 1963.
6. *Mechanics of Fluids* by G. MURPHY, International Textbook Company, Scranton, Pa., 1942.
7. *Engineering Applications of Fluid Mechanics* by J. C. HUNSAKER and B. G. RIGHTMIRE, McGraw-Hill Book Company, New York, 1947.
8. *Fluid Mechanics* by V. L. STREETER and E. B. WYLIE, 6th ed., McGraw-Hill Book Company, New York, 1979.
9. *Applied Fluid Mechanics* by R. L. MOTT, 2nd ed., Charles E. Merill Publishing Co., Columbus, Ohio, 1979.

Answers to Even-numbered Problems

CHAPTER 1

1.2 0.932 mi
1.4 32.81 ft/sec
1.6 9.2903×10^{-2}
1.8 39.77 mi/hr
1.10 28.32 litres
1.12 2.119×10^{3}
1.14 (c)
1.16 (d)
1.18 (e)
1.20 1145N
1.22 4.66 slugs
1.24 160.0(c)
1.26 10 kg
1.28 0.622 slug
1.30 3.73 slugs
1.32 20 m/s^2
1.34 2×10^{-4} m^3
1.36 Approx. 10g

CHAPTER 2

2.2 −17.7°C, −12.2°C, 10°C
2.4 $°C = \frac{20}{11}(°ARB - 20)$
2.6 Derivation
2.8 40°
2.10 251.6 kPa above atmosphere
2.12 24.5 kPa
2.14 $\rho = 0.777$ slug/ft³; $v = 0.04$ ft³/lb
2.16 $\gamma = 149.6$ lb/ft³; $\rho = 4.65$ slugs/ft³
2.18 8829 N/m³; 56.16 lb/ft³
2.20 2058.5 lb
2.22 7.01 ft
2.24 10.31 psia
2.26 0.104 ft depression
2.28 0.321 ft rise
2.30 6.3×10^{-3} m
2.32 0.55 ft/sec

2.34 1.84×10^{-2} m/s
2.36 37.6 ft/sec

CHAPTER 3

3.2 1571 psf
3.4 3659 psfa
3.6 40.7 psia
3.8 28,224 ft
3.10 19,580 ft
3.12 671.8 mm
3.14 No
3.16 191.4 kPa
3.18 2478.7 psfa
3.20 10.3 psf
3.22 0.434 psi
3.24 5.07 m
3.26 18.72 psia
3.28 1.13 psf
3.30 9405 Pa (gage)
3.32 103.9 kPa (absolute)
3.34 17.8 psia
3.36 172.7 kPa (gage)
3.38 92.5 kN
3.40 5200 lb
3.42 8986 lb
3.44 17,972 ft•lb
3.46 588.6 kN
3.48 145.5 kN
3.50 119.7 kN•m
3.52 1.49 MN•m
3.54 Derivation
3.56 156 lb/ft^3
3.58 27 011 N/m^3
3.60 643,000 lb
3.62 $\gamma_w h^2 W/2$
3.64 0.06 in.

CHAPTER 4

4.2 $V = 2.04$ ft/sec
4.4 1.89×10^{-3} m^3/s
4.6 6.31 litres/s
4.8 20 kg/s
4.10 0.849 m/s; 0.377 m/s
4.12 Derivation
4.14 1000 ft lb
4.16 25.37 ft/sec
4.18 80.2 ft/sec
4.20 38.86 ft, 388.6 ft lb
4.22 Derivation
4.24 11.58 hp
4.26 50.3 hp
4.28 833.3 ft lb
4.30 10.19 N•m/N
4.32 61.8 hp
4.34 184.4 kPa
4.36 4002 ft/sec
4.38 20.51 kJ/N

CHAPTER 5

5.2 5.1 m
5.4 39.88 m
5.6 98.2 ft/sec
5.8 94.1 psig
5.10 2741 gpm
5.12 395.3 gpm
5.14 1588 gpm
5.16 1.4×10^{-2} m^3/s
5.18 12.5 ft
5.20 8.02 ft/sec
5.22 1.25 ft^3/sec
5.24 22.2 ft/sec
5.26 66.7 ft/sec
5.28 29.1 in.
5.30 2×10^{-2} m^3/s
5.32 Yes
5.34 Yes

CHAPTER 6

6.2 248,300
6.4 Derivation
6.6 8.15×10^{-3} m/m
6.8 $V_{max} = 248$ ft/sec
6.10 9.64 psi/ft
6.12 60.7 gpm
6.14 28.1 psi
6.16 74.4 psi
6.18 156.6 psig
6.20 118.6 ft of water
6.22 36.5 psi
6.24 0.174 ft
6.26 0.58 ft of water
6.28 0.17 m

6.30 1 psi
6.32 10.5 psi

CHAPTER 7

7.2 $F \simeq (1 + R)V_2\dot{m}_A$
7.4 71.9 kN
7.6 90.7 kN
7.8 441.8 N
7.10 342.4 lb
7.12 149.6 lb
7.14 4 000 N
7.16 313.9 lb
7.18 290.7 lb
7.20 468.8 lb
7.22 112.2 lb
7.24 2.53 hp
7.26 1.44 hp
7.28 1.38 hp
7.30 30.9 hp
7.32 110.7 kPa
7.34 2068 lb
7.36 2382 lb
7.38 94.4%

CHAPTER 8

8.2 (a) $Wh/2(W + h)$
(b) $(\sqrt{3}/16)s$
(c) $W/4$
8.4 2.17 m^3/s
8.6 58 ft^3/sec
8.8 Approx. 90 ft^3/sec
8.10 $h \simeq 3$
8.12 4.47×2.24 m
8.14 17.3×8.65 ft
8.16 9.53×10^{-5}
8.18 Approx. 0.33 m
8.20 6.94×3.47 m
8.22 1.37 m
8.24 3.73 m
8.26 $q = 70$ ft^3/sec
8.28 3.22 ft
8.30 5.52 m^3/s
8.32 1.38 m^3/s
8.34 14.1 ft^3/s
8.36 1.36 m^3/s
8.38 41.9 ft^3/sec

CHAPTER 9

9.2 30.24 ft
9.4 11.37 m
9.6 1.08 hp
9.8 54.5 ft; 550 gpm
9.10 1.17 m
9.12 1287
9.14 0.296 (SI); 5091 (English)

CHAPTER 10

10.2 1321 kg/m^3
10.4 $C_D = 8.5/Re$
10.6 6.00 kN
10.8 0.023 lb
10.10 0.217 lb
10.12 2.72
10.14 5.8 N
10.16 174.5 lb
10.18 17.3 kN
10.20 4.5°
10.22 69.4 ft^2
10.24 66.8 psf
10.26 157.7 mi/hr
10.28 0.894
10.30 132.9 hp
10.32 3°
10.34 2.7°
10.36 917 ft/sec
10.38 36.1°
10.40 2170 ft

Selected Physical Data

TABLE B.1 Properties of Pipe†

Schedules, Wall Thicknesses, and Weights

Conforming to ASA Standard B36.10, 1950, for Wrought Steel and Wrought Iron Pipe

Nominal Pipe Size (in.)	Outside Diameter, D(in.)	Wall Thickness, t(in.)	Inside Diameter d(in.)	Inside Diameter d^2(squared)	Inside Diameter d^5(fifth power)	Area of Metal (in.^{2})	Internal Cross-Sectional Area in.^{2}	Internal Cross-Sectional Area ft^2	External Surface (f^2)	Moment of Inertia (in.^{4})	Weight (Pounds) of Pipe (per ft)	Weight (Pounds) of Water (per ft of pipe)
Schedule 10												
14 o.d.	14.0	0.250	13.50	182.25	448,403	10.80	143.14	0.994	3.665	255.3	36.71	62.03
16 o.d.	16.0	0.250	15.50	240.25	894,660	12.37	188.69	1.310	4.189	383.7	42.05	81.74
18 o.d.	18.0	0.250	17.50	306.25	1,641,309	13.94	240.53	1.670	4.712	549.1	47.39	104.21
20 o.d.	20.0	0.250	19.50	380.25	2,819,505	15.51	298.65	2.074	5.236	756.4	52.73	129.42
24 o.d.	24.0	0.250	23.50	552.25	7,167,030	18.65	433.74	3.012	6.283	1,315.0	63.41	187.95
30 o.d.	30.0	0.312	29.376	862.95	21,875,768	29.10	677.76	4.707	7.854	3,206.0	98.93	293.72
Schedule 20												
8	8.625	0.250	8.125	66.02	35,409	6.57	51.85	0.3601	2.258	57.72	22.36	22.47
10	10.75	0.250	10.25	105.06	113,141	8.24	82.52	0.5731	2.814	113.7	28.04	35.76
12	12.75	0.250	12.25	150.06	275,855	9.82	117.86	0.8185	3.338	191.8	33.38	51.07
14 o.d.	14.0	0.312	13.376	178.92	428,185	13.42	140.52	0.975	3.665	314.4	45.68	60.89
16 o.d.	16.0	0.312	15.376	236.42	859,442	15.38	185.69	1.290	4.189	473.2	52.36	80.50
18 o.d.	18.0	0.312	17.376	301.92	1,583,978	17.34	237.13	1.647	4.712	678.2	59.03	102.77
20 o.d.	20.0	0.375	19.25	370.56	2,643,344	23.12	291.04	2.021	5.236	1113	78.60	125.67
24 o.d.	24.0	0.375	23.25	540.56	6,793,832	27.83	424.56	2.948	6.283	1942	94.62	183.95
30 o.d.	30.0	0.500	29.00	841.0	20,511,149	46.34	660.52	4.587	7.854	5,042	157.53	286.23
Schedule 30												
8	8.625	0.277	8.071	65.14	34,248	7.26	51.16	0.3553	2.258	63.35	24.70	22.17
10	10.75	0.307	10.136	102.74	106,987	10.07	80.69	0.5603	2.814	137.4	34.24	34.96

12	12.75	0.330	12.09	146.17	258,304	12.87	114.80	0.7972	3.338	248.4	43.77	49.74
14 o.d.	14.0	0.375	13.25	175.56	408,394	16.05	137.88	0.9575	3.665	372.8	54.57	59.75
16 o.d.	16.0	0.375	15.25	232.56	824,801	18.41	182.65	1.268	4.189	562.1	62.58	79.12
18 o.d.	18.0	0.437	17.126	293.30	1,473,261	24.11	230.36	1.600	4.712	930.3	82.06	99.84
20 o.d.	20.0	0.500	19.0	361.00	2,476,099	30.63	283.53	1.969	5.236	1,457	104.13	122.87
24 o.d.	24.0	0.562	22.876	523.31	6,264,703	41.39	411.00	2.854	6.283	2,843	140.80	178.09
30 o.d.	30.0	0.625	28.75	826.56	19,642,160	57.68	649.18	4.508	7.854	6,224	196.08	281.30
					Schedule 40							
1/8	0.405	0.068§	0.269	0.0724	0.00141	0.072	0.0569	0.00040	0.106	0.001064	0.24	0.0250
1/4	0.540	0.088§	0.364	0.1325	0.00639	0.125	0.1041	0.00072	0.141	0.003312	0.42	0.0449
3/8	0.675	0.091§	0.493	0.2430	0.02912	0.167	0.1909	0.00133	0.177	0.007291	0.57	0.0830
1/2	0.840	0.109§	0.622	0.3869	0.09310	0.250	0.3039	0.00211	0.220	0.01709	0.85	0.1317
3/4	1.050	0.113§	0.824	0.679	0.3799	0.333	0.5333	0.00371	0.275	0.03704	1.13	0.2315
1	1.315	0.133§	1.049	1.100	1.270	0.494	0.8639	0.00600	0.344	0.08734	1.68	0.3744
1 1/4	1.660	0.140§	1.380	1.904	5.005	0.669	1.495	0.01040	0.435	0.1947	2.27	0.6490
1 1/2	1.900	0.145§	1.610	2.592	10.82	0.799	2.036	0.01414	0.497	0.3099	2.72	0.8823
2	2.375	0.154§	2.067	4.272	37.72	1.075	3.356	0.02330	0.622	0.666	3.65	1.454
2 1/2	2.875	0.203§	2.469	6.096	91.75	1.704	4.788	0.03322	0.753	1.530	5.79	2.073
3	3.5	0.216§	3.068	9.413	271.8	2.228	7.393	0.05130	0.916	3.017	7.58	3.201
3 1/2	4.0	0.226§	3.548	12.59	562.2	2.680	9.888	0.06870	1.047	4.788	9.11	4.287
4	4.5	0.237§	4.026	16.21	1,058	3.173	12.73	0.08840	1.178	7.233	10.79	5.516
5	5.563	0.258§	5.047	25.47	3,275	4.304	20.01	0.1390	1.456	15.16	14.62	8.674
6	6.625	0.280§	6.065	36.78	8,206	5.584	28.89	0.2006	1.734	28.14	18.97	12.52
8	8.625	0.322§	7.981	63.70	32,380	8.396	50.03	0.3474	2.258	72.49	25.55	21.68
10	10.75	0.365§	10.02	100.4	101,000	11.90	78.85	0.5475	2.814	160.7	40.48	34.16
12	12.75	0.406	11.938	142.5	242,470	15.77	111.93	0.7773	3.338	300.3	53.53	48.50
14 o.d.	14.0	0.437	13.126	172.3	389,638	18.61	135.32	0.9397	3.665	429.1	63.37	58.64
16 o.d.	16.0	0.500	15.000	225.0	759,375	24.35	176.72	1.2272	4.189	731.9	82.77	76.58
18 o.d.	18.0	0.562	16.876	284.8	1,368,820	30.79	223.68	1.5533	4.712	1172	104.75	96.93

TABLE B.1 (continued)

Nominal Pipe Size (in.)	Outside Diameter D(in.)	Wall Thickness t(in.)	Inside Diameter d(in.)	Inside Diameter d^2(squared)	Inside Diameter d^5(fifth power)	Area of Metal (in.^{2})	Internal Cross-Sectional Area in.^{2}	Internal Cross-Sectional Area ft^2	External Surface (ft^2)	Moment of Inertia (in.^{4})	Weight (Pounds) of Pipe (per ft)	Weight (Pounds) of Water (per ft of pipe)
20 o.d.	20.0	0.593	18.814	354.0	2,357,244	36.15	278.00	1.9305	5.236	1703	122.91	120.46
24 o.d.	24.0	0.687	22.626	511.9	5,929,784	50.31	402.07	2.7921	6.283	3424	171.17	174.23
Schedule 60												
8	8.625	0.406	7.813	61.04	29,113	10.48	47.94	0.3329	2.258	88.73	35.64	20.77
10	10.75	0.500¶	9.75	95.06	88,110	16.10	74.66	0.5185	2.814	212.0	54.74	32.35
12	12.75	0.562	11.626	135.16	212,399	21.52	106.16	0.7372	3.338	400.4	73.16	46.00
14 o.d.	14.0	0.593	12.814	164.20	345,480	24.98	128.96	0.8956	3.665	562.3	84.91	55.86
16 o.d.	16.0	0.656	14.688	215.74	683,618	31.62	169.44	1.1766	4.189	932.4	107.50	73.42
18 o.d.	18.0	0.750	16.500	272.25	1,222,981	40.64	213.83	1.4849	4.712	1,515	138.17	92.80
20 o.d.	20.0	0.812	18.376	337.68	2,095,342	48.95	265.21	1.8417	5.236	2,257	166.40	114.92
24 o.d.	24.0	0.968	22.064	486.82	5,229,029	70.04	382.35	2.6552	6.283	4,654	238.11	165.94
Schedule 80												
$\frac{1}{8}$	0.405	0.095¶	0.215	0.0462	0.000459	0.093	0.0363	0.00025	0.106	0.001216	0.31	0.0157
$\frac{1}{4}$	0.540	0.119¶	0.302	0.0912	0.002513	0.157	0.0716	0.00050	0.141	0.003766	0.54	0.031
$\frac{3}{8}$	0.675	0.126¶	0.423	0.1789	0.01354	0.217	0.1405	0.00098	0.177	0.008619	0.74	0.0609
$\frac{1}{2}$	0.840	0.147¶	0.546	0.2981	0.04852	0.320	0.2341	0.00163	0.220	0.02008	1.09	0.1013
$\frac{3}{4}$	1.050	0.154¶	0.742	0.5506	0.2249	0.433	0.4324	0.00300	0.275	0.04479	1.47	0.1875
1	1.315	0.179¶	0.957	0.9158	0.8027	0.639	0.7193	0.00499	0.344	0.1056	2.17	0.3112
$1\frac{1}{4}$	1.660	0.191¶	1.278	1.633	3.409	0.881	1.283	0.00891	0.435	0.2418	3.00	0.5553
$1\frac{1}{2}$	1.900	0.200¶	1.500	2.250	7.594	1.068	1.767	0.01225	0.498	0.3912	3.63	0.7648
2	2.375	0.218¶	1.939	3.760	27.41	1.477	2.953	0.02050	0.622	0.8679	5.02	1.279
$2\frac{1}{2}$	2.875	0.276¶	2.323	5.396	67.64	2.254	4.238	0.02942	0.753	1.924	7.66	1.834
3	3.5	0.300¶	2.900	8.410	205.1	3.016	6.605	0.04587	0.917	3.894	10.25	2.859
$3\frac{1}{2}$	4.0	0.318¶	3.364	11.32	430.8	3.678	8.891	0.06170	1.047	6.280	12.51	3.847

4	4.5	0.337¶	3.826	14.64	819.8	4.407	11.50	0.07986	1.178	9.610	14.98	4.976
5	5.563	0.375¶	4.813	23.16	2,583	6.112	18.19	0.1263	1.456	20.67	20.78	7.875
6	6.625	0.432¶	5.761	33.19	6,346	8.405	26.07	0.1810	1.734	40.49	28.57	11.29
8	8.625	0.500¶	7.625	58.14	25,775	12.76	45.66	0.3171	2.257	105.7	43.39	19.79
10	10.75	0.593	9.564	91.47	80,020	18.92	71.84	0.4989	2.817	244.8	64.33	31.13
12	12.75	0.687	11.376	129.41	190,523	26.03	101.64	0.7958	3.338	475.1	88.51	44.04
14 o.d.	14.0	0.750	12.500	156.25	305,176	31.22	122.72	0.8522	3.665	687.3	106.13	53.18
16 o.d.	16.0	0.843	14.314	204.89	600,904	40.14	160.92	1.1175	4.189	1,156	136.46	69.73
18 o.d.	18.0	0.937	16.125	260.05	1,090,518	50.23	204.24	1.4183	4.712	1,833	170.75	88.50
20 o.d.	20.0	1.031	17.938	321.77	1,857,248	61.44	252.72	1.7550	5.236	2,772	208.87	109.51
24 o.d.	24.0	1.218	21.564	465.01	4,662,798	87.17	365.22	2.5362	6.283	5,672	296.36	158.26
					Schedule 100							
8	8.625	0.593	7.439	55.34	22,781	14.96	43.46	0.3018	2.258	121.3	50.87	18.83
10	10.75	0.718	9.314	86.75	69,357	22.63	68.13	0.4732	2.814	286.1	76.93	29.53
12	12.75	0.843	11.064	122.41	165,791	31.53	96.14	0.6677	3.338	561.6	107.20	41.66
14 o.d.	14.0	0.937	12.126	147.04	262,173	38.45	115.49	0.8020	3.665	824.4	130.73	50.04
16 o.d.	16.0	1.031	13.938	194.27	526,020	48.48	152.58	1.0596	4.189	1,364	164.83	66.12
18 o.d.	18.0	1.156	15.688	246.11	950,250	61.17	193.30	1.3423	4.712	2,180	207.96	83.76
20 o.d.	20.0	1.281	17.438	304.08	1,612,398	75.34	238.82	1.6585	5.236	3,316	256.10	103.65
24 o.d.	24.0	1.531	20.938	438.40	4,024,179	108.07	344.32	2.3911	6.283	6,853	367.40	149.43
					Schedule 120							
4	4.5	0.438	3.625	13.15	626.8	5.578	10.33	0.0717	1.178	11.65	19.01	4.47
5	5.563	0.500	4.563	20.82	1,978	7.953	16.35	0.1136	1.456	25.73	27.04	7.09
6	6.625	0.562	5.501	30.26	5,037	10.705	23.77	0.1650	1.734	49.61	36.39	10.30
8	8.625	0.718	7.189	51.68	19,202	17.84	40.59	0.2819	2.257	140.5	60.63	17.59
10	10.75	0.843	9.064	82.16	61,179	26.24	64.53	0.4481	2.817	324.2	89.20	27.96
12	12.75	1.000	10.750	115.56	143,563	36.91	90.76	0.6303	3.338	641.6	125.49	39.33
14 o.d.	14.0	1.093	11.814	139.57	230,134	44.32	109.62	0.7612	3.665	929.8	150.67	47.57
16 o.d.	16.0	1.218	13.564	183.98	459,133	56.56	144.50	1.0035	4.189	1,555	192.29	62.62

TABLE B.1 (continued)

Nominal Pipe Size (in.)	Outside Diameter D(in.)	Wall Thickness t(in.)	Inside Diameter d(in.)	Inside Diameter d^2(squared)	Inside Diameter d^5(fifth power)	Area of Metal (in.^{2})	Internal Cross-Sectional Area in.^{2}	Internal Cross-Sectional Area ft^2	External Surface (ft^2)	Moment of Inertia (in.^{4})	Weight (Pounds) of Pipe (per ft)	Weight (Pounds) of Water (per ft of pipe)
18 o.d.	18.0	1.375	15.250	232.56	824,783	71.82	182.65	1.2684	4.712	2,499	244.14	79.27
20 o.d.	20.0	1.500	17.000	289.00	1,419,857	87.18	226.98	1.5762	5.236	3,754	296.37	98.35
24 o.d.	24.0	1.812	20.376	415.18	3,512,301	126.31	326.08	2.2644	6.283	7,827	429.39	141.52
Schedule 140												
8	8.625	0.812	7.001	49.01	16,819	19.93	38.50	0.2673	2.257	153.7	67.76	16.68
10	10.75	1.000	8.750	76.56	51,291	30.63	60.13	0.4176	2.817	367.8	104.13	26.06
12	12.75	1.125	10.500	110.25	127,628	41.08	86.59	0.6013	3.338	700.5	139.68	37.52
14 o.d.	14.0	1.250	11.500	132.25	201,136	50.07	103.87	0.7213	3.665	1,027	170.22	45.01
16 o.d.	16.0	1.438	13.125	172.29	389,670	65.74	135.32	0.9397	4.189	1,760	223.50	58.64
18 o.d.	18.0	1.562	14.876	221.30	728,502	80.66	173.80	1.2070	4.712	2,749	274.23	75.32
20 o.d.	20.0	1.750	16.500	272.25	1,222,981	100.33	213.82	1.4849	5.236	4,216	341.10	92.66
24 o.d.	24.0	2.062	19.876	395.09	3,102,022	142.11	310.28	2.1547	6.283	8,625	483.13	134.45
Schedule 160												
$\frac{1}{2}$	0.840	0.187	0.466	0.2172	0.002197	0.3836	0.1706	0.00118	0.220	0.02212	1.30	0.074
$\frac{3}{4}$	1.050	0.218	0.614	0.3770	0.08726	0.5698	0.2961	0.00206	0.275	0.05269	1.94	0.130
1	1.315	0.250	0.815	0.6642	0.3596	0.8365	0.5217	0.00362	0.344	0.1251	2.84	0.230
$1\frac{1}{4}$	1.660	0.250	1.160	1.346	2.100	1.107	1.057	0.00734	0.435	0.2839	3.76	0.46
$1\frac{1}{2}$	1.900	0.281	1.338	1.790	4.288	1.429	1.406	0.00976	0.498	0.4824	4.86	0.61
2	2.375	0.343	1.689	2.853	13.74	2.190	2.241	0.01556	0.622	1.162	7.44	0.97
$2\frac{1}{2}$	2.875	0.375	2.125	4.516	43.33	2.945	3.546	0.02463	0.753	2.353	10.01	1.54
3	3.5	0.438	2.625	6.896	124.9	4.205	5.416	0.03761	0.917	5.032	14.32	2.35
$3\frac{1}{2}$	4.0	—	—	—	—	—	—	—	—	—	—	—

4	4.5	0.531	3.438	11.82	480.3	6.621	9.283	0.06447	1.178	13.27	22.51	4.02
5	5.563	0.625	4.313	18.60	1,492	9.696	14.61	0.1015	1.456	30.03	32.96	6.33
6	6.625	0.718	5.189	26.93	3,762	13.32	21.15	0.1469	1.734	58.97	45.30	9.16
8	8.625	0.906	6.813	46.42	14,679	21.97	36.46	0.2532	2.257	165.9	74.69	15.80
10	10.75	1.125	8.500	72.25	44,371	34.02	56.75	0.3941	2.817	399.3	115.65	24.59
12	12.75	1.312	10.126	102.54	106,461	47.14	80.53	0.5592	3.338	781.1	160.27	34.89
14 o.d.	14.0	1.406	11.188	125.17	175,292	55.63	98.31	0.6827	3.665	1,117	189.12	42.60
16 o.d.	16.0	1.593	12.814	164.20	345,486	72.10	128.96	0.8955	4.189	1,894	245.11	55.97
18 o.d.	18.0	1.781	14.438	208.46	627,412	90.75	163.72	1.1369	4.712	3,021	308.51	71.05
20 o.d.	20.0	1.968	16.064	258.05	1,069,699	111.49	202.67	1.4074	5.236	4,586	379.01	87.96
24 o.d.	24.0	2.343	19.314	373.03	2,687,570	159.41	292.98	2.0345	6.283	9,458	541.94	127.15

† Data taken with permission from *Catalog* #57, the Walworth Co., New York, 1957.

‡ This column also represents the contents in cubic feet per foot of length.

§ These thicknesses are identical with those listed in ASA B36.10—1950 for standard wall pipe.

¶ These thicknesses are identical with those listed in ASA B36.10—1950 for extra strong wall pipe.

TABLE B.2 Average Properties of Tubes†

Diameter		Thickness		External			Internal				
External (in.)	Internal (in.)	BWG Gage	NOM Wall (in.)	Circumference (in.)	Surface per Lineal Foot (ft^2)	Lineal Feet of Tube per Square Foot of Surface	Transverse Area ($in.^2$)	Volume or Capacity per Lineal Foot: $In.^3$	Volume or Capacity per Lineal Foot: $Ft.^3$	Volume or Capacity per Lineal Foot: U.S. gal	Length of Tube Containing One Cubic Foot
$\frac{5}{8}$	0.527	18	0.049	1.9635	0.1636	6.1115	0.218	2.616	0.0015	0.011	661
	0.495	16	0.065	1.9635	0.1636	6.1115	0.193	2.316	0.0013	0.010	746
	0.459	14	0.083	1.9635	0.1636	6.1115	0.166	1.992	0.0011	0.009	867
$\frac{3}{4}$	0.652	18	0.049	2.3562	0.1963	5.0930	0.334	4.008	0.0023	0.017	431
	0.620	16	0.065	2.3562	0.1963	0.0930	0.302	3.624	0.0021	0.016	477
	0.584	14	0.083	2.3562	0.1963	0.0930	0.268	3.216	0.0019	0.014	537
	0.560	13	0.095	2.3562	0.1963	5.0930	0.246	2.952	0.0017	0.013	585
1	0.902	18	0.049	3.1416	0.2618	3.8197	0.639	7.668	0.0044	0.033	225
	0.870	16	0.065	3.1416	0.2618	3.8197	0.595	7.140	0.0041	0.031	242
	0.834	14	0.083	3.1416	0.2618	3.8197	0.546	6.552	0.0038	0.028	264
	0.810	13	0.095	3.1416	0.2618	3.8197	0.515	6.180	0.0036	0.027	280
$1\frac{1}{4}$	1.152	18	0.049	3.9270	0.3272	3.0558	1.075	12.90	0.0075	0.056	134
	1.120	16	0.065	3.9270	0.3272	3.0558	0.985	11.82	0.0068	0.051	146
	1.084	14	0.083	3.9270	0.3272	3.0558	0.923	11.08	0.0064	0.048	156
	1.060	13	0.095	3.9270	0.3272	3.0558	0.882	10.58	0.0061	0.046	163
	1.032	12	0.109	3.9270	0.3272	3.0558	0.836	10.03	0.0058	0.043	172
$1\frac{1}{2}$	1.402	18	0.049	4.7124	0.3927	2.5465	1.544	18.53	0.0107	0.080	93
	1.370	16	0.065	4.7124	0.3927	2.5465	1.474	17.69	0.0102	0.076	98
	1.334	14	0.083	4.7124	0.3927	2.5465	1.398	16.78	0.0097	0.073	103
	1.310	13	0.095	4.7124	0.3927	2.5465	1.343	16.12	0.0093	0.070	107
	1.282	12	0.109	4.7124	0.3927	2.5465	1.292	15.50	0.0090	0.067	111
$1\frac{3}{4}$	1.620	16	0.065	5.4978	0.4581	2.1827	2.061	24.73	0.0143	0.107	70
	1.584	14	0.083	5.4978	0.4581	2.1827	1.971	23.65	0.0137	0.102	73
	1.560	13	0.095	5.4978	0.4581	2.1827	1.911	22.94	0.0133	0.099	75
	1.532	12	0.109	5.4978	0.4581	2.1827	1.843	22.12	0.0128	0.096	78
	1.490	11	0.120	5.4978	0.4581	2.1827	1.744	20.92	0.0121	0.090	83
2	1.870	16	0.065	6.2832	0.5236	1.9099	2.746	32.96	0.0191	0.143	52
	1.834	14	0.083	6.2832	0.5236	1.9099	2.642	31.70	0.0183	0.137	55
	1.810	13	0.095	6.2832	0.5236	1.9099	2.573	30.88	0.0179	0.134	56
	1.782	12	0.109	6.2832	0.5236	1.9099	2.489	29.87	0.0173	0.129	58
	1.760	11	0.120	6.2832	0.5236	1.9099	2.433	29.20	0.0169	0.126	59

†Reproduced with permission from *Principles of Heat Transfer* by Frank Kreith, International Textbook Company, Scranton, Pa., 1958, p. 541.

TABLE B.3 Expansion Coefficient of Liquids at 1 atm Pressure and 70°F†

Liquid	$°F^{-1}$
Water	1.2×10^{-4}
Ethyl alcohol	6.21×10^{-4}
Freon 12	1.40×10^{-4}
Mercury	1.01×10^{-4}
Silicone oil	4.80×10^{-4}
Petroleum oil	4.0×10^{-4}

†Reproduced with permission from *Thermodynamics of Fluid Flow* by N. A. Hall, Prentice-Hall, Inc., Englewood Cliffs, N. J. 1951, p. 11.

TABLE B.4 Gas-Constant Values†

Substance	Symbol	M	R (ft lb/ lb_m °R)	C_p (Btu/lb_m °R) at 77°F	C_v (Btu/lb_m °R) at 77°F	K (C_p/C_v)
Acetylene	C_2H_2	26.038	59.39	0.4030	0.3267	1.234
Air	—	28.967	53.36	0.2404	0.1718	1.399
Ammonia	NH_3	17.032	90.77	0.5006	0.3840	1.304
Argon	A	39.944	38.73	0.1244	0.0746	1.668
Benzene	C_6H_6	78.114	19.78	0.2497	0.2243	1.113
n-Butane	C_4H_{10}	58.124	26.61	0.4004	0.3662	1.093
Isobutane	C_4H_{10}	58.124	26.59	0.3979	0.3637	1.094
1-Butene	C_4H_8	56.108	27.545	0.3646	0.3282	1.111
Carbon dioxide	CO_2	44.011	35.12	0.2015	0.1564	1.288
Carbon monoxide	CO	28.011	55.19	0.2485	0.1776	1.399
Carbon tetrachloride	CCl_4	153.839				
n-Deuterium	D_2	4.029				
Dodecane	$C_{12}H_{26}$	170.340	9.074	0.3931	0.3814	1.031
Ethane	C_2H_6	30.070	51.43	0.4183	0.3522	1.188
Ethyl ether	$C_4H_{10}O$	74.124				
Ethylene	C_2H_4	28.054	55.13	0.3708	0.3000	1.236
Freon, F-12	CCl_2F_2	120.925	12.78	0.1369	0.1204	1.136
Helium	He	4.003	386.33	1.241	0.7446	1.667
n-Heptane	C_7H_{16}	100.205	15.42	0.3956	0.3758	1.053
n-Hexane	C_6H_{14}	86.178	17.93	0.3966	0.3736	1.062
Hydrogen	H_2	2.016	766.53	3.416	2.431	1.405
Hydrogen sulfide	H_2S	34.082				
Mercury	Hg	200.610				
Methane	CH_4	16.043	96.40	0.5318	0.4079	1.304
Methyl fluoride	CH_3F	34.035				
Neon	Ne	20.183	76.58	0.2460	0.1476	1.667
Nitric oxide	NO	30.008	51.49	0.2377	0.1715	1.386
Nitrogen	N_2	28.016	55.15	0.2483	0.1774	1.400
Octane	C_8H_{18}	114.232	13.54	0.3949	0.3775	1.046
Oxygen	O_2	32.000	48.29	0.2191	0.1570	1.396
n-Pentane	C_5H_{12}	72.151	21.42	0.3980	0.3705	1.074
Isopentane	C_5H_{12}	72.151	21.42	0.3972	0.3697	1.074
Propane	C_3H_8	44.097	35.07	0.3982	0.3531	1.128
Propylene	C_3H_6	42.081	36.72	0.3627	0.3055	1.187
Sulfur dioxide	SO_2	64.066	24.12	0.1483	0.1173	1.264
Water vapor	H_2O	18.016	85.80	0.4452	0.3349	1.329
Xenon	Xe	131.300	11.78	0.03781	0.02269	1.667

†Reproduced with permission from *Concepts of Thermodynamics* by E. F. Obert, McGraw-Hill Book Company, Inc., New York, 1960, p. 502.

TABLE B.5 Absolute and Kinematic Viscosities of Water†

°F	$\mu \times 10^5$ (lb_f sec/ft^2)	$\nu \times 10^5$ (ft^2/sec)	°F	$\mu \times 10^5$ (lb_f sec/ft^2)	$\nu \times 10^5$ (ft^2/sec)
32	3.75	1.93	120	1.17	0.609
35	3.54	1.82	125	1.12	0.582
40	3.23	1.66	130	1.08	0.562
45	2.97	1.53	135	1.02	0.534
50	2.73	1.41	140	0.981	0.514
55	2.53	1.30	145	0.940	0.493
60	2.35	1.22	150	0.899	0.472
65	2.24	1.13	155	0.868	0.457
70	2.04	1.05	160	0.837	0.440
75	1.92	0.988	165	0.806	0.426
80	1.80	0.929	170	0.776	0.411
85	1.68	0.870	175	0.750	0.397
90	1.60	0.825	180	0.725	0.384
95	1.51	0.782	185	0.701	0.372
100	1.43	0.738	190	0.679	0.362
105	1.35	0.698	195	0.657	0.351
110	1.29	0.668	200	0.637	0.341
115	1.23	0.637	212	0.593	0.318

†Reproduced with permission from *Fluid Mechanics* by Arthur G. Hansen, John Wiley & Sons, Inc., New York, 1967, p. 486.

TABLE B.6 Surface Tension of Water Exposed to Air or Its Own Vapor†

°F	Surface Tension σ (lb_f/ft)	N/m
32	5.2×10^{-3}	7.6×10^{-2}
40	5.1	7.4
60	5.0	7.3
80	4.9	7.2
100	4.8	7.0
120	4.7	6.9
140	4.5	6.6
160	4.4	6.4
180	4.3	6.3
200	4.1	6.0
212	4.0	5.8

†Reproduced with permission from *Fluid Mechanics* by Arthur G. Hansen, John Wiley & Sons, Inc., New York, 1967, p. 486. SI units added.

TABLE B.7 Surface Tension of Various Liquids in Contact with Air†

Substance (in contact with air)	*°F*	*Surface Tension σ (lb_f/ft)*	*N/m*
Mercury	68	0.0324	4.73×10^{-1}
Benzene	68	0.00198	2.89×10^{-2}
Carbon tetrachloride	68	0.00184	2.69×10^{-2}
Glycerine	68	0.00482	7.03×10^{-2}
Ethyl alcohol	68	0.00153	2.23×10^{-2}
Methyl alcohol	68	0.00155	2.26×10^{-2}

†Reproduced with permission from *Fluid Mechanics* by Arthur G. Hansen, John Wiley & Sons, Inc., New York, 1967, p. 487. SI units added.

TABLE B.8 Properties of the Standard Atmosphere†

Altitude (ft)	**Temperature (°F)**	**Absolute Pressure (lb_f/ft^2)**	**γ (lb/ft^3)**	**Speed of Sound (ft/sec)**
0	59	2116	0.0765	1117
5,000	41	1761	0.0660	1098
10,000	23	1455	0.0566	1078
15,000	6	1194	0.0482	1058
20,000	−12	972	0.0408	1037
25,000	−30	785	0.0343	1017
30,000	−48	628	0.0288	995
35,000	−66	498	0.0238	973
40,000	−68	392	0.0189	971
45,000	−68	308	0.0148	971
50,000	−68	242	0.0117	971
60,000	−68	151	0.0072	971
70,000	−68	94	0.0045	971
80,000	−68	58	0.0028	971
90,000	−68	36	0.0017	971
100,000	−68	22	0.0011	971
150,000	114	3	9.8×10^{-5}	1174
200,000	159	0.7	2.0×10^{-5}	1220
250,000	−8	0.1	4.8×10^{-6}	1042
500,000	450	10^{-4}	3.1×10^{-9}	—

†Reproduced with permission from *Heat, Mass and Momentum Transfer* by W. M. Rohsenow and H. Y. Choi, Prentice-Hall, Inc., Englewood Cliffs, N. J., 1961, p. 523.

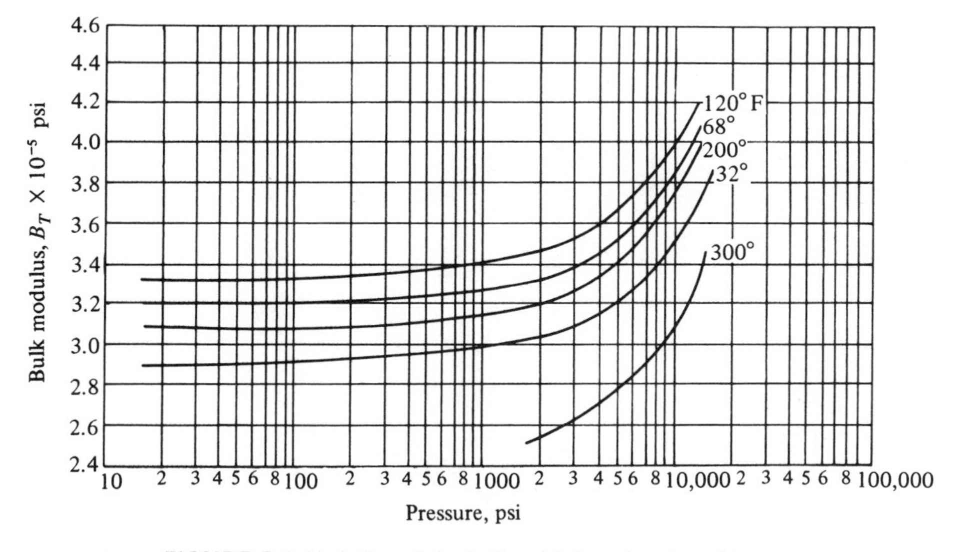

FIGURE B.1 Variation of the bulk modulus of water with pressure and temperature. (Reproduced with permission from *Fluid Mechanics* by Arthur G. Hansen, John Wiley & Sons, Inc., New York, 1967, p. 487.)

Index

B

C

D

E

W

Z